LECTURES

ON THE

PHILOSOPHY OF ARITHMETIC

AND THE

ADAPTATION OF THAT SCIENCE

TO THE BUSINESS PURPOSES OF LIFE:

WITH NUMEROUS PROBLEMS, CURIOUS AND USEFUL, SOLVED BY VARIOUS MODES; WITH EXPLANATIONS DESIGNED TO MAKE THE STUDY AND APPLICATION OF ARITHMETIC PLEASANT AND PROFITABLE TO SUCH AS HAVE NOT THE AID OF A TEACHER; AS WELL AS TO EXERCISE

ADVANCED CLASSES IN SCHOOLS.

BY URIAH PARKE.

FOURTH EDITION, REVISED AND IMPROVED BY THE AUTHOR.

"What man has done, man may do." "I WILL TRY."

PHILADELPHIA:
PUBLISHED BY MOSS & BROTHER.
No. 12 SOUTH FOURTH STREET.
1850.

T. K. & P. G. Collins, Printers.

PREFACE.

In presenting the following work to the public in a revised form, but few remarks are necessary by way of Introduction. Though not engaged in teaching, circumstances had forcibly impressed on the attention of the author, the necessity of some book adapted to the use of teachers and others desirous of studying the science of arithmetic, from principle, and tracing its uses in life. He examined accordingly every European and American publication that he supposed might meet the difficulty; but found none that seemed exactly to the purpose. *Leslie's Philosophy of Arithmetic* treats copiously on the use of counters, and the ancient modes of study, besides tracing some of the more curious properties of numbers; but there is nothing practical and life-like in it. Its scarcity, even in England, shows that it does not there meet the wants and the taste of the multitude. *Barlow's Elementary Investigation of the Theory of Numbers*, involves the necessity of an intimate acquaintance with Algebra; and though it is rigidly scientific in its development of the properties of numbers, and may suit the college, it is not sought for by the people either in Great Britain or America. Finding the ground unoccupied, and believing that a book adapted to the wants of life in our country, was necessary, the present publication was put forth, and the result fully justified the author's anticipations. The first edition was sold, and the demand continuing to increase, the work has been carefully revised, and is now offered to the public with full confidence that it will be found a useful companion to the teacher, the student, and the enterprizing reader who loves investigation.

The author has not sought to elucidate all the properties of numbers, nor to pursue a rigidly systematic course; but studying the wants of those for whom he has written, he has aimed to make a book that will be useful to them; and that will foster a spirit of investigation and study, without affecting to despise the inferior attainments with which thousands must rest satisfied. He does not seek to supersede or to build up, any particular system, but to occupy general ground, heretofore unoccupied, and to be useful. Teachers are sometimes found, that seem greatly at a loss in seeking to explain the principles of what they teach; for to know, and to be able to tell clearly what we know, are two things, not always found together; and if a teacher were learned as a NEWTON or a LARDNER, it would not benefit his pupils, unless he could communicate his knowledge. It is hoped that this book will be found to suggest some desirable modes of explanation. LOCKE says truly, "It is one thing to think right, and another thing to know the right way to lay our thoughts before others with advantage and clearness, be they right or wrong."

To the thousands of young men in our country, who are without the aid of living teachers, and yet desire to study this subject thoroughly, we trust the book before them will prove an assistant, no less valuable than to the teacher or the pupil in the school room.

If found acceptable to his countrymen, the author will have his reward.

URIAH PARKE.

ZANESVILLE, OHIO, September, 1848.

☞ It affords me pleasure to say that in the process of revision, I have been materially indebted to the assistance of Dr. SAMUEL C. MENDENHALL, a gentleman intimately acquainted with the science of Mathematics; and who has asked no living teacher for instruction in that branch of knowledge. To IMRI RICHARDS, Esq., also my thanks are due for valuable assistance.

TABLE OF CONTENTS.

LECTURE I.

The study of Arithmetic, its History, &c.

LECTURE II.

Principles of Numbering.

LECTURE XI.

Interest.

LECTURE XII.

Other uses to which a knowledge of Arithmetic is applied.

LECTURE XIII.

Proportions amongst Lines, Surfaces and Solids.

LECTURE XVIII.

Problems purely Arithmetical—Solved.

LECTURE XIX.

Questions of an Amusing Description.

LECTURE XX.

☞ Lectures 18, 19 and 20 contain solutions of a great variety of problems, that are often given as tests of ingenuity, and hence the collection will be found very desirable to persons fond of such exercises; and will be at the same time instructive. Many principles are fully explained in the course of the solutions, and different plans of solution used.

LECTURE XXI.

Unsolved Problems.

LECTURE XXII.

Arithmetical Prodigies, &c.

CONCLUDING REMARKS.

LECTURES

ON THE PHILOSOPHY OF NUMBERS.

LECTURE I.

THE STUDY OF ARITHMETIC, ITS HISTORY, ETC.

THE science of Arithmetic is of great antiquity and of much importance. It is alike indispensable to the scholar and the man of business; and must remain of primary importance through all the vicissitudes of time; for while sensible objects exist among civilized men, the science of numbering them must exist also.

Before we take up the principles and application of the science, it may be well to make some general remarks on its study, so often denounced by young persons as difficult and wearisome. We shall always find study irksome when we do not engage in it with full purpose of success; but if we direct our attention with energy and skill to any subject, and perseveringly seek to understand its principles, we seldom fail of success.

The study of Mathematics, of which Arithmetic is an important branch, has been resorted to by many philosophers, as a means of strengthening the reasoning faculties; and perhaps there is no mode more effectual than the close and connected train of thinking necessary to investigate the principles of this science. They who devote much attention to it, generally become passionately fond of the study, and often acquire great proficiency, though their general education may be very defective. Men of the highest rank have spent much time in this study and have enriched the science with many important discoveries; and who is prepared to say what timid schoolboy that is now conning his elementary rules, is not destined at some future day to make discoveries that shall carry the light of science far into the territory of ignorance and doubt?

Lord NAPIER, of Merchiston, Scotland, enriched the science of Arithmetic with one of the greatest mathematical discoveries of modern times, Logarithms; he also constructed several machines or instruments to facilitate calculation. Even the immortal Sir ISAAC NEWTON lent the powers of his mighty mind to the science of numbers, and made important discoveries.

Some persons possess by nature an astonishing aptness in calculation, being able to perform the most difficult operations in numbers, without the previous training which most minds require. But because nature has not been thus bountiful to us all, shall we fold our arms in listless despondency and do nothing?

Of a number of those persons we shall give some account hereafter. But a well balanced mind, in which the various powers are found to exist in an ordinary degree, is greatly to be preferred to one possessing some single astonishing feature, but deficient, as they generally are in others. Prodigies are seldom well adapted to the every-day affairs of life, and are generally unhappy; while the individual of strong common sense, expects nothing without labor, and applying himself diligently, shows himself sufficient for every emergency. If he cannot at once soar as high as the prodigy, in his favorite subject, neither will he be liable to fall as low in others.

One of the best mathematicians I ever knew was a weaver, who wrought daily at his loom. He commenced study late in life, but by industry and perseverance became famous for his mathematical knowledge. Indeed many of the best mathematicians of ancient and modern times have been self-taught. They have been men whose situations in youth did not afford them favorable opportunities for mental cultivation but they scorned to be discouraged; they engaged vigorously in the study of first principles, they laid broad the foundation, and in due time they were rewarded with the triumph of success.

When a young person thinks of engaging in a study that shall occupy his attention for a year or two, he is too apt to regard that time as lost to every thing else; as so much stricken from the duration of life, or to be spent in toil and privation; whereas the triumph of progress and ultimate success is a constant source of gratification, and in after life the period of acquisition seems as nothing while the possession is an inexhaustible fund of enjoyment. But if any one thinks to acquire proficiency in Mathematics without patient study of first principles, he will be disappointed, and instead of finding his way becoming more smooth and easy at every step, it will become more rugged and difficult, until lost in a labyrinth of uncertainty. There is no royal road to learning.

For the reader's encouragement I might point to our countrymen, BOWDITCH, the great navigator, and RITTENHOUSE, the great self-taught astronomer; in England to SANDERSON, blind from his infancy, yet he became not only an expert mathematician, but a professor of Mathematics; and to HERSCHEL, the astronomer, who rose from the rank of a drummer in the British army to be one of the greatest astronomers of the age; and to COBBETT, who commenced his career as a private soldier, and rose to be a member of the British parliament; and to FERGUSON, the shepherd boy who studied the stars at night while he tended his master's flock, and by perseverance rose to eminence. But I need not cross the ocean to seek examples, they abound in our own land. How many of our professional men and highest officers have been the artificers of their own fortunes! and of those who have distinguished themselves in the walks of science, how many commenced their career like BURRITT, "the learned blacksmith," under the most unfavorable circumstances!

I might name many, but it is unnecessary; let each youth cast his eye around upon his own acquaintance and see how few of those who occupy respectable stations in social community, or who enjoy the possession of wealth, were born under the smiles of fortune. It is not alone upon the high and shining mark in the list of statesmen, warriors, and men of science that the aspiring youth should look for encouragement, for if all below these were labor lost, how few would enjoy the meed of success! It is only the eagle of the strongest pinions that can reach the upper skies. The youth who aims high may fail to effect all he desires, yet he will generally effect much; while he who seeks nothing will effect nothing. When difficulties rise up in his way, let his motto be "I WILL TRY," and if he perseveringly carry out his determination, success will be his reward. The youth who is determined to succeed, must erase the word FAIL from his vocabulary.

I have remarked that the science of Arithmetic is of great antiquity, but we are not to understand that it has existed from antiquity in its present form; or containing the matter we now find in our school systems, arranged under the heads of Barter, Loss and Gain, Fellowship, and other rules. Not at all. This is a modern arrangement growing out of the application of the principles of pure science to the practical purposes of life. In the early days of this science, the art of printing was unknown, books were unknown, and even the art of writing, as it now exists, was unknown. Then the pebbles of the brook supplied the place of the pen and pencil, while a rude diagram, sketched perhaps in the sand or upon the tender bark of some forest tree, served the purpose of the geometer. As successive discoveries were made in the science, its limits were extended, and while valuable matter was added, rubbish was removed, for much time had been spent by some in idle speculations on "Magic Squares" and other matters of no practical use in life, and of little benefit to the cause of science; but as none could tell at what moment a valuable principle might be discovered, philosophers persevered in what proved to be but learned trifles. While men of science introduced principles valuable in a scientific point of view, and threw them into form for the study of youth; the wants of business were not disregarded, and hence the introduction of many rules of a practical nature, involving no new principle of science, but being entirely a practical application of the principles developed in other rules: and thus the systems of our schools have been brought to their present shape.

In the infancy of the science the power of combining numbers was limited to a very few simple operations; yet these were the groundwork, and as ages on ages have rolled on, the grand superstructure has been reared, until it has risen far above the clouds, and numbered the stars of the heavens, calculating their courses and declaring the days of their revolutions. To this result the discoveries and labors of centuries have contributed, and Arithmetic has now become an indispensable branch of education in every station of life.

The following brief sketch of its early history we copy from the *Western Academician*, where we find it suited to our purpose:

"As the arts and sciences, in their early stages must have been very imperfect, it was impossible to appreciate their value, or to predict their future importance to man. And since most of them were cultivated in some measure before mankind were qualified to record their progress, their early histories are either entirely unknown, or they are involved in doubt and obscurity, beyond which no research has been able to penetrate. This is especially the case with Arithmetic.

The Chaldeans, the Egyptians, and the inhabitants of the various parts of India, distinguished themselves at a very early period, by their knowledge of Astronomy. Their acquaintance with several periodical appearances of the heavenly bodies, some of which embraced many ages, was truly astonishing, and indicates a very advanced state in the knowledge of computation.

At this remote period of time, however, it is impossible to ascertain who were the real inventors of Arithmetic, or even to whom we are indebted for its earliest improvements. The Egyptians claim the merit of having first cultivated it; but they deem the invention too sublime to have been effected by human ingenuity, and piously ascribe it to the gods.

Some of the Greeks ascribe the invention of Arithmetic to the Phœnicians; and affirm that the first system of this science was written in the Phœnician language by AGENOR; but this appears to be without much foundation. It is highly probable that the operations of Arithmetic were improved by that commercial people; but there can be no question of their having borrowed their first ideas of it from the Egyptians.

JOSEPHUS maintains that ABRAHAM was the inventor of Arithmetic; and that his descendants carried the knowledge of it with them into Egypt. However this may be, it is certain that the Greeks copied both their alphabet and their method of notation from the Hebrews. The latter employed the first nine letters of their alphabet to represent the nine digits; and the Greeks afterwards adopted the same method, which is an evidence sufficient to determine between these two nations alone, the merit of priority in the cultivation of Arithmetic.

The Greeks are undoubtedly the first European nation among whom the subject of Arithmetic received any considerable attention. Mathematics had been cultivated to some extent when THALES appeared, (about 500 years before Christ;) but from that period may be dated the commencement of a more rapid progress. This eminent mathematician and philosopher, travelled to the East in search of information, where, no doubt, he received accessions to his knowledge of this useful science.

PYTHAGORAS, a disciple of THALES, also travelled among the Egyptians and the Indians in pursuit of knowledge, and spent twenty-two years in those countries, collecting information. Among other objects of inquiry, he gave especial attention to the science of Arithmetic. It does not appear, however, that the immediate followers of PYTHAGORAS, contributed much to the improvement of Arithmetic in its more useful branches. They devoted their almost exclusive attention to the discovery of the abstract properties of numbers, instead of trying to simplify the methods of calculation. They however, discovered many useful properties of numbers, and the common multiplication table is ascribed to PYTHAGORAS

2*

himself. One of the greatest benefits the Pythagoreans conferred on the science, was the discovery of the property of the right-angled triangle—the square of whose longest side is equal to the sum of the squares of the two other sides.

From some fragments of Grecian Arithmetic, it appears that they were not only acquainted with the operation of Addition, Subtraction, Multiplication, and Division; but also with the method of extracting the square and cubic roots, and the theory of geometrical progression. Their methods of calculation, were doubtless complicated and tedious, very unlike those of the present day; but their knowledge of the combination of numbers must have been extremely accurate as well as extensive.

About the second century of the Christian era, a system of notation was invented, called the Sexigesimal Arithmetic, of which PTOLEMY is supposed to be the author. Some traces of it are still found in the division of an hour, minute, circle, etc. The principal design of this notation was to avoid the inconvenience of the common method, especially in fractions. Every unit was divided into sixty parts, and each of these into sixty others: and in order to render the computation more simple, the progression in whole numbers was also made sexigesimal. From unity to fifty-nine the numbers were represented in the common way; and sixty, which was called *sexigesimal prima*, was denoted by unity with a dash over it; twice sixty by two units and a dash, etc., to fifty-nine times sixty, where the series was resumed, except only that sixty times sixty was denoted by unity with two dashes. When a number less than sixty was joined to a sexigesimal, it was annexed in its proper character, thus: 1′ represented sixty; 1′v sixty-five; x′ ten times sixty; x′x1 ten times sixty and ten and one, or 611, etc. Fractions were represented by placing the dash at the bottom, 1, equal one sixtieth, or at the left hand ′1 which was the same.

The notation of this system of Arithmetic is founded upon the same principle as that of the Arabian method, and differs only in the scale. The Ptolemaic by sixty, and the Arabic by ten, the dash representing the cypher. The only objection to this system, is the great number of characters necessary to its use; but this is nothing in comparison with the advantages it presents. Notwithstanding the sexigesimal Arithmetic, is ascribed to PTOLEMY, yet it is probably of Eastern origin, as the Indians of this day employ this division of time. They divide the day into sixty parts—each of these into sixty, and lastly these into sixty. They also reckon periods of sixty years as we do centuries.

The science of Arithmetic was enriched in the fourth century, by DIOPHANTUS, of the Alexandrian school; and the supposed inventor of Algebra. The time in which he lived was probably about the middle of the fourth century. He wrote thirteen books on Arithmetic, only six of which have escaped the destroying hand of time. An edition of his work was published in Paris, in 1621. The Diophantine Arithmetic was almost entirely neglected from that period to the time of the distinguished EULER; who was born at Basil, 1707.

The science of Arithmetic never received much attention by the Romans. The war-like disposition of that people being averse to the milder arts of peace. BOETHIUS, (who died about A. D. 525,) was perhaps the only mathematician of note among them. He translated the Geometry of EUCLID, and Arithmetic of NICHOMACUS.

About the middle of the seventh century, the Arabs who were a fierce and uncivilized people overran Egypt and Persia. The famous Alexandrian Library, which contained the accumulated labors of ages, and which was almost the only depository of the learning of antiquity, was consigned by them to the flames. The manners of this people, however, soon changed. In less than a century they began to cultivate the very sciences they had endeavored to banish from the earth.

About the beginning of the eighth century, they invaded the southern provinces of Spain. They carried with them the arts and sciences, and introduced into Europe the decimal scale of notation, and their admirable system of Arithmetic.

This system which is now used by every civilized nation, has all the precision we can desire, with the important advantage of conciseness and simplicity. A better scale than the decimal might possibly be adopted; but the principles of notation are incapable of improvement.

The celebrated GERBERT, who was raised to the pontifical chair, under the title of SYLVESTER II. contributed greatly to the diffusion of the knowledge of Arabian Arithmetic throughout Europe. He went into Spain himself, and acquired a knowledge of it, and returned to France, and introduced it among his countrymen, about the year 970. Soon after which, it was introduced into Britain. Though we are indebted to the Arabs for our present system of Arithmetic, that people do not pretend to have been the inventors, but acknowledge that they received it originally from India. Several manuscript copies of Arabian Treatises on Arithmetic are to be found. One of these has been preserved in the Library of Leyden, entitled "The Art of calculating according to the method of the Indians."

The difficulties experienced in the infancy of every science are great, and were not less in this than in others; and hence, much labor was wasted in speculations leading to no valuable purpose, and not a small share of ingenuity was devoted at various times to the construction of machines or instruments, designed to facilitate calculations by numbers; most of them however, exist now rather as mathematical curiosities in the cabinets of the curious, than as auxiliaries to the teacher in the school room, or the solitary student at his desk. The invention of Logarithms, so far improved the facilities of calculation, that no farther attempts were, for a long time, made to introduce machines. Recently, however, we have seen notices of further attempts at their construction, and of astonishing success.

The Greeks had their Abax, on which rows of counters were placed, consisting of pebbles, pieces of ivory, coins, &c., and from the abax of the Greeks, the Romans constructed their Abacus, which was in like manner a board on which pebbles were placed, and by their various arrangements, calculations were performed. "The use of the Abacus," says Professor Leslie "formed an essential part of the education of every noble youth. A small box or coffer called a *loculus*, having compartments for holding the *calculi* or counters, was considered a necessary appendage. Instead of carrying a slate and satchel, as in modern times, the Roman boy was accustomed to trudge to school loaded with those ruder implements, his arithmetical board, and his box of counters."

"The Greek word for pebble," says Dr. Lardner, "is *psephos*, and hence the word *psephizein*, to reckon or compute; the Latin word for pebble is *calculus*, and hence *calculare* to reckon, and our term *calculate*." *Sipher*, as appropriated to the digital characters, is an Arabic word, and introduced, says Professor Leslie, by the Saracens into Spain, and signifying to enumerate.

The form of the Abacus, as we now find it in schools, is an improvement on the original construction; the counters being made to move on wires that cross the frame. An instrument very similar to this is used by the Chinese, not only in their schools, but by their most expert accountants; it is called the Swanpan, or Schwanpan. During the middle ages, officers of the revenue in Europe, used a black cloth, on which white lines were drawn, crossing each other at right angles, as we see chess boards divided at the present day. This was called an Exchequer, and calculations were made by means of counters placed on the several squares.

Gunter's Scale, and Coggeshall's Sliding rule have lines of numbers for multiplying, dividing, &c., but they are now little used, as their accuracy is never equal to calculation, and in the hands of a careless person, or when the instrument is not well made, they cannot be relied upon at all.

BARON NAPIER, the celebrated inventor of Logarithms, contrived a machine called *Napier's Bones or Rods;* and the same gentleman contrived two other machines, for purposes of calculation, but they were complex, and though curious as well as scientific in construction, they have never been introduced into practical use.

PASCAL also invented a machine by which many combinations of numbers could be effected, but it is now found existing only as a specimen of human ingenuity. M. DE L'EPINE and M. BOITISSENDEAU, improved this machine or invented others upon the same principle. The inventions of Sir SAMUEL MORELAND, GEORGE BROWN, WILLIAM FREND, &c., followed, but like their predecessors, were never of much practical use.

For the blind, some apparatus sensible to the touch is indispensable; but that now used in the best schools for the blind, is very simple. Their slate, as they term it, is a metalic plate 8 or 10 inches square, covered with rows of cells or small apartments, (like the cells of a honey comb, only they are square,) formed by partitions crossing each other at right angles, as lines in the common Multiplication Table; and adapted to receive metalic types, on the ends of which, the figures are raised so as to be recognized by the touch. These are arranged in their elementary rules very much as figures are by those who see; but in the more advanced stages of their studies other characters are used, better adapted to the purpose of the operator, but which must be seen, and the uses explained to be well understood. After making some progress and learning the principles of numbers, the student is led by degrees to dispense with sensible characters and to conduct the process in his mind; in which the blind, having no external objects to distract their attention, become very expert. In this they are aided by having their forms of calculation adapted to a purely mental process, as is done in teaching children under the Pestalozzian system.

SANDERSON'S contrivances to enable him to perform mathematical calculations were very ingenious, and perhaps laid the foundation for the modern improvements in teaching the blind. A Mr. GREENVILLE and Dr. HENRY MOYES, both of whom like Mr. SANDERSON, were blind, contrived machines for purposes of calculation. Like the tablets used in modern schools for the blind, these machines were formed in squares with holes to receive pins, which being recognizable by the touch, and capable of various arrangement, enabled the operator to express any number at pleasure. Mr. SANDERSON was long professor of Mathematics at Cambridge, England, and Dr. MOYES pursued the occupation of a lecturer on Natural Philosophy and Chemistry, and it is said that his precision in elucidating the doctrine, even of light and colors was surprising.*

*The following extract of a letter from H. N. HUBBELL, Esq., Principal of the *Ohio Deaf and Dumb Asylum*, details some particulars relative to the mode of instructing mutes, that may prove interesting to such as have not witnessed the process.

Ohio Deaf and Dumb Asylum,
Columbus, O., Oct. 31, 1839.

U. PARKE, ESQUIRE:

Dear Sir—I received a letter from you a few days ago, requesting information respecting the manner of instructing the Deaf and Dumb in Arithmetic. I would gladly furnish you with any information in my power; I am apprehensive, however, that I cannot give such a description of the mode as will be very interesting.

It will be obvious that the mental operations of mutes in the study and use of Arithmetic, are similar to those of persons who hear and speak. They differ only in the mechanical part of the process, mutes being obliged by their necessities to lay hold of some visible symbols, corresponding in signification to the vocal sounds employed by others to express the same things; but the expression of the mute is generally most forcible, his language being natural, while that of the others is artificial and arbitrary.

It is a motto with mutes, "The hand answers the purposes of the tongue." This is true in learning and using Arithmetic, as well as in other subjects; and it is especially true in using the fingers to express numbers, for they furnish a ready means of expressing all their combinations to an unlimited extent. The ten fingers well express the *digits*, indeed the word digit signifies finger, and these were undoubtedly used by the ancients, as they are by barbarous people at the present day, to express numbers, as far as they had occasion to employ them or were capable of understanding them. The use of decimal Arithmetic by almost all nations has no force as an argument to prove that all nations had the same origin, as a celebrated author (Dr. GOOD) supposes, for it is in all cases derived from the ten fingers; and it is rather strange that such an argument should be adduced by such a man, to prove what is doubted by few.

In counting, the mute uses one hand, beginning with the thumb and proceeding to the little finger, making five; returning by touching the end of the thumb to the ends of the several fingers he makes nine. A horizontal motion with the whole hand clenched expresses ten. By a combination of these signs he expresses any number under twenty. Twenty is represented by a horizontal motion of the thumb and fore finger closed; thirty by the thumb and two fingers, and so on; a horizontal motion expressing tens. A hundred by the letter C of the manual alphabet, and a thousand by the letter M. A repetition of the thousand sign makes millions, and thus any conceivable number can be expressed with as much rapidity as can be effected by the use of speech. The ordinal numbers, first, second, third, &c., are distinguished from these by an upward perpendicular motion.

In 1831, Mr. Oliver A. Shaw, of Richmond, Virginia, invented a set of wooden figures which he called the *Visible Numerator*, and having taken out a patent for his invention, he visited different states of the Union in the character of an itinerant lecturer on Arithmetic; but though it was successful in the hands of the inventor, it has never been extensively introduced into schools. Like many other means of illustration, it probably owed much of its supposed value to the expertness of him who used it.

It consisted of a series of mahogany blocks, representing units, tens, hundreds, &c., ingeniously put up in a neat mahogany box about 9 inches wide, and a foot in length. The blocks representing units, were cubes $\frac{1}{10}$ of an inch square; the tens were ten times as large, being $\frac{1}{10}$ square and an inch long; the hundreds ten times as large as the tens, being an inch square, and $\frac{1}{10}$ thick; while the thousands were in the same ratio, being cubic inches; other blocks represented larger quantities.

From the recommendations attached to Mr. Shaw's book, it seems probable that he was able by his oral lectures to give interest to the subject, and by means of his apparatus to make

Numeration I teach thus: Instead of saying units, tens, &c., in numerating a row of figures, 6843250 for instance, I place abbreviations above them, thus:

m h ty th h ty
6 8 4 3 2 5 0

naming the figure first, which reads 6 *m*illions, 8 *h*undred, 4 for*ty*, 3 *th*ousand, 2 *h*undred, and fif*ty*. Thus with a blank over the right hand figure, ty for the tens, and the regular recurrence of ty, h, th, ty, h, m, ty, h, th, ty, h, b, (billions) I find no difficulty in enabling the learner in a few minutes to enumerate any given number.

In teaching them Addition, whatever number is to be carried from one column to another is put at the bottom of the column to which it is carried, and added to the column, and the reason why carried, of course explained to them.

In Subtraction, when the lower figure is *smallest*, they experience no difficulty. When it is *largest*, 1 is placed at the left hand of the upper figure, and 1 carried to the next lower figure; this being fully explained to them in sign language, they generally proceed with facility.

Multiplication and Division being but concise methods of performing many additions and subtractions, are learned in the same way. The elementary rules being thus acquired, their application and use in business transactions are readily explained.

Mutes, while at school, have so much to do with learning the meaning and use of language, that they devote only what time is barely necessary to the study of Arithmetic, and we find as little difficulty perhaps in teaching them whatever may be necessary in this branch of their education, as is experienced by other teachers whose pupils can hear and speak.

* * * * * * * * *

Respectfully, &c.,

H. N. HUBBELL.

some parts very plain; indeed his book is adapted to convey a very clear idea of the nature of numbers, and of our system of Notation; though like many other pieces of machinery designed to illustrate, the process is in some parts rather obscure, and far-fetched. This is often the case where formal systems of illustration are attempted; for though it is well to refer to sensible objects in our early study of numbers, it should only be for special illustration and not for the purpose of building up a general system. Such systems carried out into minutiæ are often more difficult to understand than the principle sought to be explained, and then the idea is less clear, being encumbered with the machinery of explanation.

Mr. SHAW's notion seems to have been to adopt a system of sensible or visible objects that might do for Arithmetic, what diagrams have done for Geometry; and perhaps he has done as much in that way as the nature of the subject will admit. He objects very pointedly to explaining the ratio of numbers by considering 1 in the tens' place as equivalent to 10 ones in the units' place; or 1 in the hundreds' place as equivalent to 10 ones in the tens' place, or 100 ones in the units' place; but prefers as simpler and easier to understand, that 1 in the 10's place be considered as *one*, though ten times as *large* or as *valuable* as 1 in the units' place; just as 1 dollar is as much 1 as one dime, but the 1 dollar is 10 times as great and valuable as the one dime. He thinks that usage alone leads us to commence expressing numbers at the higher denominations, and that apart from custom we might just as properly say *fifty and one hundred*, as to say *one hundred and fifty;* or the book cost *fifty cents and two dollars*, as the book cost *two dollars and fifty cents.*

There seems, however, reason for the present arrangement, apart from custom. Having a great number of units, we group them into tens, these tens into hundreds, the hundreds into thousands, &c. If none of the lower orders remain, the whole quantity is expressed in a single term of the larger class, and that is what is sought to be done; but if any remain, we express the large number first, as giving the nearest approximation in a single expression, to the whole quantity, and we then name the several overplusses in the different denominations. Suppose I ask the precise length of a street, and I am told that it is 3 inches, 1 foot, and 500 yards, I can form no approximation to an opinion until the last term is expressed; and the same may be said if I ask the price of a book, and am told that it is 12½ cents and $2; but if the leading or larger amount is first given, it gives a pretty accurate idea, which is more fully defined by what follows.

If I seek to measure a quantity of wheat, I fill my bushel measure as often as it can be filled, and when there is no longer enough to fill it again, I may use my peck, and successively my gallon, quart, pint, &c. The bushel being considered the measuring unit, while the smaller measures serve to express the overplus.

His Numerator seems principally adapted to illustrate the elementary operations of pure Arithmetic and the doctrine of Proportion; though he applies it also to the compound rules and fractions. It is easy to see that a block representing 10 would be proportionate in size to one representing 100, as a block representing 100 would be to one representing 1000.

His illustration of the extraction of roots is by blocks, and is similar to the mode used in several of our best school systems of Arithmetic.

Mr. SHAW pursued for a few years the business of a travelling lecturer, and speaks of his system having been adopted in many schools, but I have never met with it in use, nor seen more than one set of the apparatus and plates, and this was in the hands of a gentleman who had heard his lectures and at the time was pleased with the plan, but he never used it in his school. From him I purchased the apparatus with the book of lectures, and have kept them as a curiosity rather than for any practical purpose.

LECTURE II.

ON THE PRINCIPLES OF NUMBERING.

To every science is assigned certain subject matter, language to Grammar, the materials of the physical world to Chemistry, its laws to Philosophy in the limited meaning usually attached to that word, and Mathematics is the science which treats of all kinds of quantity whatever, that can be numbered or measured. That part of this great science which treats of *numbering* is called Arithmetic, while that which has reference to magnitude only is called Geometry.

Any thing which admits of increase and decrease, as surfaces, lines, weight, motion, &c., is properly called quantity; and when quantity is considered as undivided, as a quantity of land, a quantity of water, &c., we consider it in reference to magnitude, or magnitude and shape, regarding it as an undivided whole; and these are the legitimate properties for the science of Geometry to investigate. When we consider *quantity* as made up of individual and distinct parts, as ten men, a hundred books, we consider the quantity as an assemblage or multitude, and the numbering of the parts is effected by the aid of Arithmetic. Each individual of the multitude is called a unit, and the extent of the multitude is determined by the number of the units; but in determining the magnitude of undivided bulk, we must apply to it some measure of known dimension. The carpenter applies his rule to the board—the surveyor his chain to the land—while liquids are measured in vessels of known size, and the flight of time is artificially marked by seconds, minutes, and hours, or naturally by days and seasons.

While we call the unit of multitude the *natural* unit, we may call the assumed unit by which we measure magnitude, the *artificial* unit. So long as our calculations are confined to natural units, it is pure Arithmetic; and when we consider magnitude without reference to its measurement by the artificial unit, it is pure Geometry; but when our calculations respecting magnitude are based on measurement by the artificial unit, it is properly called mixed Mathematics, a denomination under which a large portion of our business calculations may be properly classed.

Algebra is a very extensive and important branch of Arithmetic, for much of Algebra may be regarded as a kind of universal Arithmetic; but as our discussions are confined at this time to first principles, it would be premature to introduce any thing on that important branch of mathematical science. It may be in place here to remark that it is amusing to hear persons of intelligence speaking of Arithmetic and Mathematics as distinct sciences; and even teachers in their bills of prices often so treat them; as though the boy who is studying his elementary rules, were not engaged in studying a portion of Mathematics; a task to him as difficult as Navigation and Astronomy will be when he is prepared for their study. Perhaps, however, it is only proper that teachers who skim the surface of Arithmetic, who teach it without reference to principle, should not degrade Mathematics by considering the arbitrary rules which they teach, as a part of that noble science. At any rate they would not be very likely to exclaim with the mathematician, who, after reading *Milton's Paradise Lost*, said to a friend, "The story 's well enough, but what does it all prove?" He could see no merit in what did not prove some mathematical proposition.

When the savage is asked the number of his family, he probably holds out to you a corresponding number of fingers; but ask him the number of his tribe, and his untutored intellect cannot frame an answer, nor his language furnish words to express so large a number. He points then to the leaves of the forest, or the hairs of his head; or tells you to count the particles of sand upon the sea shore, or the stars of the firmament.

We are so accustomed from infancy to our admirable mode of numbering, that we think it something natural; that there is a natural connexion between numbers and our mode of expressing them, and we can scarcely understand how any one can find difficulty in so simple an operation. Yet simple as it appears to us, ages elapsed before it was invented, and though now generally adopted by the civilized world, uneducated tribes are yet found, whose knowledge of numbers scarcely extends beyond the number of fingers upon their hands, and toes upon their feet.

The mode of numbering which would most naturally occur to one unskilled in the artificial science, would be to give a separate and independent name to each number, from a unit upwards as far as the series is made to extend. This we now do as far as the number ten, (and indeed as far as twelve,) but think for a moment what would be the labor and difficulty of having 10,000 distinct names to express the units or individuals in that number—and then to have 10,000 distinct characters in writing to express those numbers—and yet 10,000 is a small number compared with what we are often called upon to express. But even if we could succeed in giving a distinct name to each without reference to the rest, such would be the multiplicity of names, that we would have no distinct idea of large numbers; any more than we now have of billions, trillions, quadrillions, &c. The mind would have no resting points as it has in the present system—"it would have no landmarks in the great ocean of numbers." We would not know without reflection, larger from smaller numbers, as persons sometimes forget the order of letters in the alphabet, until repeated in succession.

One of the first steps therefore, in the science of numbering, is the adoption of a radix or basis, on which the numerical system shall be constructed. Different nations have adopted different bases, but that most extensively used, is the decimal; so called because it proceeds by tens, that number being called *decem* in the Latin language. This is our system. We commence and give a distinct name to each unit as far as *ten;* we then commence a second ten, which carries us to twenty; a name probably derived from *two* ten, as thirty, forty, fifty, &c , express three, four, five tens. We proceed thus to ten tens, which we call *one hundred;* and when by successive additions we have formed another hundred, we call it *two hundred*, and thus we proceed to *ten hundreds*, which we call *a thousand;* and a thousand of thousands we call *a million.* These tens, and hundreds, and thousands, are the "resting points," the "landmarks," which enable the mind at a glance to form a clear conception of a number composed of hundreds, or even thousands, and which if designated by an arbitrary and independent name, would be conceived by the mind faintly and indefinitely.

There is no reason why ten should have been adopted as the radix or root of the system rather than any other number, and indeed, for reasons which may be hereafter given, twelve would have been a better basis; but ten was no doubt adopted from the digits of the hands. They would be naturally resorted to in the absence of names, as they now are by persons not well acquainted with the language they are speaking, or as they are by the deaf and dumb; and the adoption of the same scale by almost all nations, civilized and savage, shows conclusively that there is a better reason for its selection than fancy or freak; and especially as the number adopted, is not for calculating purposes, as good as some others, and could not therefore have been selected for its excellence in this respect. But in its selection, which must have been antecedent to all arithmetical investigation, the computer sought only to express number, and let us see how he would be most likely to attempt this. As a sign, the fingers would express any number as high as ten, and there being no connexion between one of these numbers and another, the names assigned them would probably be independent and arbitrary, as we find them in our own language, for there is no analogy or connexion in orthography or sound between one, two, three, four, five, six, &c., to ten, but between 3, 13, 30; 4, 14, 40; &c., there is an analogy both in orthography and sound. Having expressed 10 in this way, it would be natural to express a second ten by the same *signs* somewhat varied, and by the same *words* modified so as to express the difference, as three, thirteen; four, fourteen; &c., the latter words being plainly modifications of three-ten, four-ten. The third, fourth, &c., series of tens would have their modified signs and names, and thus would the whole science of numbering resolve itself into successive periods of ten units each. It is clear that the series, by the combinations of a few names, may be made to express any number, however large.

For purposes of calculation, twelve would have been a better radix, as it admits of a greater number of divisions without remainders; and had man been a twelve fingered animal, there is little doubt but we would have had a duodecimal scale of numbers. The numbers *eleven* and *twelve* being irregular, seems to favor the idea that they are the relics of such a scale, but it is possible that the irregularity grew into use from the frequent expression of such numbers in familiar intercourse. The computation of articles by *dozens*, the division of feet and inches, the division of the circle into signs, the year into months, &c., make it probable however, that the duodecimal scale was partially used at some remote period. We have also remains of a sexagesimal scale, (by sixties,) as in degrees and minutes, and it is known that this scale, from the number of aliquot parts of which it is susceptible, has had strenuous advocates; but the number of distinct names of which the first series must consist, would be an insuperable objection to this system. Thirty has been proposed, so has one hundred and twenty, but it is not probable that the system will ever be changed; though there are those who would yet advocate the attempt.

In some nations five is known to exist as a basis, and in others fifteen and twenty, but instead of invalidating the supposition that the fingers have suggested our basis, these other bases 5, 15 and 20, the fingers on one hand, and the fingers and toes, go to confirm the position that no basis has been arbitrarily adopted. The basis 15 may have been suggested by the 3 joints of the fingers on one hand; and 20 by the joints and the ends of the fingers, as readily as by the fingers and toes, and it is more likely if there was a corresponding sign language, inasmuch as the parts of the hand could be more readily exhibited to view than the foot. The different scales will be more fully explained hereafter.

It is difficult for an educated person, or even the most ignorant person in civilized society, to realize the difference between the situation of the uneducated around him and the savage. We find persons who know very little, who cannot read, who know nothing of public men and public matters, and who seem incorrigibly stupid, yet we could scarcely find a person so ignorant as not to be able to count. Having never known the want of such ability we are scarcely able to realize the situation of one who does, nor to appreciate the value of such unthought-of advantages.

The adoption of the foregoing mode of oral counting was an

important point gained; but still it was necessary to invent some mode of expressing numbers in writing. To write them in full was not practicable in making calculations, and was at all times tedious; and to designate them severally by marks was little better, since the number of marks would require great labor in the process of reading the numbers written. This would otherwise be the simplest mode, and addition would be performed by simply adding the additional marks; as subtraction would be by striking them off. Multiplication would be nearly allied to addition; and division would consist in marking off the characters into periods equal the number in the divisor. It was found better, however, to adopt a notation corresponding to the radix of the system; and with this view the Hebrews, Greeks and Romans adopted letters from their alphabets, and this was found convenient enough to express numbers less than ten, but to express the second series of ten, and each successive ten, marks were attached to the letters, which rendered their systems complex and difficult to learn as well as use. A specimen of Roman Notation is yet seen in the letters with which the chapters in many books are numbered, but the difficulty of performing calculations with such characters caused them to fall into disuse amongst the scientific throughout Europe, very soon after the Arabian system was introduced. As late however as the tenth century, it was the system used in the study of Arithmetic as well as by the accountant. How very awkward such a system was, will be evident to any one who will attempt to add, subtract, multiply, or divide numbers so expressed. For instance let the numbers 586, 340, 619, 499, be expressed in letters according to the Roman Notation and added together.

586	will be	DLXXXVI
340	"	CCCXL
619	"	DCXIX
499	"	CCCCXCIX
——		——
——		——

Again, From	586	DLXXXVI
Take	340	CCCXL
	——	——
	——	——

Again, Multiply	586	DLXXXVI
By	340	CCCXL
	——	——
	——	——

Lastly Divide 586 by 19 i. e. XIX)DLXXXVI(

The last of these operations would be most difficult, but they are all far less convenient than the system we now use; a system which probably can never be improved in simplicity.

The modes of operation adopted by the ancients, if we may so designate such as lived hundreds of years since the Christian era, were extremely laborious. The learned Bede, of the eighth century, wrote two treatises on the science of Arithmetic, one of which was devoted exclusively to the division of numbers; showing the complicated processes, which, in consequence of their awkward notation, they were compelled to adopt. Though the Roman letters have been used for ages, they are not considered the original numeral marks. The simplest mark which was originally used to designate a unit, and still is so, was a straight stroke I, which was changed for convenience to I; when these had been repeated to ten, two strokes, crossing each other, thus X, were used, and this was afterwards represented by the letter X. For convenience half the ten mark was afterwards used to designate five, and this was changed to V. When the accountant reached 100, three straight strokes were used, thus [, (the original form of the letter C,) and being modernized in use, by rounding off the corners, that number is represented by C; while two of the strokes L, changed to L, express 50. The next power of 10, or 1000, was represented by four strokes M, and for this character the letter M furnishes a ready substitute, while half the character N, expressing 500, gradually changed to D; a much better substitute at the hands of a printer, than 5 inverted makes for the French ç: as in François for François.

These primitive characters are yet found in ancient inscriptions; the straightness of the strokes, rendering them well adapted to the purposes of the engraver. The Greek and Roman capital letters, says Professor Leslie, are more ancient than the small letters, and were originally used, like the Runic letters of modern Europe, for the purpose of the lapidary; and hence like the latter are formed principally of straight strokes. The curved and wiry forms of many alphabetical characters are probably owing to the material on which they were traced by their inventors; while the bold forms and right lines of others may be traced to the convenience of the engraver.

About the beginning of the eighth century, when the Arabs established themselves in the southern provinces of Spain, they introduced into Europe their admirable system of Notation, now used throughout the civilized world. The first writer who appears to have employed this system in calculation was JORDANUS NEMORARIUS, who wrote about the year 1230. The first treatise on Decimal Fractions was published in 1582 by

Stevinus, but they had been used in the extraction of roots and for some other purposes, for several years prior to that time. After their introduction the sexagesimal system, which had been invented by Ptolemy, or at least sanctioned and used by him, was soon abandoned.

With the Arabian system we received also the Arabian characters, or numerals; and these are now used throughout the civilized world. These characters, like hieroglyphics, represent things not words, and hence a dozen individuals, speaking as many different languages, may pursue the same arithmetical calculation, and all will understand, so long as their tongues are kept still. It is a kind of common language, which all may write, but not speak.

The great feature which gives to the Arabian system its excellence, and distinguishes it from all others, is making the value of the characters depend upon their place: an arrangement which enables the operator, with a few figures, to express the largest numbers; besides affording much convenience to the mathematician in his calculations.

The doctrine of *carrying* in the elementary rules results from this feature in our notation. When a character stands by itself, its value is simply the number it represents, as 5 or 7 will represent the number five or seven; but if we place another figure to the right of either we increase the value of the left hand figure ten fold, thus 53, 75; the character which represented five when it stood alone, becomes *fifty* when another stands at its right, and if we add another so that it shall occupy the third place, it becomes *five hundred;* and in the same way by successive removes it may be made to represent millions. It is still the same figure, but its value has changed with its place, and will fluctuate as its place does. If we desire to express a round number, as it is sometimes termed, of hundreds, or thousands, or millions, we fill all the lower denominations with ciphers, merely to fix the place of the significant figure. To express five thousand the 5 must occupy the fourth place, we therefore fill the first, second, and third with ciphers, thus: 5000; but if there be hundreds, and tens, and units, as well as thousands, then the proper figures to express them are inserted instead of the ciphers; thus: 5493. This device is apparently very simple, but it is perfectly efficient, and serves, with infinitely superior convenience, the purposes of the complex systems that had preceded it, and hence supplanted them wherever they came in contact; until the old have disappeared except in the pages of history. It may not be tedious to look a little into the history of its adoption; or rather to offer an hypothesis to show how the idea may have suggested itself to the inventor of the system; and by presenting the subject in a new light it may be made plainer than by the few remarks we have suggested.

The ancients, as well as uncivilized nations of later days, were much in the habit of using sensible objects as counters, to aid in their calculations. These were arranged on boards or tables laid off by lines, or upon the abacus. Suppose the following nine lines to represent grooves in which counters are to be placed:

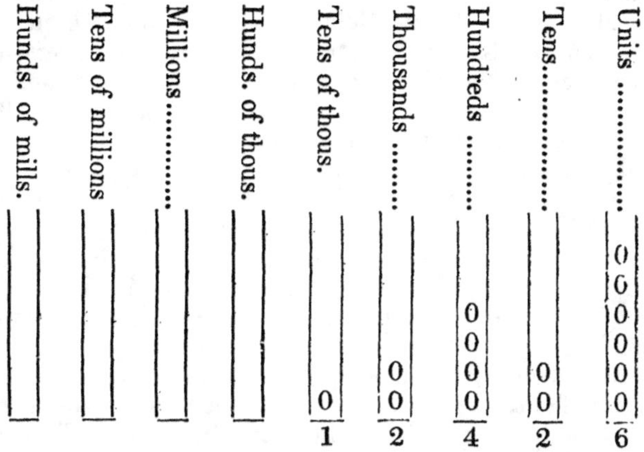

Hunds. of mills.	Tens of millions	Millions............	Hunds. of thous.	Tens of thous.	Thousands	Hundreds	Tens................	Units ....
								0
								0
						0		0
						0		0
					0	0	0	0
				0	0	0	0	0
				1	2	4	2	6

Now suppose a very large number is to be represented by counters, as in measuring grain by the bushel. We commence by putting counters into the units' groove until there are ten, as the numbers are successively announced by him who calls out the measures, and to prevent that from being filled up we take up the ten and put 1 into the next groove on the left, and as tens successively accumulate in the units' groove they are removed and 1 put into the tens' groove; after awhile there will be ten in the tens' groove, when they are to be removed and 1 put into the hundreds' groove; thus the operation is to be continued, and the tens constantly marked by putting one into the next left hand groove, until the grain is all measured, when the grooves are found to contain, as marked, viz: 1 ten thousand, 2 thousand, 4 hundred, 2 tens, and six units, or 12426 bushels. This mode of keeping "tally" where there are large numbers, would be perfectly natural, and the table of grooves and counters would be a permanent apparatus for all persons engaged in trade requiring its use; the number of grooves being increased indefinitely to the left, as the size of the numbers might require. It would not do for the ratio between the grooves, (the radix of the system,) to be too large, or the eye could not readily catch the number in each groove, neither would it do to have the ratio too small, lest the number of grooves be too great; ten is probably about right, as the operator may read the sum total from the counters in the grooves about as readily as figures could be read, nor would it be necessary to mark them with units, tens, &c., any more than it would be to mark figures so, for practice would make one familiar, as readily as the other. Here then is the principle of value according to place as fully developed as in the Arabic Notation, and it would seem but an easy step to substitute characters in place of numbers and read them as we now do, and especially in recording an amount, or in writing it for the information of others. It is true that an empty groove between such as are not so, would make the numbers read wrong, as suppose the groove of hundreds in the above was empty, it would then be 1226, which would not truly express the number; but this vacant place must be marked to preserve the proper places, and for this purpose let a cipher, which has no value, be introduced and we have 12026, the precise form used in our ordinary notation.*

Having advanced so far as to be able to express numbers in this way, the *algorithm*, or mode of calculation, would soon be discovered; though if an old mode exist, prejudices are slow to yield, and the most evident improvements are the work of time. Notwithstanding the acknowledged superiority of the Arabic Notation, business men continued for nearly 300 years after its introduction into Great Britain, to use their old

* The character representing a vacant space, 0, is variously termed zero, cipher, and nought, all signifying *nothing*, there being none of the denomination whose place is thus marked. But the name is often pronounced falsely, *aught*, which means *any thing*, as *naught*, or *nought* means *not aught* or *not any thing*, of which it is evidently an abbreviation. By this corruption the meaning is reversed and our expression falsified, for though nothing multiplied a hundred fold is still nothing, we cannot say with any regard to propriety that "aught" (any thing) however small, multiplied by 100 will be nothing, it must be something, with or without multiplication; and though nought (nothing) taken from 100, leaves a hundred still, yet if "aught" be taken there cannot be a hundred left. This may seem a criticism in a small matter, but we see this ridiculous blunder pass from the school room to the counting room, the workshop, and the office of the professional man, and thus kept up. Perfection consists in attention to accuracy. It is not sufficient that the marks of the axe and the saw be removed when the workman constructs his best furniture; he must remove every imperfection before he can bestow the proper polish. So the correct speaker must avoid every error, if he would be a fit pattern in the use of good language; and no teacher should consent to be less.

awkward Roman numerals for business transactions, and the pertinacity with which many in our own country adhered to the inconvenient currency of £. *s.* *d.*, after the adoption of the Federal money is well known. Indeed we find a disposition yet with many, to express value in that way. Such changes require at least one generation.

To understand the superiority of the Arabic algorithm, it is only necessary to compare the problems we have given, for though they may be wrought by letters, the process, especially in Division, is complex and difficult. Hence, as already remarked, mathematicians had many special rules for division, by which the operation was simplified or abridged.

By the Numeration Table as given in most school Arithmetics, enumeration is carried only to nine places, or hundreds of millions, which is far enough for all ordinary purposes; but the series may be extended almost without limit by Billions, Trillions, Quadrillions, Quintillions, Sextillions, Septillions, Octillions, Nonillions, Decillions, &c., which are terms derived from the Latin numerals as high as 10, and as each name, according to the English System of Numeration, expresses six places, there would be $10\times6=60$ places of figures, besides the units period, or 66 in all. And a series of units of one fourth that extent would be beyond the power of man to imagine. Even millions convey a very indefinite idea, and when it rises to billions, the mind can no longer grasp the number, for though we may read the expression, it is very much as we read sentences in an unknown language.

But the mind may be aided by some little calculation. We often see millions spoken of in our national expenditures, and yet even that is a large number, for if a man were to count $1500 an hour, (equal to 25 per minute,) and work faithfully 8 hours per day, it would require nearly 3 months to count a million of dollars, and if the dollars were 1 inch and $\frac{5}{8}$ in diameter and laid touching in a straight line, they would reach 136 miles; and 14 wagons carrying two tons each, would not be sufficient to convey them. Our only plan then is to group the number by imagining 1000 piles and 1000 dollars in each pile, when we can form as distinct an idea of the number of piles, as of the individuals of each pile. But when we extend even this mode to such a sum as the national debt of Great Britain, the imagination becomes bewildered. Dr. THOMPSON, Professor of Mathematics at Belfast, in Ireland, very justly remarks, "Such is the facility with which large numbers are expressed, both by figures and in language, that we generally have a very limited and inadequate conception of their real magnitudes. The following considerations may, perhaps, assist in enlarging the ideas of the pupil, on this subject:—To count a million, at one per second, would require between 23 and 24 days of 12 hours each. The seconds in 6000 years are less than one fifth of a trillion. A quadrillion of leaves of paper, each the two hundredth part of an inch in thickness, would form a pile, the height of which would be three hundred and thirty times the moon's distance from the earth. Let it also be remembered that a million is equal to a thousand repeated a thousand times, and a billion equal to a million repeated a thousand times." And if we adopt the English notation it is a million of millions. A cannon ball flies very swiftly, but were one fired at the moment that one of our national presidents takes his seat in the presidential chair, and were it to continue with an unabated velocity of 1200 feet per second, during his entire term of 4 years, including one leap year, it would not travel three millions of miles.

"We never hear," says an anonymous writer "of the 'Wandering Jew,' but we mentally inquire, what was the sentence of his punishment? Perhaps he was told to walk the earth until he counted a *Trillion.* But we hear somebody say, 'he would soon do that!' We fear not. Suppose a man to count one in every second of time, day and night, without stopping to rest, to eat, to drink, or to sleep, it would take thirty-two years to count a *Billion*, or 32,000 years to count a Trillion; even as the French understand that term. What a limited idea we generally entertain of the immensity of numbers!"

A curious practice of grouping numbers, is related of the people of Madagascar. Flacourt says, "When the people of that island wish to count a great multitude of objects, such, for example, as the number of men in a large army, they cause the objects to pass in succession through a narrow passage before those whose business it is to count them. For each object that passes, they lay down a stone in a certain place; when all the objects to be counted have passed, they then dispose the stones in heaps of ten; they next dispose these heaps in groups, having ten heaps each, so as to form hundreds; and in the same way would dispose the groups of hundreds so as to form thousands, until the number of stones has been exhausted." Nicholson, states that the native Peruvians use grains of maize, by the various arrangement of which they calculate with wonderful facility.

The following table exhibits the extended Numeration table on the English system, the periods consisting of six figures each; while that which follows exhibits the same numbers on

the French system, in which the periods consist of three figures only. Both modes of enumeration are used, and perhaps the French is now more frequently used than the English.

Units.	1 .
Tens.	1
Hundreds.	1
Thousands.	1
Tens of Thousands.	1
Hunds. of Thousands.	1
Millions.	1,
Tens of Millions.	1
Hundreds of Millions.	1
Thous. of Millions.	1
Tens of Th. of Mil.	1
Hund. of Th. of Mil.	1
Billions.	1,
Tens of Billions.	1
Hundreds of Billions.	1
Thous. of Billions.	1
Tens of Th. of Bills.	1
Hund. of Th. of Bills.	1
Trillions.	1,
&c.	

Units.	1
Tens.	1
Hundreds.	1,
Thousands.	1
Tens of Thousands.	1
Hunds. of Thousands.	1,
Millions.	1
Tens of Millions.	1
Hunds. of Millions.	1,
Billions.	1
Tens of Billions.	1
Hundreds of Billions.	1,
Trillions.	1
Tens of Trillions.	1
Hundred of Trillions.	1,
Quadrillions.	1
&c.	

We may discover by this that the systems are similar as far as hundreds of millions, which is the extent of the common table, but here the English go on to "Hundreds of Thousands of Millions," which is a million of millions, before they commence billions, while the French and Italians commence this denomination immediately after "Hundreds of Millions." The word billion is probably derived from *bis million*, second million, as trillion, quadrillion, &c., signify third million, fourth million, &c.

If it were usual to read numbers from right to left, or in more general terms, to read numbers by proceeding from the lower to the higher denominations, the result would be the same as under the present system, and we would never be compelled to stop to enumerate; but as it is not always easy for the eye to catch the whole number of figures, it becomes necessary to fix the denomination of the highest number by beginning at the units place and naming the places towards the left, preparatory to reading the amount from left to right.

To annex a cipher to the right of a number has the same effect as to multiply by 10, for every figure is thus removed one place further from the unit's place, and to annex any significant figure multiplies the number by ten, and adds the significant figure to the product. To 375 annex 0 and it becomes ten times as great, viz., 3750; and if we annex 4 the number is multiplied by 10 and 4 added, making 3754.

What would be the effect of annexing 25 to 486? *Ans.* It would be equivalent to multiplying by 100 and adding 25 to the product.

How would 835 be affected by attaching 236 to its right? *Ans.* Just as it would be by multiplying by 1000 and adding 236 to the product.

How would 836 be affected by prefixing 236 to its left? *Ans.* Just as it would be by adding 236,000 to the number.

The process of carrying in addition and the other elementary rules, is an incidental effect of this system of *value according to place.*

It is not only necessary to be able to express numbers greater than a unit, or large assemblages of units, but to express numbers representing less than a unit; or collections of such less numbers. These are called Compound Numbers, Vulgar Fractions, Decimal Fractions, &c.; though they are all properly called Fractions or broken numbers, in contradistinction to Integers or whole numbers.

Compound numbers, consist of denominations differing in value, the whole taken together expressing but one quantity; as the distance from Zanesville to Columbus may be 52 miles 4 furlongs 20 rods. Here the fractional part or distance over 52 miles is expressed in furlongs or eighths of a mile as far as they will go, and the surplus that will not make a furlong, is expressed in rods; the furlongs and rods expressing the fractional part of a mile over 52. This may be as accurately expressed in the shape of a vulgar fraction, being $\frac{9}{16}$ of a mile, or the whole distance being $52\frac{9}{16}$ miles; or it may be expressed as a decimal fraction, being .5625 of a mile, and the whole distance 52.5625 miles; but for business purposes, it is found better to express fractional parts of numbers by divisions and subdivisions, bearing distinct names, rather than to adopt the ordinary fractional form. It, for instance, suits the workman best to have his rule divided into feet, inches, &c., or feet and tenths, and so to take and express numbers; and the same may be said of the lbs. oz. drams of the grocer; the lbs. oz. dwt. grs. of the jeweler; the lbs. oz. dr. scr. of the apothecary; or the division of time into years, months, days, &c. In all these and every other case, the expression may be changed to a different fractional form if necessary in calculation.

The common mode of expressing a fractional quantity is by

two numbers, one placed above the other, with a line drawn between them, as $\frac{3}{4}$, $\frac{7}{8}$. The lower number is called the *Denominator*, because it denominates or gives name to the fraction, for if it be 4, the fractions are fourths; if 5, fifths; 6, sixths, and so on. The upper number is called the *Numerator*, because it numbers or numerates the parts expressed in the fraction.

This subject will be made plainer by considering the nature of a fraction. It means something broken, and derives its name from the Latin *frango*, to break; in Arithmetic it means a broken integer. If any thing be *cut* into parts, they are properly called *sections*, from the Latin *seco*, to cut. If we consider the unit, whether it be an apple, or a line, or any thing else capable of mechanical division, as being broken, (or cut if you choose,) into four, five, six, &c., parts, then they are 4ths, 5ths, or 6ths. Suppose we divide an apple into 8 parts and give away 5 of them, then the quantity given away will be represented fractionally as $\frac{5}{8}$: the denominator 8 giving name to the fraction (8ths) and telling into how many parts the unit is divided; and the numerator 5, numbering the parts given.

In Decimal Fractions, (so called from the Latin *Decem*, Ten,) the denominators are always tens, hundreds, thousands, &c., constantly increasing in a tenfold ratio. In Duodecimal Fractions, (so called from *Duodecim*, 12,) they are always twelves. The subject of Fractional numbers, will demand, for the purpose of detailed investigation, a distinct lecture, and needs not therefore be enlarged upon at present.

Having traced as much in detail as we think proper at present, the early efforts of mankind to perfect a system of numbering, and shown the advantages in general terms of the Arabic mode of notation; and having shown how numbers less than a unit are to be expressed, we shall in our next lecture, discuss some of *The Properties of Numbers*.

LECTURE III.

ON THE PROPERTIES OF NUMBERS.

In order to present this subject distinctly, it will be necessary first to define the terms which we shall be compelled to use; especially such as are not of very frequent occurrence.

Number has been defined by some to be "A collection of units," and so far as the definition goes, it is well enough, for every collection of units is a number; but it does not follow that every number is a collection of units. *One*, which is unity itself, has all the properties of number, but is certainly not "a collection of units." Basing their arguments on this distinction, some ancient writers contended zealously that the difference between 2 and 3 is infinitely greater than that between 1 and 2; indeed they contended that 1 could bear no ratio to 2, 3, 4, &c., since they are numbers, but one is not a number. Amended so as to include unity, the definition will answer our purpose.

Quantity is a more general term than *number*, and is that property of matter which regards size or extent; it is that property which is capable of increase or decrease. If considered as one undivided whole, it is called *magnitude;* but if made up of individuals it is called *multitude;* and it is this property of matter, either applied or in the abstract, which it is the province of Arithmetic to discuss.

Unity or a *unit*, is one of the individuals or single things of which multitude is composed, and is called *one.* Whatever measure of space is used in determining magnitude, it is called the measuring unit; and all parts of such unit are fractions. The cubic inch determines the capacity of vessels and the solidity of small bodies—the cubic foot of cord-wood, &c.—the cubic yard, of excavated earth; while the cubic mile is used to express the bulk of the globe we inhabit. A Dollar is the money unit of our Federal currency, as a Pound is of sterling or English money. All inferior denominations below the unit of the system, are fractions of the unit; and each may be considered a fraction of those above it.

An *Integer* is a unit, or any quantity considered as a whole; so called from the Latin word *integer*, signifying *whole, unbroken.*

A *Fraction* is a broken number, being part or parts of a unit.

All numbers are of course either *Integers* or *Fractions.*

A *Digit*, (from *digitus*, Latin, a finger,) means any of the single figures or characters used to express numbers, as 1, 2, 3, 4, &c. That our numerical system is based upon the fingers of the human hand is generally admitted. "The Carribbees," says PEACOCK, "call the number 10 by a name signifying all the children of the hand."

An *Abstract* number refers to no particular object, as 1, 8, 3.

An *Applicate* or *Concrete* number refers to some particular thing; as 1 man, 8 horses, 3 houses.

An *Even* number is one that is divisible by 2, without a remainder, as 4, 6, 8.

An *Odd* number is one that is not divisible by 2, without a remainder; as 3, 5, 7.

When a number can be divided by another without a remainder, it is said to be measured by it.

All numbers are either even or odd.

A *Prime* number is one that is measurable by no other number than itself, or unity, as 3, 5, 7. One number is prime to another when they have no prime factor in common, though they may or may not be absolutely prime numbers: as 8 and 9 are prime to each other, though neither is prime of itself.

A *Composite* number is one that is measurable by some other number; as 4, 6, 9, &c. It is *composed* of other numbers multiplied together.

All numbers are either Prime or Composite. The number of Prime numbers is unlimited, but by no means so numerous as the Composite; for every even number is measurable by 2, and every number ending with 0 or 5 is measurable by 5; while many others are divisible, having other terminations. In the numbers from 1 to 20, we find 9 primes, viz., 1, 2, 3, 5, 7, 11, 13, 17, 19; but in the next ten we find only 23 and 29; in the next 31 and 37; and in the next 41, 43 and 47. They occur irregularly and without any certain law, so far as can be discovered.

ERATOSTHENES, an ancient writer, used a mode of finding what numbers were prime, which from its form was called "The Sieve of ERATOSTHENES." He wrote all the odd numbers, (for except 2, the even are all composite,) as far as he wished the table to extend. He then began and dividing by 3, 5, 7, &c., he cut out such as divided without a remainder;

and when he had divided by all the prime numbers, up to the square root of the highest number in his table, and had cut out the numbers that left no remainder, all that remained in his table were of necessity prime. His table thus filled with holes, bore no faint resemblance to a sieve for domestic purposes. Though a better mode than this might be devised, there is no general and simple rule for the purpose; and it is not important there should be.

A *Perfect* Number is one that is equal to the sum of all its parts; as $6=\frac{6}{2}+\frac{6}{3}+\frac{6}{6}$: i. e. the number being divided by every integer above unity that will divide it without a remainder, the sum of the quotients will be the number itself.

In the efforts of the Pythagoreans to discover occult or hidden properties in numbers, this class was investigated, but is of no practical use whatever. The only perfect numbers known are 6, 28, 496, 8128, 33550336, and five very large numbers. A Perfect Number is necessarily composite.

Amicable Numbers are such that the sum of each is equal to the sum of all the Divisors of the others. 220 and 284 are the smallest pair of amicable numbers. 220 is divisible by 1, 2, 4, 5, 10, 11, 20, 22, 44, 55, and 110, which when added together make 284. And 284 is divisible by 1, 2, 4, 71, and 142, which added make 220.

A *Square Number* is the product of a number multiplied by itself. The number thus multiplied is called the square root of the square. 16 is a square number, being produced by squaring 4; hence 16 is the square of 4, and 4 is the square root of 16.

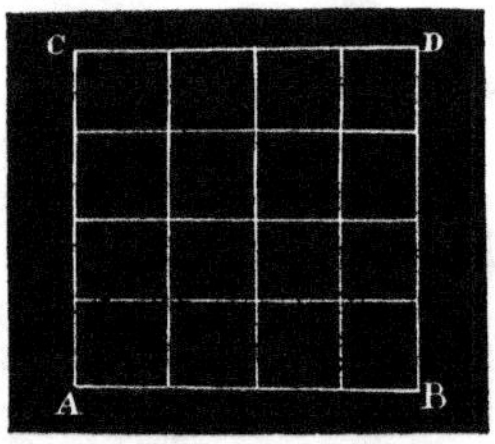

A square number may be represented by a geometrical square, one side of which will then represent the square root. The line AB, being divided into 4 parts, let it represent the number 4, then will the square ABCD, erected upon such line represent 16.

A *Cubic Number* is the product of a number multiplied by itself, and that product again by the same multiplier; thus 64 is the cube of 4, for 4 times 4=16, and 4 times 16=64. The number thus multiplied is called the cube root.

A cubic number may be represented by a square block, the length of one side being the cube root, and the area of one side will be the square of the root, which in the above illustration is 16. Numbers may be involved or raised to any power, the number raised being called the root of the resulting power. A number multiplied by itself produces the 2d power, or Square, again multiplied it produces the 3d power or Cube; another multiplication produces the 4th power or Biquadrate. This process of multiplication is called *Involution;* but if the power is given to find the root, the process is called Evolution. A *Rational* number is one whose desired root can be ascertained accurately; a *Surd* is a number, the root of which cannot be accurately ascertained; 4 is a rational number, when we seek to extract its square root, while 27 would be a surd; though rational when its cube is sought.

A *Measure* of a number, is any number that will divide it without a remainder. A common measure of two or more numbers, is any number that will divide all of them without a remainder: 3 is a common measure of 6, 9, 12 and 15, for it will divide all of them without a remainder. We might *measure* four boards respectively 6, 9, 12 and 15 feet long, by means of a measure 3 feet long; but we could not measure 16 or 17 feet with such a rule, since there would be a remainder; the length of which could not be determined with a rule 3 feet in length.

A *Multiple* of a given number is any number which the given number will measure: 8 is a multiple of 2, but not of 3. Measures are sometimes called sub-multiples.

A *Common Multiple* of two or more given numbers, is any number which may be measured by all the given numbers: 12 is a common multiple of 3 and 4.

An *Aliquot* part of a number is any measure of it; in other words any part that divides the whole without a remainder: 3 is an aliquot part of 12.

An *Aliquant* part is any part that is not a measure of the whole; 3 is an aliquot part of 9, but an aliquant part of 10; while 2 is an aliquot part of 10 but not of 9.

Ratio is the relation between numbers or quantities, expressed by the quotient of one divided by the other; as 3 is the ratio of 2 to 6.

Proportion is an equality of ratios; for when the first of three numbers bears the same ratio to the second, that the second does to the third, the three numbers are said to be proportional; as 3 : 6 : : 6 : 12. So when four numbers are such that the first is to the second, as the third to the fourth, they are called proportionals, as 3 : 6 : : 4 : 8, i. e. 3 is to 6 as 4 is to 8.

The several kinds of *Proportion* will be explained in a future lecture.

Perhaps the present will be the most appropriate time to explain the common signs used in calculation.

The sign = expresses equality, and two or more quantities having this sign between them are considered equal to each other; as $5=500 cents. It is read *equal* or *equal to*.

+ signifies addition, and is called *plus*, a Latin word signifying more. When placed between numbers they are to be added together; as 5+4=9, and the expression is read 5 plus 4 equal 9. The same mark + is used to denote a remainder, implying that there is *more*.

— signifies subtraction, and is called *minus*, the Latin for *less*. When placed between two numbers it signifies that the latter is to be taken from the former; 5—4=1.

× signifies multiplication, and is read "into" or "multiplied by," or "times," as 5×4=20.

÷ signifies division, and is read "by" or "divided by," as 8÷4=2. Division is also expressed by placing the divisor under the dividend in the form of a vulgar fraction, $\frac{8}{4}$=2.

$\sqrt{}$ signifies the square root of any quantity, $\sqrt[3]{}$ the cube root, $\sqrt[4]{}$ the fourth root, &c. $\sqrt{16}=4$.

^{2} placed after a quantity expresses the square of such number, ^{3} the cube, ^{4} the fourth power, &c. $4^2=16$, $4^3=64$, $4^4=256$.

: : : : expresses a proportion 2 : 4 : : 4 : 8.

Other terms will be explained as they occur hereafter.

We come now to treat of the properties and laws of numbers, not to search out hidden and magic properties as attempted in the infancy of the science, when its devotees imagined that even the universe was constructed in reference to the abstract properties of numbers; but to treat of such natural and accidental properties as may tend to explain the various modes of solving problems, and thus enable the attentive student to proceed understandingly.

Some properties of numbers are natural, being inherent in the nature of the subject, and existing without reference to any particular system of Notation; others are accidental, and would not exist under a different Notation. The peculiar properties of the number 9 are accidental; the division of numbers into *even* and *odd* is natural.

In order to present the most important of these properties distinctly, we shall state them in the form of separate *Propositions*, and add such explanations and comments as their importance may demand. We shall at the same time point out their practical application in the solution of problems, and show how tne common rules of Arithmetic are built upon them. This may involve considerable detail, especially as we are debarred to a great extent the aid of Algebra and Geometry, but the most important propositions can be elucidated, perhaps sufficiently for our purpose, without resorting to either. Propositions involving an amount of demonstration disproportionate to their importance, we shall not hesitate to leave for the student's future investigation, when he shall be better prepared for the task, and have leisure to look into more voluminous treatises on mathematical science. We will commence with the cardinal feature in our system of Notation.

Proposition 1.

Digits, in our system of Notation, increase in value from right to left in a tenfold ratio.

This doctrine of local value is the grand distinguishing feature of our system of writing numbers, as compared with those of the ancients. To this our system owes its efficiency and great superiority over every other, and especially in the simplicity of the algorithm or mode of calculation. It has been already remarked that this principle of value according to place gives rise to carrying in the elementary rules; and we may now show that this is true. Suppose we seek to add the annexed sums; we begin by adding up the units and they amount to 24, which is 2 tens and 4 units, the 4 units we set down and carry the 2 tens to the column of tens, by which that column comes out 22. Then as 22 tens are equal to 2 hundreds and 2 tens, the 2 tens are set down under the column of tens, and the 2 hundreds carried to the column of hundreds, by which that column is made to amount to 34 hundreds: the 4 hundreds we set down, and the 3, which are thousands, we carry to the column of thousands, which coming out 15 we set down in full; there being no column of a higher denomination in the question.

Add	3856
	3856
	3856
	3856
Sum	15424

By giving the same problem a different form we may illustrate the doctrine of Multiplication, which depends on the same principle, and produces the same result, though more briefly. It is true that in one case we call the result the *sum*, and in the other the *product;* but it is the same result, and will be readily understood by the illustrations which were given in Addition.

Multiply	3856
By	4
Product	15424

When the multiplier consists of several figures, the operation may require a remark; but it will only further exemplify the principle. In multiplying by 2, which are 2 tens we set the resulting figure under the tens, and as tens by tens will produce hundreds, the product of 5 by 2 must go into the hundreds' place; and so of the rest. We might further illustrate this operation by multiplying 3856 by 4 and then by 20, and adding the products together; which is in reality just what we have done; though if done in full, the units' place of the product of the 2 tens, or 20, would be filled with a cipher. If the multiplier were made to consist of any additional number of figures, the explanation would be made on precisely the same principle.

Multiply	3856
By	24
	15424
	7712
Product	92544

Let us now take one number from another and see whether the same principle applies. It is evident that the lower number or subtrahend is less than the upper or minuend, and may be deducted from it; but as the denominations stand, the lower cannot severally be taken from those immediately above them. We cannot take 8 from 1, 7 from 4, &c. But we can take 1 ten from the minuend, and calling it 10 units we add the 1 unit, making 11 units, from which we can take 8 units, and setting down the difference, we proceed to take 5 tens from 8 tens (the other ten having been made into units) and 3 remains. We find the same difficulty again in taking 7 hundreds from 4 hundreds, and we get over it in the same way, by taking one of the 8 thousands and converting it into hundreds, making 7 thousands and 14 hundreds.

From	38491
Take	13758
Leaves	24733

Instead, however, of considering the figure from which we borrow as 1 less, it is usual, and equally convenient, to consider the figure underneath it as increased 1; or, as we usually say, we carry 1: the operation upon the minuend being called borrowing 10. We might take away, first, the one we borrowed, and afterwards deduct the subtrahend figure; but it is as well to add them together, and take them away at once; which is just what we do in carrying. The same operation may be explained, perhaps more readily, on the principle of adding equal quantities to both minuend and subtrahend, as it is obvious we do by adding 10 units to the minuend and 1 ten to the subtrahend; or 10 tens to the minuend and 1 hundred to the subtrahend; or any other denominations. This, however, results from the same law.

In the process of Division we commence at the highest denomination, and setting down the number of times the divisor is contained in the first figure of the dividend, we reduce the remainder to the next lower denomination, and adding the number of that denomination to the result, we again divide; and so proceed to the units' place. Thus we find that 2 is contained 1 time and 1 over in 3, and the 1 over being equal to 10 of the next lower denomination, we add the 9 and have 19 which we divide, and again we have 1 over, which being equal to 10 of the next lower we add the 7 and form 17; and so we proceed to the units' place. Instead however of calling the 1 over 10 and adding it to the next figure, we may merely imagine it set to the left of the next lower, which in whole numbers amounts to the same thing. Or we may say that 2 goes into 3 tens of thousands (for 3 there represents that number) 1 ten thousand times and 1 ten thousand remains, which added to 9 thousands makes 19 thousands; this divided by 2 gives 9 thousands, and 1 thousand remains. One thousand brought to hundreds makes 10 hundreds, and 7 being added makes 17 hundreds, which we divide, and so proceed to the units' place. Both these modes of explanation depend on the principle of value according to place; and in no operation are the advantages of the system more apparent than in Division.

```
2)39756
 ------
  19878
 ------
```

In Division, as in Addition and Multiplication, we carry, and on the same principle; but as in Division we proceed from left to right, we carry 10 for 1, instead of 1 for 10, as we do when proceeding from right to left.

Long Division, as it is called, is on the same principle as Short Division, but varies in form for the convenience of finding the several remainders. It needs, therefore, no separate illustration. The process of carrying in the compound rules, fractions, &c., depends on the same principle, and may be explained in the same manner.

If we attempt to compare the simple elementary processes under our own system with the same operations wrought out by the Roman notation, as proposed in a preceding Lecture, we cannot fail to discover the great superiority of our own modes, even after making generous allowance for our want of familiarity with that system.

Proposition 2.

In any series of digits expressing a number, the value of any digit is greater than the value of all the digits on its right.

This property results also from value according to place; and that the proposition is true is obvious, for if we take the smallest digit (1) and place it on the left of the largest (9) we form 19; the 1 expresses 10 units, while the 9 expresses but 9 units; and let us add what number of nines we may, the unit will constantly retain its greater value: *e. g.* 19, 199, 1999, &c. Not only is the left hand digit higher in value than all upon its right, but the same remark applies to each digit, in reference to those on its right.

Proposition 3.

If the sum of the digits in any number be a multiple of 9, the whole number is a multiple of 9.

This is one of several peculiar properties of the number 9, all arising from its being just one less than the radix of our system of notation, and hence the highest number expressed by a single character; and these properties will belong to the highest number so expressed in any system. We might go a step further in reference to this property, and say that it belongs to any number that will divide the radix of the system, and leave one as a remainder.

If we carefully examine the genesis of numbers, we must see that, so far as the number 9 is concerned, this is an accidental property, resulting from our scale of notation. We constantly express each successive number from unity to 9, inclusive, by a digit of greater value than any preceding it; but when we pass 9 we express the next number, 10, by a unit and a cipher. The number is one greater than 9, and the sum of its digits is 1. Eleven is 2 greater, and the sum of its digits is $1+1=2$. Thus we proceed, the sum of the digits constantly expressing the excess over 9, until we reach 18, or twice 9. Nineteen is 1 and 9, and it is one over twice 9. 20 is 2 over twice 9, and the sum of its digits is 2. The same course continued to millions, would but produce the same recurring result. Nine is 1 less than 10; twice 9 are 2 less than 20; 3×9 are 3 less than 30; and so on; and hence the 1 of 10, 2 of 20, 3 of 30, &c., come just in the proper place to keep up the excess above 9 and its multiples. If the multiples of 9 did not constantly fall at each product, one further behind the corresponding multiples of 10, the 2 of 20,

3 of 30, &c., would not fall in the right place to keep up the regular order of the series.

Let us try 8. We stumble at the threshold, for we cannot get over 9; and if we could, the 1 of 10 does not express the excess of that number above 8. The sum of the digits will be 1 less than the excess, and the sum of the digits, 1+6, expressing 16 or twice 8, will not make 8, as the digits expressing 18, or twice 9, make 9. At each multiplication the sum of the digits will be 1 less, 1+6=7; 2+4=6; 3+2=5; 4+0=4; 4+8=4 over 8; 5+6=3; 6+4=2; 7+2=1; 8+0=0; &c., &c. So we might trace other numbers, and find some pervading law in each; but in none would we find the law that the sum of the digits of the several products, equals the figure multiplied, or is uniformly a multiple of it, except in 9 and 3.

Let us try 9 and 3. Add together the digits of the several multiples of 9, viz: 18, 27, 36, 45, 54, 63, 72, 81, 90, 99, 108, &c., indefinitely, and you find they make 9; or multiples of 9. So the sums of the digits arising from multiplying 3, which is a measure of 9, are multiples of 3, *e. g.* 6, 9, 12, 15, 18, 21, &c.

Proposition 4.

If the sum of the digits in any number be a multiple of 3, *the number is a multiple of* 3.

The same reasoning applied to the number 9 to show the correctness of the preceding proposition, will show the correctness of this. Ten, the sum of whose digits is 1, is 1 over 3 times 3; 11, the sum of whose digits is 2, is 2 more than 3 times 3; 12, the sum of whose digits is 3, is a multiple, &c., &c.

Proposition 5.

Dividing any number by 9 *or* 3, *will leave the same remainder as dividing the sum of its digits by* 9 *or* 3.

This proposition follows as a matter of course from the two next preceding it; and we shall adduce no other proof of its correctness. Like the former, it is an accidental property of the highest number expressed by a single digit in any system, and of all its factors. If 9 were the basis of our system, these properties would belong to 8, 4 and 2; if 8, then 7 only, since 7 has no factors; and if 7 were the basis, then 6, 3 and 2 would possess these properties; and if 12 were the basis, then 11 only would possess such properties; for it would in that case be expressed by a single digit, and would be the highest number so expressed. Twelve would be written with a unit and a cipher as 10 now is; and 11 being prime, it would be the only number that would divide the radix of the system, and leave 1 as a remainder.

As early as 1657, Dr. Wallis, of England applied this principle to prove the correctness of operations in the elementary rules of Arithmetic; and the practice has been continued to the present time. The operation is performed thus:

We cast the nines out of each number separately, and set the excess on the right. We then cast the nines out of the sum total 305160, and also out of the sum of the excesses 8+1+8+7, and they are equal: both being 6, and we hence infer that the work is right. To cast out the nines, the number may be divided by 9; but a better way is to add the digits together in each number, rejecting 9 whenever it occurs, and carrying forward only the excess. Thus 7 and 8 are 15; 9 being rejected, we carry 6 to 6 is 12; rejecting 9, we carry 3 to 4=7; the number carried in each place is the excess over 9; and where 9 occurs it is passed over.

Add	79864=7
	32075=8
	83214=0
	61840=1
	48167=8
	305160=6

In Subtraction cast out the nines from the minuend and subtrahend, and also from the remainder. If the excess in the remainder is equal to the difference of excesses in the minuend and subtrahend, the work is right.

Here as we cannot take 8 from 6, we take from 9 and add 6; the result, 7 agrees with the excess above 9 in the difference of the numbers.

From	6894321=6
Take	2960864=8
Leaves	3933457=7

In Multiplication, find the excesses in the factors, and if the excess in the product of these two excesses equals the excess in the product of the factors the operation is correct.

Multiply	48756=3
By	245=2
	243780
	195024
	97512
	11945220=6

3	2
6	6

This is often called proving by the cross; and instead of placing the excesses after marks of equality, they are placed in the angles of a cross as on the right hand of the above operation.

In Division cast the nines out of the divisor, dividend, quotient, and remainder; then to the product of the excesses in the divisor and quotient, add the excess in the remainder, and cast the nines out of the sum, and if the excess equal that in the dividend the work is right.

```
                      Excess in Divisor               0
                      Excess in Quotient              8
    27)465                                            —
       ——             Product of Excesses             0
        17+6          Add Excess of Remainder         6
        —                                             —
                      Excess in Dividend            6=6
                                                    ———
```

Hence the work is right, the excesses being equal.

It is proper to remark that this mode of proof is liable to much objection. If the figures become transposed, or if mistakes are made that balance each other, the work will prove right when it is wrong. The work will, however, never prove wrong when it is right. In the product above, you may transpose the digits as you please, the work will prove, since the excess is the same whatever is the order of the digits; and ciphers may always be omitted. Or if mistakes balance each other, as if instead of 91145 it be 83216. The excess here will be the same and the work will prove, though not a figure is right.

Proposition 6.

If from any number the sum of its digits be subtracted, the remainder is a multiple of 9.

```
From  31416                          For,
Take     15=3+1+4+1+6            31416÷9=3490+6
      ————                           15÷9=   1+6
    9)31401                      ———————  ————
      ————                       31401÷9=3489
      3489 times 9, diff.        ————————————
      ————
```

As the remainder on dividing the given number by 9 will be just the same as on dividing the sum of its digits, (prop. 5,) it is obvious that the difference must be an even multiple. The given number is 6 more than 3490 times 9; the sum of digits is 6 more than 1 time, hence their difference is 3489 times.

Or we may illustrate the proposition in this way. Separate the number into its constituent parts:

$$31416=\begin{cases}30000\\1000\\400\\10\\6\end{cases}$$

If we divide these several numbers by 9, the remainders will be the same as the significant figures which they contain, viz, 3, 1, 4, 1, 6, as is obvious for the reasons already given.

Now it is plain that if these numbers were severally diminished by their remainders, they would be multiples of the divisor leaving such remainder, viz; 9. Thus:

30000÷9=	3333	and 3	over.
1000÷9=	111	" 1	"
400÷9=	44	" 4	"
10÷9=	1	" 1	"
6÷9=	0	" 6	"

30000— 3=29997	These are severally multiples of 9.
1000— 1= 999	
400— 4= 396	
10— 1= 9	
6— 6= 0	

31416—15=31401=3489 times 9 as before.

It needs no proof to show that if the remainders taken separately from the parts leave multiples, the sum of the remainders taken from the sum of the parts will leave a multiple.

Proposition 7.

The difference between a given number, and the digits composing such number reversed or any how arranged, is always a multiple of 9.

The difference for instance between 7425, and any arrangement you can make of the same figures is a multiple of 9.

From	7425	7425	7425
Take	5247	5724	2457
	9)2178	9)1701	9)4968
	242	189	552

This is based on the same reason as the preceding; for whether you take the sum of the digits or transpose the digits, it is the same in effect. The excess over an even multiple being the same as in the given number, the difference must necessarily be an even multiple.

A practical application is sometimes made of this principle by a person setting down two rows of figures for subtraction, but being careful to have the figures of the subtrahend and minuend the same, though differently arranged. One figure of the remainder is then stricken out, and the puzzle is to restore it without seeing the minuend and subtrahend. It is done by taking such number as will make the remainder a multiple of 9.

Here if 5, 4, or 1 be erased, any one may restore it; but if the 9 or 0 be removed he cannot know whether a 9 or a cipher should be supplied, as either will make the number a multiple of 9.

From	7354681
Take	1864537
	5490144

The mode we have adopted in explaining the four last preceding propositions, appears to us plain and sufficiently satisfactory.

In addition to the use of these properties as modes of proof, they are the key to many numeral puzzles and amusing questions; and hence the care we have bestowed in explaining the principle. What has been said may be a sufficient explanation of the following article on the "*Wonderful Properties of the Number Nine*," which we copy from a periodical of the day.

"Multiply 9 by itself or any other digit, and the figures of the product added will be 9.

Take the sum of our numerals 1+2+3+4+5+6+7+8+9=45, the digits of which 4+5=9. Multiply each of these digits by 9, and their sum will be 405; which added 4+0+5=9; and 405÷9=45, also a multiple of 9.

Multiply any number, large or small, by 9 or 9 times any digit; and the sum of the digits of the product will be a multiple of 9.

Multiply the 9 digits in their order, 1 2 3 4 5 6 7 8 9 by 9, or any multiple of 9 not exceeding 9 times 9, and the product, except the tens' place, will be all the same figures; while the tens' place will be filled with 0. The significant figure will always be the number of times 9 is contained in the multiplier.

27, or 3 times 9, will produce all 3s; 4 times 9 all 4s.

Omit 8 in the multiplicand and the product will be all the same digits, the 0 having disappeared."

```
 123456789
        18=9×2
 ---------
 987654312
123456789
----------
2222222202
----------
```

To a superficial observer the above results may seem accidental, but investigation will show that they all flow from the laws and principles we have laid down; and that a much longer list might be made of apparently simple and detached facts, but really of results flowing from well established laws *There are no unaccountable properties in numbers;* and in the common acceptation of the word accidental, as implying a happening unexpectedly or by chance, opposed to that which is constant, there are no accidental properties. The word accidental, as applied by mathematicians to numbers, means rather as in Logic, not essential, but rather incidental to something else. In this sense the properties of matter, as hardness, softness, color, &c., are called accidents.

While the number 9 has some peculiar properties from being the next below the radix of the system, the number 11 has some peculiarities from being next above the radix. Among these are the following: "If from any number the sum of the digits standing in the *odd* places be subtracted, and to the remainder the sum of the digits standing in the *even* places be added; then the result is a multiple of 11." Again, "If the sum of the digits standing in the *even* places be equal to the sum of the digits standing in the *odd* places, or differ by 11 or any of its multiples, the number is a multiple of 11."

As these however are of no practical utility, we shall not discuss them.

The number 7 has also some peculiarities, but we shall name only one, as they are useless. *If a number be divided into periods of three figures each, beginning at the units' place, when the difference of the sums of the alternate periods is a multiple of 7, the whole number is a multiple of 7.*

Here 862—428 is a multiple of 7, and so is 907—382, therefore the whole number is a multiple of 7.

```
7)382,907,428,862
  ---------------
   54,701,061,266
```

The division of numbers into *Even* and *Odd*, seems to arise from considering them in pairs. The following facts growing out of this division will be readily understood.

The sum of two even numbers is even, and so is their difference: 8+4=12, 8—4=4.

The sum of an odd number of odd numbers is odd; but the sum of an even number of odd numbers is even; 3+5+7=15; 3+5=8 and 5+7=12.

An even and an odd number being added together, or one subtracted from the other, the result will be odd: 8+3=11, 8—3=5.

If a number has 0, 2, 4, 6 or 8 in the units' place, it is divisible by 2, and is consequently even.

No odd number can be divided by an even number without a remainder.

If an odd number measure an even one, it will also measure the half of it. 7 measures 42, and therefore measures 21, the half of it.

LECTURE IV.

ON PRIME AND COMPOSITE NUMBERS. MEASURES, MULTIPLES, ETC.

One of the most important classifications of numbers is into *Prime* and *Composite.* The two classes include all numbers, and the consideration of their peculiarities enables us to understand distinctly some important practical applications of the science. It is obvious that a composite number may be represented by its prime factors, so connected as to imply multiplication; but a prime number having no factors, cannot be so expressed. $6=2\times3$, but 7 cannot be so expressed, unless we say $7=1\times7$, which does not divide the number into factors. $12=2\times2\times3$ or $2^2\times3$, $324=2^2\times3^4$, or $2\times2\times3\times3\times3\times3$.

There is no mode of determining by mere inspection, whether a number is certainly prime, though we may at a glance determine that many are not prime, and by other calculations the field of search, within which such numbers are found, may be reduced to very narrow limits. No *even* number, above two, can possibly be prime, for they are all measured by 2. All numbers ending in 5 or 0 are measured by 5, and many other numbers of other terminations are composite; as 21, 27, 33, 39, &c. Prime numbers therefore must always end in 1, 3, 7 or 9; but it is not true that all numbers ending in 1, 3, 7 or 9 are prime. Every prime number above 3 is either one greater or one less than some multiple of six, for six being even, all its multiples will be even, and any number 2 greater or 2 less will also be even: if 3 greater or 3 less, it will be a multiple of 3: if 4 greater or 4 less, it will again be even; but if 5 greater or 5 less, it will be again within 1 of some greater or less multiple of 6, and *may be* prime. In the same way we may show that every prime number is 1 greater or 1 less than some multiple of 4; but it is not necessary to pursue this subject further, as no formula has ever been devised that will produce only prime numbers.

In order therefore to determine whether a number is prime, the most certain and expeditious mode, perhaps, would be to see if it be odd and does not end in 5, and if so to divide by 6, and if the quotient either falls short or exceeds by 1, an even multiple of 6, it *may be* prime. To determine the fact positively, divide by all the prime numbers less than the square root of the number, except 2 and 5, and, if none will measure evenly, the number is prime. It is evident that you need not divide by 2, for if 2 were one of a thousand prime factors, the product would not be odd, and of course not prime; and if it end in 5, it will be measurable by 5, and of course not prime; and as factors exist by pairs, one exceeding and the other falling short of the mean or square root, if there is not a factor *less* than the square root, there cannot be one *greater*.

To ascertain the prime factors of a Composite number, divide it by 2 if practicable, and repeat the same operation on the quotient, and so on until the final resulting quotient cannot be measured by 2; this will determine how often 2 enters as a prime factor into the number. Then treat the quotient last found in the same manner, only using 3 as a divisor; and so on by each succeeding prime number, until the resulting quotient is known to be prime, or your divisor equals the square root of the original number. Or we might say—divide the number successively by all the prime numbers less than its square root, continuing the division of the quotient each time by the divisor, so long as no remainder occurs; but the former mode is preferable. Required the prime factors of 84.

2	84
2	42
3	21
7	7
	1

Answer. 2, 2, 3 and 7; or better expressed, 2^2, 3 and 7.

It might be thought that a different result would be produced by dividing in different order, but this is not true; if the prime factors enter into the composition of the number, they will come out unchanged; and we commence with 2 as being the least prime number, and ascend in the scale of primes, only for the sake of system. Our first division shows that $84=2\times42$; the second that $42=2\times21$; the third that $21=3\times7$; and the fourth that 7 is prime. But though 21 is not measured by 2, may not a future resulting quotient be measured by 2; especially where there are many divisions? Certainly not; for we first determine how often 2 entered into 84 as a constituent prime factor, and it cannot afterwards be developed in a factor. We must distinguish between a *prime* and a *composite* factor of a number. 2, 3 and 7 are prime factors of 84; but 42, 28, 21, 12, 6 and 4 are composite factors.

If we know the prime factors of a number, we may readily by multiplying them together, determine all the possible composite factors; or, in other words all the divisors which the number admits of; for no number can be measured except by its prime factors; and their various products by each other.

Try by how many numbers 84 may be divided evenly. The prime factors, as determined above, are 2^2, 3^1 and 7^1; hence,

1	3	7	21
2	6	14	42
4	12	28	84

It is measured in the first place by 1, 2 and 2^2, or 2 times 2=4. It is then measured by 3 times each of these,=3, 6 and 12. And, again, it is measured by 7 times the first column,= 7, 14 and 28; and 7 times the second,=21, 42 and 84; and it can be measured evenly by no other integral number.

This operation is in accordance with the following rule, as will be obvious on close inspection. "Increase the exponent of each factor 1, and multiply the several sums together, the product will express the whole number of divisors." The exponents are 2, 1, which increased a unit will be 3, 2, 2; and $3\times2\times2=12$, the whole number of divisors that 84 can have, that will leave no remainder. Hence, if we had a walk to measure, 84 rods long, we might have 12 different measures to do it with, viz. 1, 2, 4, 3, 6, 12, 7, 14, 28, 21, 42 or 84 rods in length.

How many divisors or aliquot parts has the number 4725?

3	4725 ...
3	1575 ...
3	525
5	175
5	35 ...
7	7 ...
	1

The prime factors are therefore 3^3, 5^2, 7^1, and increasing these exponents each a unit, and multiplying the sums together we have $4\times3\times2=24$, the number of divisors of which 4725 is susceptible.

The converse of this operation, or the finding a numbei that shall have a given number of divisors, will be spoken of before we close.

The most frequent application of the doctrine of prime and composite numbers is in the calculation of fractional quantities, and in the solution of questions involving common measures, common multiples, &c., of quantities.

The following are such of the properties of prime numbers as may aid us in future investigations.

Proposition 8.

If two numbers be prime to each other, then will their sum and difference be prime to each of them.

Five and sixteen are prime to each other; therefore 11 their difference, and 21, their sum, are each prime to 5 and 16.

Proposition 9.

If two numbers are prime to each other, then will their sum and difference be prime to each other, or only have 2 as a common measure.

Twenty-one and 11 are the sum and difference of 5 and 16, and they are prime to each other. 8 and 2 are the sum and difference of 3 and 5, and they are divisible only by 2.

The truth of Proposition 9 may be thus shown. Let a and b represent the numbers, then if $a+b$ and $a-b$ have any common measure, their sum has the same measure. But a and b being prime to each other, $2a$ and $2b$ can have no common measure except 2; and if either a or b is odd, their sum and difference will be odd; and not being measured by 2 will necessarily be prime.

The following propositions might all be proved in a similar manner, but this will serve as a specimen of this mode of demonstration.

Proposition 10.

The product of any set of prime numbers, is prime to the product of any other set of entirely different prime numbers.

The product of $3\times5\times7=105$, is prime to the product of $11\times13\times2=286$, or any other set of prime numbers, not including 3, 5, or 7. Hence if two numbers have no prime factors in common, their product is their least common multiple: $105\times286=30030$, which is the least common multiple of 105 and 286. The reason is obvious.

Proposition 11.

If there be two numbers prime to each other, the product of neither by a third integral number, can be divisible by the other.

You cannot, for instance, multiply 11 by any whole number, except 5, that will make a product divisible by 5; nor 4 by any whole number, except 11, that will make a product divisible by 11.

Proposition 12.

The product of two different prime numbers cannot be a square.

This is evident; for the two different prime factors are the only integral divisors of the number, and being unequal, neither of them can be the root.

The following are the leading properties of Composite numbers.

Proposition 13.

No number can have more than one set of prime factors, and their general product will be the given number.

The prime factors of 60 are 2, 2, 3, 5; and it can have no other set. It may have other factors as 6×10, 4×15, &c., but they are all resolvable into 2, 2, 3, 5.

Proposition 14.

All composite numbers that measure any given number must have the same prime factors as such number.

The prime factors of 60 are $2\times2\times3\times5$. It is divisible also by the following composite numbers, 4, 6, 10, 12, 15, 20, and 30, but these are all resolvable into the prime factors we have named.

Proposition 15.

If a number be a common multiple of all the prime factors of two or more numbers, it is a common multiple of such numbers also.

Sixty is a common multiple of 2, 3, and 5, into which the following composite numbers, 4, 6, 10, 12, 15, 20, and 30, may be resolved; and therefore it is a common multiple of such composite numbers also.

Proposition 16.

If a number be not measurable by all the factors of a proposed number, neither is it measurable by such proposed number.

If 60 were not measurable by all the factors of 30, neither would it be by 30.

Proposition 17.

If a number measure another, it will measure every product of such other.

Five measures 60, and will therefore measure any and every multiple of 60; for if 60 is 12 fives, twice 60 will be twice 12 fives, or 24 fives, &c.

Proposition 18.

The common measures of two or more numbers, are either the prime factors, which they have in common, or some products of them, one by another; and the greatest common measure is the product of all the prime factors they have in common.

Take 60 and 84. The prime factors of 60 we have found to be 2, 2, 3, 5; and those of 84 we will seek thus:

$$84 \div 2 = 42$$
$$42 \div 2 = 21$$
$$21 \div 3 = 7$$
$$7 \div 7 = 1$$

By dividing 84 and the quotients found, by the prime numbers, 2, 3, &c., successively, we find it resolved into $2 \times 2 \times 3 \times 7$. Any prime factor common to both, will necessarily be a common measure of both; and so will any products that can be formed in both: 2, 2, 3, are common to both; 4, 6, 12 may be formed in either, and will also be common measures. Fifteen may be formed in the primes of 60, but not in those of 84; 21 in those of 84, but not in 60; neither 15 nor 21 is therefore a common measure. This might be neatly explained mechanically.

Proposition 19.

If a number measures two other numbers, it will also measure their sum, and their difference.

Four measures 12 and 20, it will therefore measure 32, their sum; and 8, their difference. The difference between 3 times 4, and 5 times 4, is 2 times 4; and if to 3 times 4, we add 5 times 4, the result will be 8 times 4.

Proposition 20.

If a common measure of two or more numbers be not the greatest common measure, it will be a measure of such greatest common measure.

We found, at prop. 18, the common measures of 60 and 84 to be 2, 3, 4, 6, and 12; the last named number being the greatest possible; and each of the others measures of it.

Proposition 21.

If there be three numbers, the first of which is a measure of the second, and the second of the third, the first will be a measure of the third.

Let there be 5, 10, 30. Five is a measure of 10, and 10 is a measure of 30; therefore 5 is a measure of 30. Thirty is 3 times 10, and 10 is twice 5, therefore 30 is 3 times twice 5, or 6 times 5.

Proposition 22.

If the product of two numbers be divided by any factor common to both; the quotient will be a common multiple of the two numbers.

$6\times4=24$; and if we divide 24 by 2, (which is a factor common to both 6 and 4,) the quotient 12, will be a common multiple of 6 and 4.

Proposition 23.

The least common multiple of two or more numbers is their product divided by their greatest common measure; or what amounts to the same thing, the product of all their prime factors not common to any two or more of them.

The greatest common measure of 8 and 20 is 4, the least common multiple is $8\times20\div4=40$; and $40=2\times2\times2\times5$ the prime factors of 8 and 20.

The principle of common measures and common multiples is practically applied in operating on fractional quantities; and likewise in the solution of some descriptions of problems. In reducing fractions to their lowest terms, we seek the greatest common measure of the numerator and denominator, by which both are to be divided; and for the aid of memory, the rule may be thus versified:

> "The greater by the less divide,
> The less by what remains beside,—
> The last divisor still again,
> By what remains—till 0 remains,
> And what divides and leaveth 0,
> Will be the common measure sought."

Let $\frac{21}{35}$ be reduced to its lowest terms.

Seven "divides and leaveth 0" and is therefore the greatest common measure of 21 and 35, and 7)$\frac{21}{35}$($\frac{3}{5}$, the lowest terms of the fraction.

```
21)35(1
   21
   --
   14)21(1
      14
      --
       7)14
         --
         2+0
```

That 7 is *a* common measure is evident, for it divides both 21 and 35, and that it is *the greatest* common measure may be thus shown: In the operation we perceive that 7 measures itself and measures 14, it therefore, (by proposition 19,) measures their sum, 21, which is one of the numbers to be divided; and measuring 14 it will also measure their sum, 35, which is the other number to be divided; it therefore must be a common measure. It also is the greatest, for if there be a greater measure of 21 and 35, it must also measure 14, their difference; and measuring 21 and 14, it must also measure 7, their difference. That is, a greater number must measure a less, which is absurd. 7 is therefore the greatest common measure.

We may explain this principle also by lines.

Let AB represent a line 35 rods long, and CD another 21 rods long; then as the common measure cannot be greater than CD, cut off from AB the length CD as often as it can be applied, which here is but once, A*a*, and 14 remains; then apply this 14 to the shorter line CD and it measures once C*c* and 7 rods remain: try whether this piece of 7 rods will measure the remainder *a*B, and we find it will without a remainder; it is therefore a common measure. To do this properly no numbers should be applied; and it might be illustrated very handsomely and expeditiously, by taking two strips of wood or paper, and cutting off alternately from the remainders of each, until one remainder would measure another. The length of the measure thus found, if the strips were respectively 21 and 35 inches long,

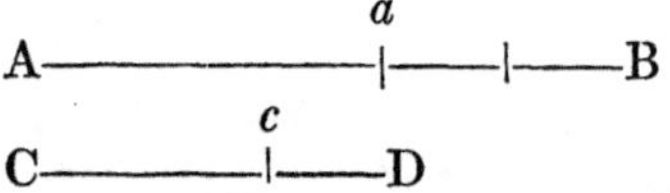

would prove to be 7 inches. The reason must be plain from what is said of numbers. We would recommend that this experiment be tried.

Let us try to reduce $\frac{105}{286}$ to lower terms, they being the number named under proposition 10. We find they run down to a unit.—The numbers therefore are prime to each other.

```
105)286(2
    210
    ---
     76)105(1
         76
         --
         29)76(2
            58
            --
            18)29(1
               18
               --
               11)18(1
                  11
                  --
                   7)11(1
                      7
                      -
                      4)7(1
                        4
                        -
                        3)4
                          -
                          1+1
```

This principle is also applicable in the solution of problems.

1. A gentleman owns a prairie 320 rods long and 180 rods wide, and wishes to lay it off into the smallest practicable number of precisely square fields. What will be their size and number?

Twenty we find is the largest number that will divide both side and end without a remainder, the fields cannot exceed 20 rods square; and 320÷20 =16, the number of rows of fields across; and 180÷20= 9, the number of rows the other way. Then 16×9=144, the whole number of fields.

```
180)320(1
    180
    ---
    140)180(1
        140
        ---
         40)140(3
            120
            ---
             20)40(2
                40
```

2. My garden is 430 feet long, and 320 broad, and I wish to enclose it with a fence, the panels of which shall be all of equal length. What is the greatest length I can use?

Proceeding as before, we find 10 to be the greatest common measure, and of course the greatest length that can be used.

If there were several unequal sides to the garden, or in other words, if we desire to find the greatest common measure of several numbers, we commence by finding the greatest common measure of any two numbers, then of this measure and a third; and then of the new measure and a fourth; and so on until all the numbers are included. It follows of course that the probability of finding a large common measure decreases as the number of quantities increases; indeed if there are many numbers there is little probability of any number being a common measure.

The word "common" occurs very frequently in discussing certain properties of numbers; as when we speak of "common measures," "common denominators," &c. It is in this construction used to mean belonging to more than one—possessed by several. So we say "Light and air are common property," "A and B own the lot in common," "The Ohio common schools," meaning such as all have a common interest in; not such as are of inferior grade.

To find the least common multiple of two or more numbers, we arrange the numbers in a line and proceed to divide by any *prime* number that will divide two or more of them without a remainder; and continue the operation on the quotients and the remaining numbers, until no two of them have a common measure: then multiply the several divisors, and undivided quotient figures, and any undivided numbers that may remain, and their product will be the least possible common multiple. This operation is in effect finding the several prime factors of the numbers.

Required the least common multiple of 6 and 9?

3	6	9
	2	3

Then $3\times2\times3=18$, the least common multiple of 6 and 9, or the least number that both will divide without a remainder.

The *natural* common multiple, if we may so express it, of two or more numbers, is their product; which in the above case would be 54; and the *least* is the same product divided by all the factors such numbers have in common; or in other words by their greatest common measure.

Required the least common multiple of 4, 9, 6, 8?

3	4	9	6	8
2	4	3	2	8
2	2	3	1	4
	1	3	1	2

Then $3\times2\times2\times3\times2=72$, the least common multiple.

We might in this solution have divided by 4 instead of by 2 twice, and the result would not have been affected; but we have no assurance when we divide by composite numbers that the result will be the least common multiple. Take for example the following; in which we seek the least common multiple of 12, 25, 30, 45.

6×5×2×5×9=2700, which appears to be the least common multiple.

6	12	25	30	45
5	2	25	5	45
	2	5	1	9

Let us now use prime numbers only:

3×2×5×2×5×3=900, the true least common multiple.—This difference arises from the fact that 45 is a multiple of 3, one of the factors of 6, but is not a multiple of 6. The conclusion then is that though the result when composite divisors are used *may be* the least common multiple; when prime numbers are used the result *must be* the least. It is never the least, unless the numbers multiplied are prime to each other: in the above 6 and 9 have a common measure, 3; and 2700÷3=900 the least common multiple.

3	12	25	30	45
2	4	25	10	15
5	2	25	5	15
	2	5	1	3

To show still farther that composite numbers cannot be relied upon, let us use another set.

15×3×2×2×25=4500,—still worse than the first.

15	12	25	30	45
3	12	25	2	3
2	4	25	2	1
	2	25	1	1

To abridge the process we may cancel any number that is a measure of another that is to be divided. In the numbers 4 9 6 8, we may reject the 4 altogether; because it is a measure of 8 and we may reject 2 in the line of quotients, because it also is a measure of 8; for let the multiple of 8 be what it may, both 4 and 2 will measure it.

3	4	9	6	8
		3	2	8

If we consider a number and any multiple of it, as resolved into their constituent prime factors, we shall find in the latter all the factors contained in the former. So if two numbers be not prime to each other, they must contain factors common to both; and these being rejected, the multiple must be the least, since no factor is retained that is not necessary to produce one of the given numbers. But if two numbers are prime to each other, then they possess no common factor; and as there is none such to reject, their product will be the least common multiple.

The purpose to which this calculation is most frequently applied, is finding the least common denominator in the addition and subtraction of fractional quantities; though where the quantities are small, an expert accountant will readily determine the least common multiple by inspection. Any one will perceive at a glance that the least number into which 2, 3, and 4 will divide without remainder is 12; and with a little practice much larger numbers are readily determined.

We might illustrate the doctrine of common multiples mechanically, as we did that of common measures; but we pass on to an example or two showing its direct application to the solution of problems.

Suppose 63 galls. to fill a hogshead, 42 a tierce, and 35 a barrel, what is the smallest quantity of molasses that being first shipped in hogsheads, then reshipped in tierces, and then again reshipped in barrels, shall always just fill the vessels without defect or redundancy?

7	63	42	35
3	9	6	5
	3	2	5

$7 \times 3 \times 3 \times 2 \times 5 = 630$ Galls. Ans.

What is the least whole number that being divided by 2, 3, 4, 5, 6, 7, 8, 9 and 10, will leave as remainders 1, 2, 3, 4, 5, 6, 7, 8, and 9 respectively?

The least common multiple of the above divisors is 2520; and it is obvious that if we diminish this number one we shall have the required result, viz. 2519, which will answer the conditions of the question.

Sometimes one or more of the numbers have fractions annexed, in which case the mixed number must be brought to an improper fraction, and the numerator alone used in the calculation. Thus—find the least common multiple of $3\frac{1}{3}$, $6\frac{2}{3}$ and 30 that shall be a whole number.

$3\frac{1}{3}$, $6\frac{2}{3}$ and $30 = \frac{10}{3}$, $\frac{20}{3}$ and 30, and using the numerators only we have

5	10	20	30
2	2	4	6
	1	2	3

and $2 \times 3 \times 2 \times 5 = 60$, the least common multiple.

To divide by a fraction we are taught to multiply by the Denominator, and divide by the Numerator, being just the reverse of multiplication. If therefore any number is a multiple of the numerator it is a multiple of the fraction; and if the fraction be in its lowest terms, it will be the least multiple. In the above example, 10 is the lowest integral number that $3\frac{1}{3}$ will measure, and it can measure no other integral number, that is not a multiple of 10. So 20 is the least whole number that $6\frac{2}{3}$ will measure, and it can measure no whole numbers but 20 and its multiples. If therefore we bring the annexed fractions to their lowest terms, and then the mixed numbers to improper fractions, the least common multiple of the numerators must be the least common multiples of the fractions.

On the same principle, we may find the least common multiple of two or more fractional quantities, so as to give integral quotients, by reducing the fractions to their lowest terms, and taking the least common multiple of the numerators.

Required the least common multiple of $\frac{3}{4}$, $\frac{6}{7}$ and $\frac{4}{5}$, so that the quotients may be whole numbers.

3	3	6	4	Numerators.
2	1	2	4	
	1	1	2	and $2\times2\times3=12$ Ans.

$$\text{Proof}\begin{cases}12\div\frac{3}{4}=16\\12\div\frac{6}{7}=14\\12\div\frac{4}{5}=15\end{cases}$$

We have shown how the number of divisors of which a number is capable may be determined; it is sometimes required to determine the least number that shall have a given number of divisors. This involves the same principle as finding the number of divisors in a given number.

Required the least number that shall have 30 divisors.

$30=2\times3\times5$, which according to the principles already explained, are the exponents increased 1, of the factors of the required number, the exponents therefore are 1, 2, 4, and as that number will be least when the number having the greatest exponents is least, and so on in order, therefore taking 2, 3, 5, (the least primes) as the factors, and applying the exponents as explained, we have

$2^4\times3^2\times5^1=720$. The number required.

Had we taken $5^4\times3^2\times2^1$ the result would have been 11250, and $5^4\times3^1\times2^2$ gives 7500.

The reason is obvious enough from what has been said. In this case, as in seeking the number of divisors, 1 and the number itself are both included.

We may sometimes save ourselves trouble in calculating, by observing certain simple facts, the reason of which will be obvious without explanation. For instance—

If a number has 0, 2, 4, 6 or 8 in the units' place, it is measurable by 2; but

No odd number can be divided by an even number without a remainder.

If the two right hand digits of a number be a multiple of 4, the whole number is a multiple of 4. 724 is a multiple of 4, because 24 is.

If the three right hand digits of a number be a multiple of 8, the whole is a multiple of 8.

If any number have 0 or 5 in the units' place, the number is a multiple of 5.

To the above others might be added, but as they will readily occur to the attentive student, we shall omit them here; lest more important matter be displaced.

LECTURE V.

PROPERTIES OF FRACTIONAL NUMBERS.

THE subject of our present lecture is Numbers in their fractional or broken form, as contrasted with Numbers in their whole or integral form. All are familiar with the present mode of expressing fractional quantities, and we have not space to dwell on the modes adopted in times past; but after stating a few propositions we shall pass on to a discussion of the subject in general, which will probably be more interesting than discussing the several propositions individually.

PROPOSITION 24.

The value of a fraction does not depend on the size of the numbers by which it is expressed, but on their ratio to each other.

Whether we consider an integer as divided into two parts, two hundred, or two thousand, is not important, if in the first case we can claim 1 of the 2 parts; in the second 50 of the 100; or in the third 500 of the 1000, we shall in either case have one half. From this proposition the following results as a matter of course.

Proposition 25.

Multiplying or dividing both terms of a fraction by the same number does not alter its value.

If the terms of $\frac{4}{6}$ be multiplied by 2, they become $\frac{8}{12}$ and if divided by 2 they become $\frac{2}{3}$; but the value is not altered, for $\frac{2}{3}$, $\frac{4}{6}$, and $\frac{8}{12}$ are the same thing in value. The ratio between the numerators and denominators remains unaltered.

Proposition 26.

If the numerator be increased or diminished, and not the denominator, the value of the fraction will be increased or diminished.

If instead of multiplying, as just stated, both terms by 2, we multiply only the numerator, $\frac{4}{6}$ becomes $\frac{8}{6}$; and had we divided that term only, $\frac{4}{6}$ would have become $\frac{2}{6}$. The unit would be divided as before, but the number of parts taken would vary.

Proposition 27.

If the denominator be increased without altering the numerator, the value of the fraction will be diminished; but if diminished the value will be increased.

Let the denominator of $\frac{4}{6}$ be multiplied by 2, and it becomes $\frac{4}{12}$, but if divided by 2, it becomes $\frac{4}{3}$. From propositions 26 and 27, it is obvious that we may multiply a fraction by dividing its denominator or multiplying its numerator; and we may divide it by dividing the numerator or multiplying the denominator.

Proposition 28.

Multiplying the numerator of a fraction has the same effect on its value, as dividing the denominator, and vice versa.

$\frac{4}{6}$. Here to multiply 4 by 2 will have the same effect as dividing 6 by 2; and multiplying 6 by 2 has the same effect as dividing 4 by 2: and it will obviously be so in all cases.

Fractional numbers have in them much that is interesting when properly understood, and an intimate knowledge of their principles is a most efficient weapon in the hands of the expert arithmetician; for many operations that involve no fractions of integers, involve nevertheless, the use of the same principles.

Fractions, as we remarked in a former lecture, are of various kinds, including what are usually called compound quantities and decimals, as well as the more common form known as *Vulgar* Fractions; a name equivalent to common or usual fractions. The word vulgar is now generally used in a different and rather opprobrious sense; but its original meaning was nothing more than *common, not unusual;* and if the word common were not used in numbers in a still different sense, as a common measure, common multiple, common denominator, &c., it might be well to designate this class of fractions as common fractions. The present mode of expressing a vulgar fraction is by two quantities placed one above the other with a dash drawn between them, thus, $\frac{7}{8}$; the lower figure being called the denominator, because it denominates or gives name to the fraction, and the upper the numerator, because it numbers or numerates the parts taken. We may consider that the denominator tells into how many parts the integer is divided, and the numerator numbers the parts taken, as in the expression $\frac{7}{8}$ of an apple, we suppose the apple to be divided into 8 parts and 7 of them are the $\frac{7}{8}$; had it been $\frac{7}{9}$ then the apple would have been cut into 9 parts; and had it been $\frac{7}{10}$ the apple would have been cut into 10 parts; the denominator constantly giving name to the fraction which is called eighths, ninths, or tenths, as the unit or entire thing is divided into 8, 9 or 10 parts. A fraction is said to be in its lowest terms, when the numerator and denominator are prime to each other.

Or we may consider the numerator as expressing so many integers and the denominator as telling how they are to be divided. Thus $\frac{7}{8}$ would imply that 7 apples are to be divided into 8 parts, and the value would be the same as before. This would be $\frac{1}{8}$ of 7, instead of $\frac{7}{8}$ of 1. This view of the case accounts more plainly than the other for forming fractions of remainders in division, by placing the divisor underneath the remainder; and also for the mode of expressing division by making the dividend the numerator of a fraction and the divisor the denominator, thus if we would express that 8 is to be divided by 2, we may adopt either of these forms, 2)8($8 \div 2$, or $\frac{8}{2}$, which latter form would be an ordinary improper fraction, and reducing it to its proper value, would require the very operation proposed; viz, *division.* So in all cases we may consider the numerator as the dividend and the denominator as the divisor; or in other words that the numerator expresses some number of units, and the denominator expresses the manner of their division. In the expression $\frac{7}{8}$ of an apple, we may imagine that we divide an apple mechanically into 8 parts and take 7 of them, as we have already supposed; or that we have 7 apples, and take $\frac{1}{8}$ from each; or could we blend the whole into a mass as we would 7 pounds of butter, then that we should take $\frac{1}{8}$ of the whole; for $\frac{1}{8}$ of 7 pounds would be just the same as $\frac{7}{8}$ of 1 pound.

The nature of fractions may be illustrated by drawing a line and dividing it into parts to represent fractional numbers; and in the same way the nature of compound fractions may be shown, as $\frac{1}{2}$ of $\frac{1}{3}$, thus: Divide the line into three equal parts, which will represent thirds, then take the $\frac{1}{2}$ of one of these thirds, which will be $\frac{1}{2}$ of $\frac{1}{3}$, and as each of the 3 thirds will make two halves, the whole line will make 6 such parts, and hence $\frac{1}{2}$ of $\frac{1}{3}$ is $\frac{1}{6}$ of the unit. In the same way one of the halves of a third may be subdivided, say into 4 parts, and then one of the parts will be $\frac{1}{4}$ of $\frac{1}{2}$ of $\frac{1}{3}$, and as there would be 6 such parts to be so divided into fourths, each such fourth would be a twenty fourth part of the whole; and so parts of parts might be taken without limit. The reason of the rule for changing such fractions of fractions into fractions of a unit would hence seem plain.

In the same way complex fractions as $\frac{3\frac{1}{2}}{5}$ or $\frac{3}{5\frac{1}{2}}$ may be illustrated. In the first the line or unit is divided into 5 parts or fifths, and $3\frac{1}{2}$ of such parts are taken; in the last the line or unit is divided into $5\frac{1}{2}$ parts, and 3 of them constitute the fraction. The first expression would be equal to $\frac{7}{10}$ and the last $\frac{6}{11}$, for (by Proposition 25,) the value of a fraction is not altered by multiplying both numerator and denominator by the same factor, or dividing by the same divisor, and here to reduce the complex fractions to simple ones we have multiplied both terms by 2. So the expressions $\frac{3\frac{1}{2}}{5\frac{1}{2}}$ would mean that the line be divided into $5\frac{1}{2}$ parts and that $3\frac{1}{2}$ of them be taken, and will be equal to $\frac{7}{11}$. These forms of fractions are not very common, but they are perfectly natural, and it is vain to attempt to learn the rules and principles of operating upon fractions unless the value and principle of every form of fraction are distinctly understood. $\frac{3\frac{1}{2}}{5}$ lb. of butter would be $\frac{1}{5}$ of $3\frac{1}{2}$ lbs., or $3\frac{1}{2}$ fifths of 1 lb. We have already shown that the value of a fraction is not affected by increasing or lessening the terms, so long as both are changed in the same ratio; we will only add that if we divide a unit, or the line before referred to, into 8 parts and take 4 of them, we shall have $\frac{1}{2}$, just as clearly as if we had divided the line into only 2 parts and taken 1 of them; or into 4 parts and taken 2 of them; or into 16 parts and taken 8 of them; or into 32 parts and taken 16 of them; or 11 parts and taken $5\frac{1}{2}$ of them. As we double the number of parts that the unit is divided into, each part is only half as large, but then at the same time the number received is doubled. Every child knows that if 1 dollar be divided between two, each will have one 50 cent piece; or two 25s; or 4 "levies;" or 8 "fips;" according as the change is made in halves, quarters, eighths, or sixteenths of a dollar; nor will he object to the diminution in the value of each piece, so long as the number of pieces increases in the same ratio in which the size decreases. The rule for reducing a fraction to its lowest terms is based on this principle.

The terms or numbers in which fractions are expressed, are frequently large and inconvenient to operate upon, and are not at once understood, for though every one knows at a glance what $\frac{1}{2}$ means, it might require some calculation to determine the value of $\frac{327}{654}$, or other large numbers; and to reduce them to a more convenient, as well as more intelligible size, we are instructed to "*Divide both terms by any number that will divide both without a remainder;*" and the larger such divisor is, the better, since the resulting fraction will be proportionately less. Suppose in the course of calculation a fraction comes out $\frac{156}{312}$ of a yard, I discover that 2 will reduce this unwieldy expression, making it $\frac{78}{156}$ which is better, but still inconvenient, and I divide by 2 again, making $\frac{39}{78}$, which is still better; but by dividing by 3, I get $\frac{13}{26}$ yet smaller; and lastly dividing by 13, I get $\frac{1}{2}$; so that the large expression $\frac{156}{312}$ of a yard, is resolved into *half of a yard*. The fractions found are all equivalent to each other, but had the greatest common measure been found as explained in the preceding lecture, only a single division would have been necessary. Thus $156)\frac{156}{312}(\frac{1}{2}$. The mode of finding the greatest common measure, and the reason of the mode, were there explained.

The value of a fraction may be expressed by means of the quotients produced in finding the greatest common measure; but of this hereafter.

The counter operation to dividing by a common measure, and thus reducing the amount of the terms to such as are less, is multiplying both terms by a common multiplier, *i. e.* by the same number; a process that would increase the size of the numbers, but would in no wise affect the value; it would give a greater number of pieces to make the "change," but they would be individually less, and the aggregate would be the same in value as before the number was increased—it would be giving 8 "fips" instead of one fifty cent piece. This operation occurs in changing fractions to such as have a common denominator, and also in cases where it is desirable to change the fraction to one equivalent in value, but of a larger numerator or denominator, either of which operations may become necessary in the process of calculation.

The class of numbers known as improper fractions have the numerators equal to, or greater than the denominators, and hence are equal to, or greater than a unit; as $\frac{4}{4}$, $\frac{7}{3}$.

Numbers are often expressed for division by placing the divisor underneath the dividend, instead of using the division mark ÷; thus $\frac{312}{18}$. In division of integers as treated of amongst the elementary rules, the operation is often imperfect, the quotient not showing the true number, but only the number of times the divisor is contained in that multiple of itself next less than the dividend; the difference between the multiple and the dividend remains undivided, and is called the *remainder*. Take for instance the number we have given for illustration, $\frac{312}{18}$.

```
18)312(17
   18
   ---
   132
   126
   ---
     6
```

Here by division we find, not how often 18 is contained in 312, but how often 18 is contained in 306, which is 17 times; the 6 remaining over, an undivided part of the dividend; neither does the young learner usually attach any definite idea to the remainder; and when he is directed to form a fraction of the remainder by putting the divisor under the remainder for a denominator, he too often regards it as a mode of getting clear of the remainder, rather than an arrangement expressive of division. In the above problem 6 remains, under which 18 being put, we have $\frac{6}{18}$, implying that 18 was contained in the dividend 17 times and 6 remained over, which is also to be divided into 18 parts.

It is difficult to understand how the number 6, abstractly considered, is to be divided into 18 parts, but suppose you consider the dividend as miles or some other real quantities, then there are 6 miles over, and there can be no more difficulty in understanding that the space of 6 miles is to be divided into 18 parts, than that 6 miles are to be divided into 2 or 3 parts. The quotient will then be 17 units and 6 units yet subject to division. Or we may consider $\frac{312}{18}$ as equal to $\frac{306}{18}+\frac{6}{18}$, the $\frac{306}{18}$ we can express in an integral form, but the $\frac{6}{18}$ we cannot; we therefore leave it in a form that implies that it is yet to be divided. To make it more convenient we may reduce it to $\frac{1}{3}$, for dividing 6 miles into 18 parts would be just the same as to divide 1 mile into 3 parts; and in either case, if we express the remainder in yards, the result will come out $586\frac{2}{3}$ yards. The name *improper* is no doubt derived from the fact that with the form of a fraction, the value is greater than any fraction of an integer; while a *proper* frac tion is always less than an integer.

It may be in place here to remark that the division of one number by another may be regarded in a two fold light; first, as that operation by which we discover how many times the divisor is contained in the dividend; and, secondly, as that process by which the dividend is resolved into as many equal parts as there are units in the divisor. According to the first view division is clearly impracticable where the divisor exceeds the dividend, since it is not contained at all in the dividend; but under the second aspect we may effect the division as well when the divisor exceeds the dividend as when it does not; though the parts will have less than a unit in each. We generally obtain our first notions of numbers from integers, hence it costs an effort to see how a number is increased by division, or rather how the quotient can exceed the dividend. From the same cause we take it for granted that a number must be increased by being multiplied; but a little thought will make it plain that the quotient is less than the dividend, equal to it, or greater than it, according as the divisor is greater than a unit, just a unit, or less than a unit: and the same being reversed applies equally in Multiplication. The definitions in Multiplication and Division given in our school books are not adapted to convey a perfect idea of the subject; as they do not embrace fully the fractional operations. Multiplication is not always repeating a sum *a given number* of times, for it may be less than a time; and when the divisor exceeds the dividend, you cannot be said to find "how often one number is contained in another," since the less cannot in strictness be said to contain the greater. In that case however the second view just named strictly applies, and we may imagine the dividend however small to be divided into any number of parts, though each part should be less than the thousandth part of a unit.

We may furthermore regard the dividend as a composite number, equal to the product of two numbers, one of which we have, and the other of which we seek to find. The known number is our divisor, the number sought is the quotient. Or the dividend may be regarded as the area of a rectangular parallelogram, and the divisor as one side, the quotient will be the other.

From one end of a field 25 rods wide, it is required to cut off 10 acres: what must be the length of the part cut off?

10 acres=1600 rods, and 1600÷25=64 rods. The Ans.

We are sometimes called upon to say what part one number is of another. What part of 75 is 25?

One is $\frac{1}{75}$ of 75, it is one part of 75, and 25 being 25 times as much as 1, must be $\frac{25}{75}$, which being reduced will be $\frac{1}{3}$. In the same way it is easy to show that we have only to form a fraction by placing the given number as a numerator, and the number of which it is a part as the denominator, and the fraction may then be reduced to its lowest terms.

What part of 13 is 6? One is $\frac{1}{13}$, and 6 is 6 times as much, or $\frac{6}{13}$.

What part of a foot are 8 inches? One inch is $\frac{1}{12}$, 8 inches are $\frac{8}{12}$ or $\frac{2}{3}$.

What part of a dollar are 23 cents? One cent is $\frac{1}{100}$, 23 cents are $\frac{23}{100}$.

What part of a pound Avoirdupois are 9 ounces? One ounce is $\frac{1}{16}$, and 9 ounces are 9 times as much, or $\frac{9}{16}$.

The reverse of this operation is to find the value of a fraction of any denomination in terms of a lower.

What is the value of $\frac{9}{16}$ of a pound Avoirdupois? $\frac{1}{16}$ of a pound Avoirdupois is one ounce, and $\frac{9}{16}$ being 9 times as much will be 9 ounces. Or the operation is as well explained in all cases, and much better in some, by considering the fraction as the one sixteenth part of 9 pounds, instead of the $\frac{9}{16}$ of one pound; for had it been the $\frac{9}{15}$ or indeed any thing else than the $\frac{9}{16}$, we could not have given so readily the value of one part, though we might then have set down 1 lb. 0 oz. 0 dr. and divided by 15, by which we would have got 0 lb. 1 oz. $1\frac{1}{15}$ dr. as the $\frac{1}{15}$, and this multiplied by 9 would give 0 lb. 9 oz. $9\frac{3}{5}$ dr., as the $\frac{9}{15}$. But the operation would be simpler to make 9 lb. 0 oz. 0 dr. the dividend, and 15 the divisor. But even adopting this form, we may consider that the $\frac{9}{16}$ is still the $\frac{9}{16}$ of 1 lb., rather than $\frac{1}{16}$ of 9 lbs., and that we have multiplied the 1 lb. 0 oz. 0 dr. by 9 before dividing, just as we are in the habit of multiplying by the numerator and dividing by the denominator in taking fractional parts of numbers. As if we would take $\frac{2}{3}$ of 200, it is rather easier to multiply by 2 and then divide by 3, than to take $\frac{1}{3}$ first, which would give $66\frac{2}{3}$, and multiply the fraction $\frac{2}{3}$ by 2, and yet this is a very simple case. Where the denominator will divide the given number without a remainder, it is generally best to divide first, and then multiply by the numerator, but not where dividing first would create an awkward fraction.

How many cents are $\frac{23}{100}$ of a dollar? $\frac{1}{100}$ is 1 cent, hence $\frac{23}{100}$ are 23 cents.

How many inches are in $\frac{2}{3}$ of a foot? Here we cannot so readily tell what $\frac{1}{3}$ is, since 3 is not the number of inches in a foot; but we can set down 1 foot 0 in. and divide by 3 and thus we find there are 4 inches in 1 third, and hence twice $4=8$ in $\frac{2}{3}$; or what is still better and must produce the same

3)2 ft. 0 in.

Ans. 8 in.

If fractions are to be added together, the first circumstance to be considered is whether they are all of the same denomination; that is, all similar parts of a unit, and hence of the same name or denomination, as they will be if their denominators are alike. Suppose it be required to add $\frac{3}{16}$, $\frac{1}{16}$, $\frac{8}{16}$, and $\frac{5}{16}$ together, it is easy enough to see that these are all similar parts, and $3+1+8+5=17$ such parts, or $\frac{17}{16}=1\frac{1}{16}$.

But suppose the fractions were $\frac{3}{4}$, $\frac{1}{2}$, $\frac{3}{16}$, and $\frac{5}{8}$, we may add the numerators as before, and they will come out 17, but what are they? they are neither 17 halves, fourths, sixteenths, nor eighths. We might just as well add 1 qr. 20 lb. 12 oz. 13 dr. into one mass, and we would be just as much at a loss to find a name for the resulting sum, in one case as in the other. But to get clear of the difficulty we may bring all the weights to drams, and the fractions to sixteenths, being the lowest denominations mentioned, and hence all may be brought to them without fractional parts. Thus:

1 qr. (25 lbs.)	$=6400$	$\frac{1}{2}=\frac{8}{16}$
20 lbs.	$=5120$	$\frac{3}{4}=\frac{12}{16}$
12 oz.	$=\ 192$	$\frac{5}{8}=\frac{10}{16}$
13 dr.	$=\ \ 13$	$\frac{8}{16}=\frac{8}{16}$
	11725 dr.	$\frac{38}{16}=2\frac{6}{16}=2\frac{3}{8}$ the sum.

From this it is evident that when we wish to add fractions of various denominations, or any numbers of various denominations, we must bring them all to the same name; and the principle is just the same in fractions as in whole numbers.

In order to bring any number of fractions to similar ones having a common denominator, we must find a common multiple of the denominators, and for convenience it is best to find the *least* common multiple, for by so doing, the new fractions will be the least possible. The rule for finding this multiple and the reason of the rule were explained in Lecture 4.

Let us add together $\frac{1}{2}$, $\frac{2}{3}$, $\frac{3}{4}$ and $\frac{4}{5}$. Result $2\frac{43}{60}$.

Different persons adopt different forms, but the principle is the same in all. In the first place we may with a little practice generally find the least common multiple *by inspection; i. e.* without any formal calculation; and should the multiple not be the least it will in no wise affect the value of the result, though it will the size of the numbers. In this case it is easy to see that 2, 3, 4 and 5 will divide 60 evenly, and will do so with no less number. But let us calculate.

Here we find $\frac{1}{2}=\frac{30}{60}$, $\frac{2}{3}=\frac{40}{60}$, $\frac{3}{4}=\frac{45}{60}$, and $\frac{4}{5}=\frac{48}{60}$, making in all $\frac{163}{60}$ or $2\frac{43}{60}$. In order to find the new numerators, after finding the common denominator, we multiply the common denominator by the numerator and divide by the denominator of each given fraction; or divide by the denominator and multiply by the numerator.

2)2, 3, 4, 5,

1, 3, 2, 5, and 5×2×3×2=60
Then, 60 common denom.

$\frac{1}{2}=30$
$\frac{2}{3}=40$
$\frac{3}{4}=45$
$\frac{4}{5}=48$

$\frac{163}{60}=2\frac{43}{60}$. Answer.

Perhaps the correctness of this cannot be shown in any way more readily than by laying down the position that any fraction may be changed to another having any given denominator, by increasing or diminishing the numerator in the same ratio, for a new numerator. This in effect we have already shown, by proving that the value of a fraction is not altered by multiplying or dividing the terms by any number whatever so that both are changed in the same ratio: *i. e.* multiplied or divided by the same number.

Here the first fraction is $\frac{1}{2}$, and we wish it changed to another, having a denominator 30 times as large, we then multiply the numerator by 30 for a corresponding numerator.

The next number is $\frac{2}{3}$, we divide the common denominator and find it 20 times as large, we therefore multiply the numerator by 20, to find a corresponding new numerator.

Or you may prove the statement by proportion, As 3 (the given denominator) : 60 (the proposed denominator) : : 2 (the given numerator) : 40 (the new numerator.)

In this way both numbers are increased in the same ratio, and while that is the case the value will not be altered.

Instead of this form some prefer the following:

Find a common denominator by multiplying together all the denominators; and then, to find the several numerators, multiply each numerator into all the denominators except its own; thus:

This mode does not admit of the least common multiple being sought, and hence the new fractions are seldom in the lowest terms that the condition of having a denominator in common will admit. In

$\frac{1}{2}\times\frac{2}{3}\times\frac{3}{4}\times\frac{4}{5}=120$ com. dem.

$$\left.\begin{array}{l}1\times3\times4\times5=60\\2\times2\times4\times5=80\\3\times2\times3\times5=90\\4\times4\times3\times2=96\end{array}\right\}\text{Numerators.}$$

the present instance it is twice as large as necessary. This is often a decided inconvenience.

The following form, still different from both we have given, is also used by some. Add together $\frac{7}{10}$, $\frac{11}{12}$, and $\frac{4}{9}$.

$$\begin{array}{l|lll} 2 & 10 & 12 & 9 \\ \hline 3 & 5 & 6 & 9 \\ \hline \end{array}$$

$$5\times2\times3\times3\times2=180 \text{ c. den.}$$

$$\left.\begin{array}{r} 10 \\ 12 \\ 9 \end{array}\right\} 180 \left\{\begin{array}{l} 18\times\ 7=126 \\ 15\times11=165 \\ 20\times\ 4=\ 80 \end{array}\right\} \text{Numerators.}$$

These modes differ in form from the first, but the principle is the same, resolving itself at last into the fact, that both terms of each fraction are multiplied by the same factors, and hence the value is not affected. Here we divide the common denominator, 180, by the several denominators to find their ratio to it, that the corresponding numerators may be increased in the same ratio.

The same preparation is necessary in *Subtraction* as in *Addition*, for we can no more take the difference than the sum of dissimilar quantities; but having changed them to such as are similar the sum or difference may be taken as readily as if the numbers were integers. The denominators serve the same purpose as different names and a table of their value in the compound rules.

Multiplication of Fractional quantities does not require this identity of denomination, neither does Division; but if they are alike it will not affect the result.

Multiply 150 by $\frac{1}{2}$.

In the absence of any rule, we have here two factors, either of which may be considered the multiplier, so that the product will be 150 times $\frac{1}{2}$=150 halves=75 units; or $\frac{1}{2}$ of 150=75. If 150 were multiplied by 1, the product would be 150, and if by the half of 1, the product must be half as much.

We are generally instructed to multiply by the numerator and divide by the denominator; let us try an example.

Multiply 150 by $\frac{2}{3}$.

$$\begin{array}{r} 150 \\ 2 \\ \hline 3)300 \\ \hline 100 \text{ Answer.} \\ \hline \end{array}$$

Here we multiply by 2 and the product is 300, but as the multiplier was really $\frac{1}{3}$ of 2, so the product will be $\frac{1}{3}$ of 300=100. Thus in any other case, as the multiplier or fraction is equal to the numerator divided by the denominator, so will the product be equal to the product of the numerator divided by the denominator.

By the nature of fractional notation, if as already stated, we increase the numerator in any ratio, in the same ratio will the value of the fraction be increased, for we thus increase the number of parts without diminishing their value; but if the denominator be multiplied, the value of the fraction will be diminished in the same ratio in which the denominator is increased, for the value of the fractional parts is thus decreased without increasing their number. But if both terms be increased or decreased in the same ratio, the value will not be altered, for as the number of parts increases or diminishes, the value of each part changes inversely in the same ratio, and thus the value of the expression is unchanged. If I take $\frac{3}{8}$ and multiply the numerator by 2, the product is $\frac{6}{8}$, the number of parts being doubled while they are eighths still; but if I multiply the denominator by 2, the result is $\frac{3}{16}$, the number of parts not being increased, while the value is decreased one half, *i. e.* from eighths to sixteenths; but if both terms be multiplied by 2, the result will be $\frac{6}{16}$, which is neither more nor less than $\frac{3}{8}$.

Again, if we divide the denominator it is equivalent to multiplying the numerator, for it increases the value of the parts without diminishing their number; and on the other hand if we divide the numerator the value is reduced, for the number of parts is lessened without increasing their value; but if both be divided, as we have before shown, the ratio between the terms is unaffected and so is the value. Thus taking $\frac{2}{8}$, if we divide the numerator by 2, it becomes $\frac{1}{8}$, and if we divide the denominator by 2 it becomes $\frac{2}{4}$ or $\frac{1}{2}$, but if both be divided it is $\frac{1}{4}$, the same in value as $\frac{2}{8}$.

These positions being understood, it is easy to see that if we wish to multiply a fraction, we may either multiply the numerator or divide the denominator; and if we would divide a fraction we may either divide the numerator or multiply the denominator, and the same effect will be produced.

Hence, when we multiply two fractions, we multiply the numerators together for a new numerator, and the denominators together for a new denominator, and this is just what we did in multiplying 150 by $\frac{2}{3}$; it is multiplying by the numerator of the multiplier and dividing by the denominator; for multiplying the denominator of the multiplicand is the same as dividing the numerator. Multiply $\frac{4}{8}$ by $\frac{3}{4}$.

$$\frac{4}{8}\times\frac{3}{4}=\frac{12}{32}=\frac{3}{8}$$

Multiplying $\frac{4}{8}$ by the numerator 3, gives $\frac{12}{8}$ which we may divide by 4 and we will have $\frac{3}{8}$, or we multiply 8 by 4 which makes $\frac{12}{32}$, the same in value as $\frac{3}{8}$.

Division is the reverse of Multiplication, and as in the latter we multiply by the numerator and divide by the denominator, so in the former we multiply by the denominator and divide by the numerator; or we may divide the numerator and denominator of the dividend by the numerator and denominator of the divisor, and the result will be the same.

Divide $\frac{6}{8}$ by $\frac{2}{4}$.

We may say 2 into 6 three times, for a numerator, and 4 into 8 twice, for a denominator, making $\frac{3}{2}=1\frac{1}{2}$. Answer.

Or, inverting the divisor and multiplying, $\frac{6}{8}\times\frac{4}{2}=\frac{24}{16}=1\frac{1}{2}$. That this is the true quotient may be thus shown: $\frac{6}{8}=\frac{3}{4}$, and $\frac{2}{4}=\frac{1}{2}$; then $\frac{1}{2}$ will be contained $1\frac{1}{2}$ times in $\frac{3}{4}$, for $\frac{1}{2}$ will be contained 1 time in $\frac{1}{2}$, and half a time in $\frac{1}{4}$, and $\frac{1}{2}+\frac{1}{4}=\frac{3}{4}$, the dividend.

As in multiplying by a proper fraction, the product is always less than the multiplicand, for the multiplicand is taken less than once, so in dividing by a proper fraction the quotient is always greater than the dividend, for the divisor is less than a unit, and must be contained oftener than a unit would be. This result is contrary to the result in whole numbers, for when a number is taken more than once, it is increased; and if divided by more than a unit, it is diminished; the fixed point in both operations is unity.

But let us look into the reason of both the above modes and see why they are identical.

There is no difficulty in seeing that dividing the numerator and denominator will give the correct quotient; for we thus find a number that multiplied by the divisor will produce the dividend, and the object of division is finding such a number.

Divide $\frac{15}{16}$ by $\frac{3}{4}$.

We are here required to find how often $\frac{3}{4}$ are contained in $\frac{15}{16}$, or in other words, we are required to find a number that multiplied by $\frac{3}{4}$ will produce $\frac{15}{16}$. If we divide 15 by 3 we obtain such a number for the numerator, and if we divide 16 by 4 we obtain such a number for the denominator, and this must be always the case in principle, whether the numbers divide evenly or not; but if one is not a multiple of the other a complex fraction will be the result. If we seek to divide $\frac{5}{19}$ by $\frac{2}{3}$, the result by this mode of calculation will come out $\frac{2\frac{1}{2}}{6\frac{1}{3}}=\frac{15}{38}$: for in order to reduce the complex fraction to a simple one, we must multiply both terms by some number that will clear both of fractions, and 6 is the least number that will do that, it being the least common multiple of the two denominators 2 and 3.

But if that mode of operation be correct, then any mode which is equivalent to it, must also be correct; and we have already shown that to multiply the numerator of a fraction by any number is the same as to divide the denominator by that number, therefore multiplying the numerator of the dividend 5, by 3, the denominator of the divisor, and the denominator 19, by 2, has the same effect, and must produce the same result, as dividing the contrary terms; and by this mode we avoid complex fractions.

Another view may be taken of this subject. Suppose I seek to divide 360 by $\frac{3}{4}$, or what is the same, by the fourth part of 3. I proceed first to divide by 3, and produce 120, but as I was to have divided by only the fourth part of 3, my divisor was 4 times too great, and consequently my quotient is 4 times too little; I must therefore multiply the quotient by 4 to get the true quotient. In order to generalize this operation it is only necessary in dividing by any number, to divide by the numerator and multiply by the denominator; or to get clear of remainders, we may multiply first by the denominator, and afterwards divide by the numerator, as it can make no difference in the result, which operation is first performed.

For convenience, the terms of the divisor are sometimes reversed, and then the operation becomes one of multiplication, but this is not necessary, as they may be multiplied crosswise; thus, $\frac{5}{19}\div\frac{2}{3}=\frac{15}{38}$ or by reversing $\frac{5}{19}\times\frac{3}{2}=\frac{15}{38}$, the same as before.

When a remainder occurs in dividing a mixed number by an integer, multiply the denominator of the annexed fraction by the remainder and add in the numerator, for a new numerator, and then multiply the denominator of the annexed fraction by your divisor for a new denominator; the new numerator and new denominator united fractionwise, will give the fractional part of your quotient.

Divide $125\frac{3}{4}$ by 3.

$$\begin{array}{r} 3)125\frac{3}{4} \\ \hline 41\frac{11}{12} \\ \hline \end{array}$$

Here we divide 125 and find 2 over, which we multiply by 4, changing them to 8 quarters and adding the 3, we have 11 quarters, to be divided by 3, which we may do by division, making $\frac{3\frac{2}{3}}{4}$, or by multiplying the denominator 4 by 3, for it will be remembered that dividing the numerator and multiplying the denominator have the same effect.

In stating questions in the Rule of Three in Fractions, the numbers are arranged just as they are in whole numbers, and then the terms of the dividing number are inverted, when the several numerators are multiplied together for a numerator.

and the several denominators for a denominator. This is strictly multiplying the 2nd and 3rd terms together and dividing by the first.

Where fractions are very large in their terms, it is sometimes difficult to form an opinion at once of their value; and modes of calculation have been devised to show very nearly their value in fractions of smaller terms, but it is not thought necessary to introduce them here, as the same may be effected by a little consideration; thus $\frac{315}{456}$ is about $\frac{3}{4}$, as we may see by dropping the units and tens of each term, an operation equivalent to dividing each by 100: but it is nearer to $\frac{31}{45}$ or $\frac{3}{4\frac{1}{2}}$. In this way a pretty accurate idea may, at a glance, be formed of any fractional expression however large.

It has been remarked that the fractional principle is frequently applied to questions that involve no fractional numbers, though fractional parts of the whole quantity are considered; the following are solved on that principle.

1. Thomas gave away $\frac{3}{4}$ of all the apples in his hat, and had 5 left; how many had he at first?

If he gave away $\frac{3}{4}$ he had $\frac{1}{4}$ left, and this one fourth was 5, hence 4 fourths, or the whole quantity would be 20.

2. A person spent $\frac{1}{3}$ and $\frac{1}{4}$ of his money, and had \$60 left; how much had he at first?

He spent $\frac{1}{3}+\frac{1}{4}=\frac{7}{12}$, and hence had $\frac{5}{12}$ left=60. Then if $\frac{5}{12}$ was \$60, $\frac{1}{12}$ was \$12, and $\frac{12}{12}$ was \$144.

In questions of this kind the whole quantity is considered a unit, and the fractions are of such quantity, not of numbers. The principle is very important in solving questions by analysis, without reference to any formal rule, and a large proportion of the questions usually solved by Single Position, may be solved by this means. We shall take occasion to present other examples hereafter.

We shall now close this subject by adding two propositions, that may be sometimes useful.

Proposition 29.

If to any number a fractional part of itself be added, and from the resulting sum such fractional part of the sum be deducted that the original number shall be left; then the latter fraction will be found to have the same numerator as the former, and the denominator will be the sum of both terms of the former.

If we add $\frac{2}{3}$, we must subtract $\frac{2}{5}$; if we add $\frac{2}{5}$, we must subtract $\frac{2}{7}$, &c. To 12 add $\frac{2}{3}$ of 12, and we have 20; and from 20 deduct $\frac{2}{5}$ of 20 and we have 12 again. This principle is rather an important one in solving some problems. Take the following for instance:

If to my age there added be,
One half, one third, and 3 times 3;
Six score and ten the sum will be;
What is my age? pray show it me.

This may be solved by Position, but the process will be less simple than either of the following.

3 times 3=9, and 9 taken from "6 score and ten" leave 121, which by the question is 1 age, + $\frac{1}{2}$ an age, + $\frac{1}{3}$ an age=$1\frac{5}{6}$. Then, As $1\frac{5}{6}$: 1 : : 121 : 66 *Ans.* *Or,* As we must add $\frac{1}{2}+\frac{1}{3}=\frac{5}{6}$ to the unknown number to make 121, we must by the above proposition, deduct $\frac{5}{11}$ of 121 to leave the unknown number, which we are seeking; $\frac{5}{11}$ of 121=55; and 121—55=66 *Ans.* as before. If we deduct the fixed number "3 times 3," and operate with 121 instead of 130 we may perform the solution by Single Position; otherwise we must resort to Double Position.

That this theorem is correct may be readily shown: Suppose to a number I add its $\frac{2}{3}$, then I must deduct $\frac{2}{5}$ to restore the first number; for the number itself has 3 equal parts or thirds, to which 2 more being added, there are 5 just such parts, hence what were thirds in the first instance become fifths; so that we add $\frac{2}{3}$ and subtract their equivalent $\frac{2}{5}$ to restore the original number.

Proposition 30.

If from any number we subtract a fractional part, and then we desire to add such part of the remainder to the remainder as will make the original sum, the part to be added will have the same numerator, and a denominator equal to the difference between the first numerator and denominator.

From 15 take $\frac{2}{5}$ of itself, and to the remainder we must add $\frac{2}{3}$ of itself to make 15 again: 15—$\frac{2}{5}$ of 15=9, and 9 +$\frac{2}{3}$ of 9=15. The reason of this is obvious from proposition 29.

LECTURE VI.

DECIMAL FRACTIONS, CONTINUED FRACTIONS, ETC.

It has been already shown that any fraction may be changed to another fraction of equal value, having a given numerator or denominator, by changing the other term of the fraction in the same ratio; thus, we may change $\frac{3}{4}$ to a fraction having 6 for a denominator, by increasing the 3 in the same ratio as it would be necessary to increase the 4 in order to make 6 of it. This we may do by the rule of three, thus:—As 4 : 6 : : 3 : $4\frac{1}{2}$ the new numerator, the fraction being $\frac{4\frac{1}{2}}{6}$. Or we may change it to a fraction having 6 for a numerator, thus:—As 3 : 6 : : 4 : 8, the new denominator, and the fraction would be $\frac{6}{8}$.

If we were to reduce all our fractions to a common denominator, it would be useless to express that denominator to every number, as it would be understood that they were all of a kind, all fourths, fifths, sixths, &c., as the case might be. Duodecimals are an instance of this, the several periods being expressed as whole numbers and called seconds, thirds, fourths, &c., as they diminish in value from the units' place. The numbers in our compound rules are expressed as whole numbers, but the name attached to them is as expressive a "denominator" as a number could be if placed under a line. 8 inches is the same in value as $\frac{8}{12}$ of a foot, the value being as decidedly marked in one case as in the other. But there is only one common denominator that could be adopted, which would supersede the necessity of dividing and otherwise operating as in the compound rules, and this would also more effectually than any other dispose of the complex fractions which will be found more or less troublesome in every such system.

From the higher places towards unity, our numbers decrease in value in a tenfold ratio, and if we were to extend the same ratio of decrease onward indefinitely beyond the units' place, we would have a mode of expressing numbers less than a unit, that would have many advantages. Small numbers might have 10 as a denominator, such as are larger might have 10 times 10, and as the numerators would increase, let the denominators increase in the same tenfold ratio. Such a mode of expressing fractional quantities would present many advantages in facility of calculation, being as much superior to the common modes, as our Federal money is superior to the old currency.

In order to change vulgar or common fractions to this form, the rule of stating already explained in this lecture might be adopted, or for greater convenience we might simply multiply the numerator by 10, 100, 1000, or whatever number might be found necessary, and divide the product by the denominator, for a new numerator; and as all the fractions would be tenths, or hundredths, or thousandths, there would be no necessity for writing down the denominators as they would always be 10, 100, 1000, 10000, &c., according as the numerator might consist of one, two, three, or more figures. Such a system as this was invented by M. PURBACH, of Austria; and I need scarcely add, if the student has attended to the subject, that it is our common system of *Decimal* Fractions. From unity we proceed upward by tens, hundreds, thousands, &c. of units, and downward by tenths, hundredths, thousandths, &c. of a unit; we may therefore consider unity as the centre of our numerical system, with an infinite ascending series above it, and an infinite descending series below it.

Simple as this seems to us in theory, and convenient as it is in practice, the invention is of recent date. The natives of India, though the inventors, perhaps, of our notation, are still ignorant, says Professor LESLIE, of its application to fractions. Below the place of units they change the rate of progression, and descend by continued bisections, in halves, fourths, eighths, sixteenths, &c.

In changing vulgar fractions to decimal ones, the principle we have advanced is always applied, though it is simplified to a mere addition of ciphers to the numerator and division by the denominator; the division if practicable being continued by the addition of ciphers, until no remainder occurs; and then the decimal fraction is read by supposing the denominator to consist of a unit with as many ciphers as there are figures in the numerator.

Or this principle may be explained by considering that the addition of three ciphers is equivalent to multiplying by 1000, and by doing so our quotient will be a thousand times too much, but by cutting off 3 places we in effect divide the quotient by 1000, and thus bring it to its proper size.

Change $\frac{3}{8}$ to a decimal.

$$8)3.000$$
$$.375$$

Here 3 ciphers are added, or supposed to be added, before the division comes out without a remainder, and the result is $\frac{3}{10}$, $\frac{7}{100}$, $\frac{5}{1000}$, or $\frac{375}{1000}$, the denominator may or may not be written; but it is more convenient not to write it; and to mark the decimal character of the expression by placing a point on its left. The addition of ciphers to the right of the decimal, has no effect upon its value, for as the number of places in the numerator is thus increased in a tenfold ratio, the number in the denominator is increased in the same ratio, and the value will in that case be unaffected. But ciphers on the left *decrease* the value in a tenfold ratio by increasing the denominator without increasing the numerical amount of the numerator: thus, $3=\frac{3}{10}$, $.03=\frac{3}{100}$, $.003=\frac{3}{1000}$, &c. They change the position of the significant figures with regard to the units' place, as the addition of ciphers to the right of a whole number does; but placing ciphers on the left of a whole number has no such effect, any more than placing ciphers on the right of a decimal.

But in changing vulgar fractions to decimals, it is not always practicable to continue the division until there is no remainder; for instance $\frac{1}{3}$ would produce .333 &c. forever, and $\frac{2}{3}$ would produce .666 &c., $\frac{5}{7}$ would produce .71428571 &c., the same figures recurring again in the same order forever. These are called *Repeating and Circulating Decimals;* of which we shall speak hereafter. But though such decimals always leave a remainder, and do not therefore express perfectly the value of the vulgar fraction, they are for ordinary purposes sufficiently exact after being extended to a few places of figures, as every place to which they are extended brings the value nearer to the truth. .3 is $\frac{1}{30}$ less than $\frac{1}{3}$, .33 is $\frac{1}{300}$ less than $\frac{1}{3}$, .333 is $\frac{1}{3000}$ less, and so on at every step getting 10 times nearer the truth, without any possibility of ever quite reaching it. But if the figures are dissimilar, as in reducing $\frac{5}{7}$ above, to a decimal, the ratio of approximation is not precisely tenfold, though it does not vary materially from it.

Decimal fractions are generally called for brevity sake *decimals*, and sometimes the whole subject is called *decimal arithmetic*; but the word decimal is often applied to our entire system of notation, to distinguish it from such as are based on some other radix. For decimal is derived from the Latin *decem*, ten, and only means proceeding by tens, or having reference to tens, and is hence used to distinguish ours from the *binary*, the *ternary*, and other systems having two, three, &c., for their bases.

To change a decimal fraction to the form of a vulgar fraction, place the proper denominator under it, but still as long as the denominator is 10, or any power of 10, the fraction is a decimal one; but reduce it to its lowest terms or to any form that changes the denominator from a power of 10, and it loses that character.

Before passing to Addition, &c., it may be well to consider a little further the nature of decimal expressions, for much of the difficulty found in numbers results from want of clearly understanding first principles.

We may read all the decimal expression together, and that is the usual mode, or we may call the first figure tenths, the second hundredths, the third thousandth, the fourth is generally called so many *ten thousandths*, but it would perhaps be rather more accurate to call them so many tenths of a thousandth; and instead of calling the fifth place hundred thousandth, it should be called hundredth of a thousandth. The same criticism may be applied to other forms of expression. We may say $\frac{3}{8}=\frac{375}{1000}$, or $\frac{3}{8}=\frac{3}{10}+\frac{7}{100}+\frac{5}{1000}$, and the meaning will be the same. So we may read a mixed number, 25.13, as $25\frac{13}{100}$, or disregarding the point we may call the expression $\frac{2513}{100}$, and the value will be the same.

In arranging decimal numbers for addition, the general rule applies that numbers of a like denomination must be placed under each other, so that tenths and hundredths, as well as units, tens, and hundreds, shall be in vertical columns, and thus add together; and to effect this we are generally instructed to regulate the numbers by the points; for if they be in a vertical column the numbers will all be right. And as the value of all the columns, decimal as well as integral, increases from right to left in a tenfold ratio, the addition is the same as in whole numbers. If one number has more decimal places than another, you may, if it will make it any plainer to you, imagine the vacancies filled with ciphers, for coming on the right of decimals they will not affect their value; but this is in truth not necessary, as the numbers of like kind will add together if properly placed; though by adding ciphers to make the numbers even, the entire denominator of each number will be common to all, and it was shown under the head of Vulgar Fractions, that if fractions have a common denominator, the numerators are to be added together and placed over it for the sum total, and the same principle applies here. We may however consider each column as having a common denomintator, for the numbers are all tenths, hundredths, &c., and then it is not necessary to fill out the numbers to make a common denominator for all.

In Subtraction the same reasoning applies.

In Multiplication the only peculiarity that distinguishes it from multiplication of whole numbers is in placing the point. It is necessary to point off as many decimal places in the product as are in both the factors, and that this must be so will be seen from changing the numbers of a problem into the form of vulgar fractions.

Multiply .375 by .54. $\frac{375}{1000}\times\frac{54}{100}=\frac{20250}{100000}=.20250.$

We may drop the cipher after 5, since it is only dividing both terms of the fraction by 10, and does not affect its value.

Multiply 4.25 by 3.5. $\frac{425}{100}\times\frac{35}{10}=\frac{14875}{1000}=14.875.$

It is here evident that the product of the denominators will always contain just as many ciphers as there are figures in the factors; and that dividing by it will cut off precisely as many figures as there are ciphers in the factors.

The law of placing the point in Division is just the reverse of that in Multiplication, and results from the relation between the two rules.

Divide .20250 by .54

```
.54).20250(.375
     162
     ---
      405
      378
      ---
       270
       270
```

As the dividend is always equal to the product of the divisor and quotient it must have just as many decimal places, and hence the rule "Point off as many places in the quotient as are in the dividend more than are in the divisor;" and that number must be had, even if it is made by prefixing ciphers.

Or, the operation may be explained on the principle that fractions having a common denominator may be divided into each other by dividing the numerator of one by the numerator of the other. We will divide 43.047 by 2.53698. The number of decimal places will be equalized by annexing two ciphers to the dividend: this being done, and the decimal point removed, which will not affect the proportion of the numbers, the process of division will be as follows:—

```
253698)4304700(16
       253698
       -------
       1767720
       1522188
       -------
        245532
```

The integral part of the quotient is 16, and the remainder 245532, which becomes a vulgar fraction by having the divisor placed under it, or a decimal by adding ciphers and continuing the division; for the addition of ciphers is but multiplying by the denominator of the new fraction, and the principle of this has been already fully explained.

In treating of the reduction of vulgar fractions to their lowest terms, by means of the greatest common measure of the terms, we showed how the common measure may be found, and how by its use, fractions involving large numbers may be changed to equivalent ones expressed in smaller numbers. The lowest terms of a fraction may be determined also by means of the quotient figures arising in the process of finding the greatest common measure.

Reduce the fractional expression $\frac{27}{36}$ to its lowest terms.

```
27)36(1                        27(1                      27(1
   27             Or, 27)———————           Or, 27)——————— = 3/4
   ——                          36(1                      36(1 1/3
    9)27(3                     27
      27                       ——
      ——                        9)27(3
                                  27
                                  ——
```

We assume the less number, which in proper fractions is the numerator, as a divisor; and in such numerator it is contained once of course; and in the above fraction it is contained in the denominator once and 9 over; which remainder is contained three times in the divisor; showing that it is one third of it, or that the numerator is contained in the denominator $1\frac{1}{3}$ times. The quotients then express the value of the fraction as $1\frac{1}{3}$, and multiplying both terms by 3 to clear it of complexity, we have $\frac{3}{4}$, the value of the fraction in its lowest terms.

The quotients are the denominators of a complex fraction, that may run into a *Continued Fraction,* the several numerators being units. A Continued Fraction is indeed only a complex fraction extended to several places; and this mode of reducing a fraction is the same in principle as the ordinary mode of reducing to the lowest terms; only that the numbers assume a complex form, from not dividing evenly.

Reduce $\frac{77}{175}$ to the lowest terms.

```
77)175(2
   154
   ———
    21)77(3
       63
       ——
       14)21(1
          14
          ——
           7)14(2
             14
```

Here the process runs farther before it terminates, but the principle is the same. Neither have we placed the units over the quotient figures, but they obviously belong there, for as 77 is contained 2 times only in 175, it is about the half of it; but there is a remainder of $21=\frac{21}{77}$. This remainder is about the $\frac{1}{3}$ of 77, but with a remainder that is contained once in the divisor, and 7 over; which is just half its divisor, for it is contained in it just two times. These several quantities may be arranged thus:—

$$\tfrac{77}{175}=\tfrac{1}{2}+\tfrac{1}{3}+\tfrac{1}{1}+\tfrac{1}{2}$$

$$\text{Or } \tfrac{77}{175}=\cfrac{1}{2+\tfrac{1}{3}+\tfrac{1}{1}+\tfrac{1}{2}}$$

$$\text{Or } \tfrac{77}{175}=\frac{1}{2\tfrac{3}{11}}$$

Here the numerator is a unit, while the denominator is expressed fractionally in a continuous form. To condense this continued fraction we may begin at its conclusion $\frac{1}{1+\frac{1}{2}=1\frac{1}{2}}=\frac{2}{3}$; then $\frac{1}{3\frac{2}{3}}=\frac{3}{11}$; and $\frac{1}{2\frac{3}{11}}=\frac{11}{25}$, the value of the fraction; just as found by dividing $\frac{77}{175}$ by 7, the greatest common measure of the terms.

Reduce the expression $\frac{27}{34}$ to its lowest terms.

```
27)34(1
   27
   --
    7)27(3
      21
      --
       6)7(1
         6
         -
         1)6(6
           6
           -
```

$$\text{Hence } \tfrac{27}{34}=\cfrac{1}{1+\cfrac{1}{3+\cfrac{1}{1+\cfrac{1}{6}}}}$$

Here it is found that the numbers are prime to each other.

Let us see how the fractions will work back $\frac{1}{1\frac{1}{6}}=\frac{6}{7}$; $\frac{1}{3\frac{6}{7}}=\frac{7}{27}$; $\frac{1}{1\frac{7}{27}}=\frac{27}{34}$, which is the original expression.

Or we may effect the summation by a *direct* process, though the reason is, perhaps, not quite so obvious. It is however sufficiently clear, for the process of division constantly approximates to the true value of the expression, the quotients forming a series that taken all together gives the precise value *in the lowest terms.*

Let us resume the fraction $\frac{77}{175}$. Dividing both terms by 77 we have $\frac{1}{2+}$, the remainder showing the denominator to be a little more than 2, and of course the value rather less than $\frac{1}{2}$.

Take in another term, and we have $\frac{1}{2\frac{1}{3}+}=\frac{3}{7+}$, which is nearer the truth than $\frac{1}{2}$, which would be $\frac{3}{6}$. Embrace the next period and we have $\frac{1}{2+\frac{1}{3+\frac{1}{1+}}}=\frac{4}{9+}$, which is yet nearer, but not accurate, for there is still a remainder, which the next step will embrace, and we have the perfect fraction $\frac{1}{2+\frac{1}{3+\frac{1}{1+\frac{1}{2}}}}$, which being condensed, gives $\frac{11}{25}$, as before.

$\frac{77}{175}$ then are about $\frac{1}{2}$, still nearer $\frac{2}{7}$, yet nearer $\frac{4}{9}$, and just equal to $\frac{11}{25}$. The terms of the series are alternately less and greater than the true value, but converge rapidly to it. One half is $\frac{3}{50}$ too much; $\frac{3}{7}$ is $\frac{2}{175}$ too little; $\frac{4}{9}$ is $\frac{1}{225}$ too much; and $\frac{11}{25}$ is just enough.

Before passing on to Circulating Decimals, we shall consider another series or two, all tending to prepare the way for a clear understanding of the decimal series.

The sum of the geometrical series $\frac{1}{2}+\frac{1}{4}+\frac{1}{8}+\frac{1}{16}+\frac{1}{32}$, &c., extended indefinitely onward, constantly approximates to the value of a unit, without however exactly reaching it; and of course without the possibility of ever passing it. It is analogous to two lines approaching forever without coinciding.

It is obvious however that the series may be made to approach nearer in value to unity, than any assigned difference; and we may consider the value of such a series, when extended to infinity as *a unit.* But when so extended, the extreme extension, or last term, is considered as diminished to nothing. The greatest term then is $\frac{1}{2}$ and the least 0, and the ratio 2; and having the extremes and ratio, to find the sum of the series, the rule is—"Divide the difference of the extremes by the ratio less 1, and the quotient increased by the greater term, will be the sum of the series."

$\frac{1}{2}-0=\frac{1}{2}$ difference of extremes; and $\frac{1}{2}\div 1$ (the ratio less 1)$=\frac{1}{2}$. Then $\frac{1}{2}+\frac{1}{2}=1$, the sum of the series. In the same way we may show that $\frac{1}{3}+\frac{1}{9}+\frac{1}{27}+\frac{1}{81}$ &c.$=\frac{1}{2}$; that $\frac{1}{4}+\frac{1}{16}+\frac{1}{64}$ &c.$=\frac{1}{3}$; that $\frac{1}{5}+\frac{1}{25}+\frac{1}{125}$ &c.$=\frac{1}{4}$, and so with other similar series. We might indeed commence with any number, and increase by any ratio, and the series would tend to some fixed amount, beyond which it could not pass, but towards which it would forever approximate; and which consequently we would regard it as equal to.

$$\tfrac{3}{4}+\tfrac{1}{2}+\tfrac{1}{3}+\tfrac{2}{9}+\tfrac{4}{27}\ \text{\&c.}=\tfrac{9}{4}=2\tfrac{1}{4}$$

$$\tfrac{3}{4}+\tfrac{9}{16}+\tfrac{27}{64}\ \text{\&c.}\qquad =3$$

$$\tfrac{2}{5}+\tfrac{4}{25}+\tfrac{8}{125}\ \text{\&c.}\qquad =\tfrac{2}{3}$$

$$\tfrac{5}{3}+1+\tfrac{3}{5}+\tfrac{9}{25}\ \text{\&c.}\qquad =4\tfrac{1}{6}$$

$$\tfrac{1}{10}+\tfrac{1}{100}+\tfrac{1}{1000}+\tfrac{1}{10000}\ \text{\&c.}=\tfrac{1}{9}$$

We may regard this then as another mode of expressing fractional quantities, by expanding them into infinite series.

But in addition to both the foregoing, there is yet another mode of expressing fractional quantities, that will require still closer investigation—*i. e.* expressing them in a series of other fractions, whose denominators shall decrease from unity in a tenfold ratio Thus $\frac{7}{8}=\frac{8}{10}+\frac{7}{100}+\frac{5}{1000}$. $\frac{15}{16}=\frac{9}{10}+\frac{3}{100}+\frac{7}{1000}+\frac{5}{10000}$.

We cannot sum up these series as we did some of the former, because the terms are not strictly a geometrical series, for though the denominators increase in a geometrical ratio, the numerators are regulated by no fixed law. We must therefore add the terms according to the ordinary rule for adding fractions, if we would find their sum.

It is not however, every fraction that can be expanded into such a series; and be limited in extent, while expressing the precise value of the fraction. The following would form infinite geometric series, that may be summed up as such.

$$\tfrac{1}{3}=\tfrac{3}{10}+\tfrac{3}{100}+\tfrac{3}{1000},\ \text{\&c., forever.}$$

$$\tfrac{1}{9}=\tfrac{1}{10}+\tfrac{1}{100}+\tfrac{1}{1000},\ \text{\&c., forever.}$$

The following would not terminate but still could not be summed as the above; for the ratio between all the terms is not the same.

$$\tfrac{1}{7}=\tfrac{1}{10}+\tfrac{4}{100}+\tfrac{2}{1000}+\tfrac{8}{10000}+\tfrac{5}{100000},\ \text{\&c.}$$

$$\tfrac{1}{13}=\tfrac{0}{10}+\tfrac{7}{100}+\tfrac{6}{1000}+\tfrac{9}{10000}+\tfrac{2}{100000},\ \text{\&c.}$$

All fractions having *in their denominators*, no other prime factors than 2 and 5, the prime factors of our scale of notation, will form a limited series. All multiples of $\frac{1}{9}$, and consequently of $\frac{1}{3}$, will form a geometrical series in which the same numerators will be repeated continually, while all others will form series whose denominators will increase in a uniform ratio, but their numerators will not; neither do they remain as the same. The reader can scarcely have failed to recognise in this series, our common decimal system. The numbers that can be changed to a finite series, are our ordinary decimals; while such as run into infinite series, are called *Repeating* or *Circulating Decimals.* All circulating decimals, if taken by their full periods, form geometrical series.

$\frac{1}{3}$ and $\frac{1}{9}$ produce *Single Repetends,* the figures being all alike.

$\frac{1}{11}$, and some others, produce *Compound Repetends,* the figures recurring alternately.

If other numbers precede the repeating ones, it is called a *Mixed Repetend,* as .1666, &c., is a *Mixed Single Repetend;* .378123123 &c., is a *Mixed Compound Repetend.*

Similar Repetends are such as consist of the same number of figures, and begin at the same place. *Dissimilar* are the reverse.

Similar and *Conterminous Repetends* are such as begin and end at the same distance from unity.

Perfect Repetends are such as have in the circle of repetition, as many places, save one, as there are units in the denominator. Sevenths circulate in six places, and consequently produce perfect repetends.

So far as regards the value and properties of decimals that terminate, we have heretofore considered the subject pretty fully, and we shall now take up such as circulate, and examine the laws by which they are governed.

For the sake of brevity it is usual to designate circulates by placing a period over the repeating figure, if it be a single repetend; and a period over the first and last figures, if the repetend consists of several places; thus $.\dot{6}$, instead of .666 &c., $.\dot{3}2\dot{4}$ instead of .324324324 &c.

As 1 with as many ciphers annexed as there are places in the decimal, is the proper denominator of a decimal, so a series of nines of as many places as there are figures in the circulate, is the proper denominator of a circulate. This may be readily inferred from what has been said on the sum of the series $\frac{1}{10}+\frac{1}{100}+\frac{1}{1000}$, &c., $=\frac{1}{9}$. $.\dot{6}$ then is $\frac{6}{9}$ instead of $\frac{6}{10}$, and $.\dot{3}2\dot{4}$ is $\frac{324}{999}$. If the circulate be mixed, as $.32\dot{1}4285\dot{7}$, then the part that does not circulate (.32) has the common decimal denominator, (100) while the circulating part has a series of nines, with as many ciphers annexed as there are finite places. The sum will then be correctly expressed $\frac{32}{100}+\frac{142857}{99999900}$. This is obvious, $\frac{1}{6}=\frac{1}{10}+\frac{6}{9}$ of another tenth, or $\frac{6}{90}$ of a unit. That is $\frac{1}{6}=\frac{1}{10}+\frac{6}{90}$, or $\frac{1}{6}=\frac{15}{90}$. If the circulate embrace integral numbers, as $\dot{2}.3\dot{7}$, the denominator will be nines, and the numerator must have as many ciphers added, as there are integral places. $\dot{2}.3\dot{7}=\frac{2370}{999}$, the reason of which is obvious enough, since $.\dot{2}3\dot{7}=\frac{237}{999}$, and $\dot{2}.3\dot{7}$ is ten times as great; we therefore add a cipher to the numerator to increase the value ten fold.

In order to determine whether a fraction will produce a terminating decimal, we may analyze the denominator, and see whether there is in it any prime factor but 2 and 5. If not, the decimal will terminate. *But how soon?* Every number is made up of its prime factors, either multiplied simply, or in certain powers. Ten is the product of the simple factors 2 and 5; $20=5\times2^2$; $40=5\times2^3$ and $50=5^2\times2$. Having reduced the fraction to its lowest terms, and determined the highest power of either factor in the denominator, the equivalent decimal will contain just as many places as there are units in the exponent of such power. $20=5\times2^2$, therefore any number of 20ths, if in their lowest terms, will resolve into decimals of two places, $\frac{17}{20}=.85$. $\frac{18}{20}$ can be reduced to $\frac{9}{10}$. $50=5^2\times2$ will give two places; $80=5\times2^4$ will require four places; 125 $=5^3$ will require three places; while $64=2^6$ will extend to six places, viz. .140625.

Perhaps in practice it would be best to divide by 5 as often as possible, and if the last quotient be not a unit, then divide it by 2 as often as possible, and if in dividing by either 5 or 2, the final resulting quotient be a unit, the fraction will terminate after as many places of figures as there are divisions by 5 or 2, whichever was contained the oftenest. By trying the foregoing numbers, the operation will be obvious.

But if in such division there is still a quotient not divisible by either 5 or 2, then some other prime factor enters into the number, and the decimal will begin to circulate after as many places, as there were 5's or 2's used, whichever was used oftenest.

To determine how many places of figures will be in the repetend, after dividing by 5 and 2 as often as possible, let the quotient of the last division be used as a divisor for a series of 9's, and just so many nines as must be used before nothing remains, just so many places the repetend will contain.

Will $\frac{33}{625}$ produce a finite decimal?

5	625
5	125
5	25
5	1

Yes, it will terminate in 4 places, for it involves 5^4.

Will $\frac{7}{12}$ terminate?

$$\begin{array}{r|l} 2 & 12 \\ 2 & 6 \\ & 3 \end{array}$$

And as 3 divided into 9 leaves no remainder, there will be a single repeating figure, after two finite places, as indicated by 2 being included twice as a prime factor.

$$12)\underline{7.00000}$$
$$.58333$$

Five and 8 are finite places, but the 3 will repeat forever.

What form will $\frac{19}{112}$ assume?

$$\begin{array}{r|l} 2 & 112 \\ 2 & 56 \\ 2 & 28 \\ 2 & 14 \\ & 7 \end{array}$$

5 will not divide.

The prime factors are 2, 2, 2, 2, 7; there will therefore be 4 finite places, and

$$7)\underline{999999}$$
$$142857,$$

shows that the repeat will be in periods of six places; and the quotient by dividing the nines gives the figures of the repetend. It is obvious that this must be so, and that as the decimal from $\frac{1}{7}$ will begin to repeat as soon as 1 occurs as a remainder, (for that will give the dividend with which the operation commenced) the series of nines being 1 less, than a unit with ciphers annexed, there must at that point be no remainder.

A circulate will occupy the same number of places, whatever may be the numerator of the fraction from which it is produced.

Every decimal must terminate or circulate before there are as many places as there are units in the denominator. The reason of this, and other facts that have been stated, will be given before we close this lecture.

In order to add or subtract circulates with perfect accuracy, they should be made similar and conterminous, for which purpose the least common multiple of the number of places may be ascertained, and all the decimals extended to the same number of places; the denominator of the repeating portion being regarded as a series of nines; and of the finite portion as decimals. In subtraction the right hand place of the minuend must be regarded as one less than the figure represents, as the denominator is 1 less. To multiply and divide, it is best to change the circulates to vulgar fractions, else the true result cannot be produced.

We shall now close the subject with a few propositions illustrating the principles involved.

Proposition 29.

All vulgar fractions, when changed to decimals, will either terminate or circulate.

Proposition 30.

If the prime factors of the denominator of any vulgar fraction, when in its lowest terms, be 5 and 2, or either of them, and no other, the resulting decimal will terminate.

Proposition 31.

If the prime factors of the denominator be 5 and 2, or either of them, the decimal will terminate, when the number of places shall equal the units, in the exponent of the highest power of 5 or 2 involved.

Proposition 32.

If the denominator contain neither 5 nor 2 as a prime factor, then the decimal will circulate from the commencement.

Proposition 33.

No circulate can contain as many places as there are units in the denominator of the vulgar fraction from which it is produced.

Proposition 34.

If the denominator of any vulgar fraction in its lowest terms, contain either 5 or 2, combined with any other prime factor, or factors, the circulate will be preceded by as many finite places as the powers of 5 or 2 involved, would indicate, according to Proposition 31.

Proposition 35.

Repeating and Circulating Decimals constantly approximate to the true value of the vulgar fraction from which they are produced.

Proposition 36.

The proper denominator of a finite decimal is a power of 10 whose index is equal to the number of places in the numerator of the decimal.

Proposition 37.

The proper denominator of a circulating decimal, is a series of nines equal to the number of places in the numerator.

The foregoing propositions, taken in connection with what has been said in the preceding and present lectures, embrace all that need be said on the subject of vulgar and decimal fractions; and it now remains only to show the truth of some of those propositions, so far as the same are not obvious from what has been already said. When speaking of fractions, we generally mean the form designated as vulgar fractions, but the principles are just as true of duodecimals and all other compound quantities, the inferior denominations being regarded as fractions of the unit of the system. Proposition 29 might be inferred from what follows, but we wish to call the reader's attention distinctly to the fact, that if a fraction does not terminate, it must run into a circulate, and cannot run on beyond a certain limit, in places that show no fixed law.

Propositions 30 and 31 may be taken together, and will be found to involve an important law of numbers; and for convenience we may consider them together. If the denominator be 10 or any of its powers, it is obvious that it will be a decimal in principle, whether it has the ordinary form of one or not. $\frac{7}{10}$ is as much a decimal as .7; and $\frac{77}{100}$ as .77. If the denominator be 10, it will give the simple factor 5×2; if 100, then $5^2\times2^2$; if 1000, then $5^3\times2^3$; and so on forever, for as these numbers are powers of 10, they are the products of similar powers of 5 and 2, the prime factors of 10. If a number contains 10 once, it must contain 5 and 2, the prime factors of 10, once. If it contain the square of 10, it must contain the square of 5 the square of 2 times, or *vice versa*, for each pair of such divisors is equivalent to 10. And as the square of 10 gives 3 places, the cube 4 places, &c., and the numerator must contain one place less, the number of places in the decimal, or numerator, will be just equal to the exponent of the power of 10, or its prime factors, involved. So much then for cases in which each factor is involved to the same power. Indeed, all such fractions are decimals already.

But if one factor be involved to the 2d, and the other to the 3d power, or if to any other different powers, why should the decimal terminate? And if it does terminate, why should the number of places equal the exponent of the highest power involved?

So far as the lowest power extends, it forms, with corresponding powers of the other factor, a series of tens, and will produce as many ciphers in the denominator of the given fraction, as there are such similar powers involved. Hence, some authors say, "Divide by 10 as often as possible, and then by 2 or 5 until the quotient is a unit, and the number of divisions will show the number of places that will be in the decimal." This is the same in principle as the above, for the number of divisions by 2 *or* 5 (of course you cannot divide by 2 *and* 5, if the dividend will not contain 10) will be according to the excess of the highest power above the lowest involved.

Change $\frac{1}{20}$ to a decimal.

$20=5\times2^2$, or $5\times2\times2$. Hence the decimal will contain two places. Or $20\div10=2$; and $2\div2=1$.

Here there are two divisions and the decimal will contain two places. The 5 and one of the 2 s$=10$, and will require one decimal place, while the other two will require another place; and it would require but the same if the additional factor were 5, for either will divide 10 without a remainder. If the square of either remained, two places would be necessary, for either 2^2 or 5^2 will require the addition of two ciphers to any significant digit, to divide evenly; and in the same way the cube of either will require three places; and this is equally true, whether the numbers stand by themselves, or are involved with others.

Or we may consider that if the powers of the factors are alike, the denominators must be 10, 100, 1000, &c., being a unit, with ciphers annexed. If either factor be of a power higher than the other, the ciphers will not be altered, but the significant figures will be. If 2 or 5 be the single factors, they will require one cipher to be added to the significant figure; if 2^2 or 5^2, they will require two places; 2^3 or 5^3 will require three places, &c.

We may make this plainer by adopting the doctrine of proportion. What is the decimal equivalent to $\frac{3}{5}$?

As 5 : 3 : : 10.

$$\begin{array}{r} 10 \\ \hline 5)\overline{30} \\ \hline 6 \\ \hline \end{array}$$

Had the denominator of the given fraction been 5^2or 2^2, the third term must have been 100; that is, two ciphers must have been added. It is evident that there must be as many ciphers as there are pairs of factors, or powers of single factors; each additional power of either requiring an additional place. It may be necessary to remark that whatever number of ciphers must be added to obtain a complete finite decimal or a complete circulate, when the numerator of a fraction is a unit, the same number will be necessary with any other numerator. $\frac{1}{25}$ will require two places in its decimal form, (.04) then will any other number of 25ths require two places.

Propositions 32 and 33 may be noticed together.

It is very clear that as only ciphers are added, whenever a remainder occurs either similar to the numerator, which forms the significant portion of your dividend, or to any previous remainder, the same dividuals, the same quotient figures, and the same remainders must recur continually, and thus the quotient will become a succession of circulating periods. It is evident that every possible remainder must occur before the quotient will reach as many places as there are units in the divisor. Take $\frac{1}{7}$ for instance.

$$7)\underline{1.0000000}$$
$$\underline{.1428571}$$

Here the division continues to six places, before a remainder occurs of 1; but as soon as it does occur, the next dividual is 10, and the repetition commences. The remainders are 3, 2, 6, 4, 5, 1; being every unit less than the divisor.

But how is it that no remainder will in any case recur, until the numerator of the fraction has been reproduced? Why for instance in changing $\frac{1}{7}$ to a decimal, as above, might not some remainder occur, similar to some previous remainder, before 1 occurs, which it does at the sixth place? Take any other number of 7ths, and the same result will be produced. But if 2 or 5 had entered into the composition of the denominator, the circulation would have commenced at the second place; if either of those numbers squared, then at the 3rd place; and thus onward.

Let us change $\frac{3}{7}$ to a decimal.

$$7)\underline{3.000000}$$
$$.428571 \text{ \&c.}$$

Taking this special case as an example, let us suppose that the circulation commences at the sixth place, by a remainder of two occurring, then will the period 28571 be a circulate, and 4 be a finite decimal. But by the supposition of the case, the divisor (7) is prime to 10, (for it is divisible by neither 2 nor 5, the prime factors of 10,) and by a principle advanced in our fourth lecture, "If there be two numbers, prime to each other, the product of neither by a third integral number, can be measured by the other." But if the number .4 be finite, then 7 and 30 are not prime, but have a factor in common; which by the principle just cited they cannot have,—the supposition is therefore absurd, and .4 is not finite. In the same way, we may show the absurdity of supposing any other portion of the decimal expression finite; and what is shown of 7 may be shown of any other divisor prime to 2 and 5.

The decimal, therefore, must circulate from the commencement, and within a period whose number of places is less than the given denominator.

By Proposition 34, we are taught that if the denominator of the fraction contain 5 or 2, either in their prime or higher powers, there will be a corresponding number of finite places, after which the decimal will circulate according to the law of the other prime factors, as though they occurred alone.

The question may be asked why the finite portion should come first in the result, and this we cannot better answer than by taking an example as a subject of remark.

Change $\frac{1}{35}$ to a decimal.

We will separate the denominator into the factors 7 and 5; and proceed thus:

```
7)1.000000000000
  5).142857142857 &c.
    .02857142857 &c.

5)1.000000000000
  7).200000000000
    .028571428571 &c.
```

It has been already shown that the division by 5 will terminate, but that the division by 7 will circulate; and it only remains to show that the finite places will be the first of the series. It is obvious from the above that the 5 affects but one place, and that is the first of the series; and in the same way it may be shown of any other composite denominator. If 2 or 5 square enters, then it will affect two places, &c. It may not be improper to remark that similarity of remainders is not all that is necessary; for the residue of the dividend must be similar. If we divide first by 5 it will affect but one of the ciphers; if by 5^2, it will affect two, &c.; and the same remark applies if 2 be used.

The truth set forth in Proposition 35, is obvious from what has been said on the subject of series. Proposition 36 has been fully elucidated; and it was shown in the early part of this lecture, that a series of tenths, decreased by a common ratio of $10 = \frac{1}{9}$. That is $\frac{1}{10} + \frac{1}{100} + \frac{1}{1000} + \frac{1}{10000}$, &c., to infinity $= \frac{1}{9}$; and this is strictly analogous to a circulating decimal. Perhaps in strictness, circulating decimals should be called nonary fractions, as their denominators are nines instead of tens.

We might pursue the investigation farther, but it is believed that the foregoing will be found to embrace all that is useful or important; and he that is curious in such matters has enough to arouse his mind to the subject.

LECTURE VII.

PROPORTION.

THIS word, like many others that are applied technically as well as in common discourse, is freely used by every one, while few examine it with scientific care. In general conversation we say that a building, a piece of furniture, or an animal is in proportion or otherwise, according as the parts are properly adapted in size and place in reference to each other, agreeably to nature's standard, or general custom if it be a work of art. Nature is exceedingly uniform in her works, seldom changing her standard, but pursuing her old and beaten track. With the works of man it is far otherwise, for as whim and taste change, the laws of proportion change also, except when nature's works are looked to as a standard; and we are so constituted that use will reconcile us to what at first seemed monstrous. At one time we see our coats reaching down far towards our feet, and then the tailor flourishes his shears, and, inch by inch, the mandates of fashion bid him abridge the length of the garment, until it reaches the opposite extreme, and in the language of Burns, becomes "*unco scanty.*" The latitude as well as longitude changes, but if we would have our garments in proper *proportion*, as adjudged by connoisseurs in such matters, they must conform to the laws of fashion, for the time being. Still there is a natural fitness in things, and extremes are soon abandoned; while good taste is gratified when the means used seem adapted to effect the end designed. The orders of architecture established by the ancient Greeks, are still the standards of correct proportion, because none are found better adapted to the purposes for which they are intended. The massive Tuscan and the Doric are placed where strength seems necessary to support the superstructure, while the light Ionic and Corinthian, with their long and tapered columns, support the lofty portico of lighter structure. There is a natural fitness between the end and the means. Rearing a mighty pillar to support a trifling weight, would be as absurd as loading a cannon to shoot a fly. But our business is not with the word in this wide sense, but rather in its technical meaning.

The geographer constructs his map of a country by adopting his scale and then reducing every part of the territory in the same ratio. He lays down his plan so that when his work is done all may be in proportion, and the little map be the great country in miniature. If the scale is a tenth of an inch to a mile, then measure what part of the map you may, a tenth of an inch on the map will be equivalent to a mile upon the land, or an inch upon the map, to 10 miles upon the land.

The relation which one line or number bears to another, in regard to magnitude, is called its *ratio*; thus if one line be 3 times as long as another, or one number three times as great as another, the ratio is as 1 to 3, and if several numbers thus increase as 1, 3, 9, 27, 81, &c., we say they increase in a triple or three fold ratio.

The first number is called the *Antecedent*, the second the *Consequent*. In expressing *ratio*, the French make the antecedent the *denominator*, and the consequent the *numerator* of a fraction; while the English place the *antecedent* as *numerator*, and the *consequent* as *denominator*. In expressing the ratio of 8 to 4, the French would say it is $\frac{4}{8}$ or $\frac{1}{2}$, the English $\frac{8}{4}$ or 2. Both modes of expression are met with in American books. The former mode expresses the multiplier necessary to make the antecedent equal to the consequent; the latter the multiplier that would make the consequent equal to the antecedent.

So also ratio is sometimes expressed by the mark of division, $4 \div 8$ or $8 \div 4$, instead of $\frac{4}{8}$ or $\frac{8}{4}$.

It is plain that various numbers may have the same *ratio* to each other, thus 2 has the same ratio to 4, that 3 has to 6, for 2 is half of 4 and 3 is half of 6 ; and when four numbers have this equality of ratio, they are said to be proportional : *i. e.* equality of ratio is *proportion.* Numbers thus proportional are arranged for calculation thus, 2 : 4 : : 3 : 6, and they are read, *As* 2 *is to* 4 *so is* 3 *to* 6; or *As* 2 *are to* 4, *so are* 3 *to* 6.*

The word proportion is often used as synonymous with ratio, as when I speak of the proportion of 4 to 8, 3 to 9, &c., but this is incorrect, we compare 2 numbers by their *ratio ;* and if between two pairs of numbers, the ratio is equal, we say they are in proportion ; but if the ratios are not equal, the numbers are not proportional. We cannot properly say 2 : 4 : : 3 : 9, for a proportion does not exist ; 3 is a third of 9, while 2 is half of 4; but we might say 2 : 4 : : 3 : 6, for that is true.

A little attention to the distinction here drawn between Ratio and Proportion, may save the young student from the quandary of Dr. ——'s pupil, who was party to the following dialogue.

Dr. Pray, sir, what is ratio ?

Pupil. Ratio sir? ratio is proportion.

Dr. And what is proportion ?

Pupil. Proportion sir ? proportion is ratio.

Dr. And pray sir, what are they together ?

Pupil. Excuse me, I can answer but one at a time.

The doctrine of ratio and proportion is the basis of the Rule of Three, a rule formerly deemed so important by mathematicians and business men as to be called the *Golden Rule.* The numbers of the statement abstractly considered are proportionals, and their application shows that they should be proportional, for it is natural that as one quantity is to any other quantity, so will be the price of the one to the price of the other, if the rate be the same ; or as any quantity is to its price, so is any other quantity to its price.

If 3 *pounds of butter cost* 30 *cents, what will* 9 *pounds cost?* To find the value of the 9 lbs., we have but to consider that

* The practice of using the singular form of expression in reading proportion is very common, but its propriety is doubtful, for the numbers 2, 4, &c., certainly express plurality of idea ; and especially so when used applicately. *As* 2 *yards is to* 4 *yards, so is* 3 *dollars to* 6 *dollars,* sounds very much like doing violence to one of the leading rules of syntax ; and yet some object to the expression *As* 2 *are to* 4 *so are* 3 *to* 6, as pedantic. Neither does it seem clear that there is less plurality in an abstract than in an applicate number, that 3 is less a plural number than 3 men or 3 houses. In the above the singular is used in deference to custom, It is true that we sometimes consider even a large number as a unit, or simply as a number, but it does not seem that the size of numbers can be thus compared, unless we consider numbers as representatives of magnitude and not as made up of many distinct parts.

as 9 is three times as much as three, so 9 lbs. will cost 3 times the price of 3, and 3 times 30=90; or as 3 lbs. cost 30 cents, 1 lb. will cost $\frac{1}{3}$ as much, viz, 10 cents, and 9 lbs. will cost 9 times 10=90 cents. Here then we have a proportion, As 3 lbs. : 9 lbs. : : 30 cts. (the price of 3 lbs.) : 90 cts. (the price of 9 lbs.)

Or, As 3 lbs. : 30 cts. (the price of 3 lbs.) : : 9 lbs. : 90 cents, (the price of 9 lbs.) But it is evident that the rate must be the same, or the proportion would not exist; for if the last lot of butter cost 12$\frac{1}{2}$ cents per pound, it would amount to 112$\frac{1}{2}$ cents, and the ratio between the first quantity and its price, would be different from the ratio between the last and its price; and without equality of ratio there can be no proportion.

Every child would perceive that such a question as the following would be absurd and ridiculous, for want of connection between the antecedents and consequents: If 3 lbs. of butter cost 30 cents, what are 20 horses worth? Or, If wheat is worth one dollar per bushel, how far is it from Wheeling to Columbus?

Our proportion must be between similar things at similar prices; and not between dissimilar things or things at dissimilar prices; though it is true we might estimate dissimilar things at the same rate, and then our calculation would apply. For instance "If 3 lbs. of butter cost 30 cents, what will 9 lbs. of lard come to at the same rate?" is a fair question.

The colon : denotes that two numbers are compared in reference to the ratio between them, and the double colon : : denotes equality of ratio; but instead of using the double colon, the parallel mark of equality is used by the Germans, thus 3 : 30=9 : 90, implying that the ratio of 3 to 30 is equal to the ratio of 9 to 90; and when we proceed to express the ratio of each pair in a fractional form, this mark seems peculiarly appropriate, thus $\frac{3}{30}=\frac{9}{90}$, which is literally true.

The proportion to which we have alluded has reference to the quotient of the antecedent divided by the consequent, or *vice versa*, and in deducing one number of the series from another, we multiply or divide by the ratio, according to the requirement of the case. This relation of numbers is called *Geometrical Proportion*, and where the numbers of a series increase or decrease from multiplying or dividing by a common ratio, it is called a *Geometrical Progression.* Thus the numbers 1, 3, 9, 27, 81, &c., form a geometrical progression increasing by the common ratio 3; and 81, 27, 9, 3, 1, is another series decreasing in the same ratio. We often meet with questions involving this progression, and indeed in most arithmetics they are classed under a rule called Geometrical

Progression; though they may all be solved so far as any practical purpose is concerned without any special rule. The following question is of this class:

"A gentleman, whose daughter was married on new year's day, gave her a guinea, promising to triple it on the first day of each month in the year; pray, what did her portion amount to? *Ans.* 265720 guineas."

It is here plain that we may write down the months in order and set the payments received opposite, and then find their amount; thus—

January	1
February	3
March	9
April	27
May	81
June	243
July	729
August	2187
September	6561
October	19683
November	59049
December	177147
Total	265720

In treating this subject it is usual to divide it into a number of cases, according to the parts given and the parts required. In this problem we have the first term (1 guinea,) the ratio, (3) and the number of terms, (viz., 12, the months in a year,) to find the sum of the series; and with less data we could not find the sum of the series, neither could we find the last term, 177147. But if we had the first and last term and the number of terms, we could find the ratio, or we could find the sum of the series. So we may successively consider certain parts given and others required, until we make up 10 cases; but for every ordinary purpose the whole may be dispensed with, though few rules of Arithmetic contain more important principles than this does. Hence to the speculative mathematician it is indispensable, but for his use its principles are fully set forth in Algebra.

The following questions may be solved like the above, though they could be done much more briefly by some special rules.

1. A young fellow, well skilled in numbers, agreed with a rich farmer to serve him 10 years, without any other reward than the produce of one wheat corn for the first year, and the annual produce to be sowed from year to year, till the end of the time. What is the produce of the last sowing, allowing the increase but in a tenfold ratio, and what will it amount to at $1.25 a bushel, allowing 7680 grains to a pint?

Ans. 10000000000 grains, worth $25431.30.8+

2. A man travelled 252 miles: the first day he went 4 miles and the last 128, and each day's journey was double the pre-

ceding one. How many days was he performing the journey? *Answer*, 6.

Many more questions might be given, but it is not necessary to introduce them here; neither shall we now take up the laws of Geometrical Proportion, but we will pass to the other mode of comparing numbers, viz., by their differences; first, however, remarking that a sum of money at compound interest, in its increase at stated times, forms a geometrical series, of which the amount of $1 for the time between the additions of interest, forms the ratio.

In the series we have considered, the numbers were increased and decreased by *multiplication* and *division*, but it is plain that a series may be formed by the constant *addition* or *subtraction* of a common difference; thus: 1, 4, 7, 10, 13, 16, &c., is a series increasing by the common difference 3; and if we reverse it, it will be a series decreasing by the same common difference. This is called Arithmetical Proportion, or Arithmetical Progression, and may be extended to ten cases, according to the parts given and the parts required.

Such questions as the following fall under this rule:

1. How many strokes does a regular clock strike in a year? *Answer*, 56940.

2. A body falling by its own weight, not resisted by the air, would descend in the first second about 16 feet 1 inch; in the next second through 3 times that space; in the third through 5 times, in the fourth through 7 times, &c. Through what space, at the same rate of increase, would it fall in a minute? *Answer*, 57900 *feet*.

The first term; the last term; the common difference, sometimes called the arithmetical ratio; the number of terms; and the sum of the series, are the five parts usually given or sought in this rule; any three of which being given the other two can be found. The name Arithmetical Progression, as well as Arithmetical Proportion, is objected to by many, and the term *Equi-different Series* preferred. Prof. Thompson says, "these names, Geometrical Proportion and Arithmetical Progression, for continued proportionals and equi-different quantities, are highly improper. Series of both kinds belong equally to Arithmetic and Geometry. The appellations Arithmetical Progression and Geometrical Progression, should be entirely disused as tending to impress false ideas on the mind respecting the nature of the quantities. The term proportion is applied, if possible, still more improperly to equi-different quantities, as this term is always expressive not of *equality of differences*, but of *equality of ratios*. The latest and best continental writers have accordingly rejected these terms, and substituted more appropriate ones, calling them by the names above given, or others of similar import, such as *progressions by differences* and *progressions by quotients*. With respect to the name *continued proportionals*, applied to the second kind of quantities, it may be observed that besides its being perfectly expressive of the nature of such quantities, it has long been thus applied in works on Geometry, and it is equally applicable in arithmetic."

It need scarcely be remarked that the increase is much less rapid in a series of this kind than in a Geometrical one, but there are relations between the series that make it desirable to speak of their properties together.

In both series the first and last terms are called the *extremes*, and the intermediate the *means*, and if there be 3 terms in the series, as 2, 4, 8, the product of the extremes will be equal to the square of the mean, $2\times8=16$ and $4^2=16$; but if it be an equi-different series, the sum of the extremes will be double the mean: thus, 2, 4, 6; $2+6=8$ and $4+4=8$; *hence to find the geometrical mean between two numbers we multiply them together and take the square root, while to obtain the arithmetical mean, we add them together and take half.*

If there be four terms in a geometrical series, the product of the extremes is equal to the product of the means; and if there be any number whatever the product of the extremes is equal to the product of any two terms equally distant from the extremes. Thus as 2 : 4 : : 8 : : 16; here $16\times2=32$ and $4\times8=32$; and if we extend the series, thus 2, 4, 8, 16, 32, 64, the products of 64×2, 32×4, 16×8 will all be equal, and had we carried the series to an odd number, the square of the middle term would have been equal to the other products.

If the series be equi-different the *sum* of the extremes will be equal to the sum of the means, or *double* the middle term if odd, as may be seen by inspecting the following series, where 16 will be the sum of each pair: 2 4 6 8 10 12 14. Thus we find that addition and subtraction in one correspond with multiplication and division in the other.

The law of the geometric series, that the product of the extremes is always equal to the product of the means, (which we shall presently explain) gives rise to the mode of solution in the Rule of Three, when treated as a rule of proportion; viz: multiplying the 2nd and 3rd terms together and dividing by the first, to find the 4th term or answer. In every statement of four proportionals the 2nd and 3rd are the means, the first and 4th the extremes, and by the rule, we multiply the means together and divide by the first term or given extreme for the other extreme; which is the number sought. The product of the means is in effect the product of the extremes, and if we divide the product of any two numbers by one of the numbers, the quotient will be the other. In the Rule of Three we state the question, the 2nd and 3rd terms of which are the two means that are to be multiplied together, the first term is the given extreme and the fourth term the extreme sought.

If two yards of cloth cost $5 what will 6 yards cost?

As 2 yds. : 6 yds. : : $5
6
2)30
The answer, $15

Here we have one extreme (2) given, and we want the other. We have no direct mode of finding it, but we have two numbers given us, and by multiplying them together, we shall find a number equal to the product of the given extreme by the unknown one, and in all cases the product of two factors being divided by one of them, will give the other, as we have already shown. You will perceive that the proportion is never perfect until the answer is read as the fourth term:

As 2 yds. : 6 yds. : : $5 : $15.

This mode of stating the Rule of Three has become very general within a few years, and it is found that the young mind readily catches the idea, for it is not difficult for even a child to see that 2 bears the same ratio to 6, that 5 bears to 15; for 2 is a third of 6 and 5 is a third of 15. And it is reasonable in a business point of view that two yards should be proportionate to 6 yards, as the price of 2 to the price of 6. It is about as plain as the little girl's reply to Miss Pepper's niece, when asked her relationship to her playmate Mary, who was her aunt, but she did not wish to say so. Turning to the inquirer she stated the relation, "*as Miss Pepper is to you, so is Mary to me.*"

This mode of stating compares things of like kind, and not such as are unlike. By the old mode we would have said

As 2 yds. : $5 : : 6 yds. : $15.

Thus comparing yards and dollars, which involves an absurdity; and this may appear more plain by varying the question.

If I give a cabinet-maker 2 hogs for a bureau, how many must I give for 3 bureaus?

bureau,	hogs,	bureaus,	hogs.
As 1 :	2 : :	3 :	6

It seems difficult to understand the ratio between a hog and a bureau; though we may readily enough understand that 2 hogs will bear the same ratio to 6 hogs that 1 bureau does to 3 bureaus—each is a third of the other: but a hog is not half a bureau, though two hogs might pay for one.

Another important advantage resulting from this mode of statement is, that the distinction between Direct and Inverse Proportion is dispensed with.

In the proportionals we have considered, the increase or decrease is onward in direct or straight forward proportion; and hence its applicability to estimating value as applied in the Rule of Three and to all other purposes where "more requires more" or "less requires less." If I buy goods, the more I buy the more I must give for them, and the less I buy the less I shall be required to give for them. The more work I have to do the more time it will take, and the less, the less time.

But then there are many cases wherein this *Direct* or straight forward proportion, as between cause and effect, does not apply. In doing work for instance, the more hands are employed, the less time will be necessary; and the fewer hands the more time. Or suppose I wish to buy a coat, the less the width of the cloth is, the greater must be the length; and the greater the width the less length is necessary. In all such cases the proportion is inverted; it is "more requiring less," or "less requiring more;" and by the old mode of statement a peculiar mode of solution was used. Instead of multiplying the second and third terms and dividing by the first, the first and second terms were multiplied and the product divided by the third. Here the 1st and 2nd are the means of the proportion and the 3rd the given extreme. An example may make this plain.

I have a board 72 inches long and 18 wide, but I wish to obtain another board that shall be 24 inches wide and be just equal to the above, how long must it be?

inches	inches	inches
As 18 :	72 : :	24

It is here plain that as the new board is wider than the old one, it must have less length to make just the quantity; but if I adopt the rule of direct proportion, the length will be increased in the same ratio as the width, for multiplying 24 by 72 and dividing the product by 18 will give 96 as the length, and such a board would contain $24 \times 96 = 2304$ sq. inches, instead of 72×18=1296 sq. in. I must then multiply the first and second, viz: 18 by 72 and divide the product by 24, by which I get 54 inches as the length; and that this would give the proper length we may see (disregarding proportion) for multiplying 18 by 72 I get 1296, which must also be the product of the length and breadth of the new board, for it is to contain the same, and this product divided by 24, one of the factors, must give the other.

As to the scientific nature of *Inverse Proportion*, it is when the first term of the series is to the third, as the fourth to the second, and *vice versa:* thus 2, 9, 6, 3, are inversely proportional, for 2 is to 3 as 6 to 9, or 2 to 6 as 3 to 9. This is sometimes called *reciprocal* proportion. If we arrange the numbers in regular order so as to form a direct proportion, they will be As 2 : 6 : : 3 : 9, As 6 : 2 : : 9 : 3 and here it is plain if we had the 1st, 2d and 3d, to find the 4th we must multiply the 1st and 2d of the inverse statement, (which are the *means* of the proportion,) and divide by (6) one of the extremes, for (3) the other. Numbers inversely proportional, may be made directly so, by change of arrangement: and *vice versa.* Let us examine a problem involving the principle of inverse proportion.

If 4 men build a house in 12 days, how long should 6 men be engaged in doing a similar job?

	men	days	men	days
As	4 :	12 : :	6 :	8

Here the first term, (4 men) is to the 3rd, (6 men) as the 4th, (8 days) is to the 2nd (12 days) 4 and 12 being the extremes of the series, 6 and 8 the means, hence 4 and 12 must be multiplied together and the product divided by 6 (one mean) to find 8, (the other.)

In explaining this calculation without reference to the doctrine of proportion we multiply 4 by 12 to find the number of days' work there would be in the job for one man, and find the number to be 48, and dividing this by 6, we find 8 day's work for each man. We do not multiply 12 days by 4 men, but simply multiply 12 days by 4, because 4 men would do 4 times the work of one man, or as much as one would do in 4 times 12 days; or we may consider that 4 men are multiplied by 12 to show how many men could do the work in one day.

Some state the Rule of Three by using *If* instead of *As*, and so arranging the numbers that they make the same sense as in the proposed question. This mode keeps up the distinction between Direct and Inverse Proportion, but is readily applied and explained.

	men		days		men	
If	4	:	12	: :	6?	Ans. 8 days.

This may be read, If 4 men (do the work in) 12 days (in what time will) 6 men do it? and being Inverse Proportion we multiply together the 1st and 2nd terms and divide by the 3rd.

We may also read the *general* statement with If. Thus, If 6 (were) 4 (what would) 12 be?

Instead of multiplying the 2nd and 3rd terms together and dividing by the first, we may in any case produce the fourth term by multiplying or dividing the third term, by the ratio of the first to the second. This is obvious, for the 4th term must be as many times greater or less than the 3rd as the 2nd is than the 1st, in order to have the same ratio, and to constitute a proportion.

As 18 inches : 72 inches : : 24 inches : 96 inches

Here instead of multiplying 24 by 72 and dividing by 18, we see at once that the ratio of 18 to 72 is 4, and 4 times 24 are 96. The same course may be adopted in any problem, but the ratio found by dividing the second by the first would often be fractional, and thus increase the labor of the process. Thus $2\frac{1}{2} : 7\frac{1}{3} :: 10$ would work badly in this way; but as $2\frac{1}{2} : 7\frac{1}{2} :: 10$ would work handsomely.

When however the statement is made according to the general rule, the numbers become directly proportional, instead of inversely so; thus:

```
       men   men  days
    As  6  :  4 : : 12
                     4
                    --
                  6)48
                    --
                     8 days.
                    --
```

If four numbers be proportional they may be variously arranged and will be proportional still, or they may all be multiplied or divided by the same numbers, and the results will be proportional; and hence the mode of abridging in the Rule of Three, by dividing, for as one number is to another, so is the half or any given part of one to the half or any given part of the other. Thus—

	2	:	6	: :	5	:	15
By Inversion	6	:	2	: :	15	:	5
or	5	:	15	: :	2	:	6
Alternately	2	:	5	: :	6	:	15

Or Compoundedly, As 2 : (2+6=)8 : : 5 : (5+15=)20
Or As 2 : (6—2=)4 : : 5 : (15—5=)10
Or As (2+6=)8 : (6—2=)4 : : (5+15=)20 : (15—5=)10
Or multiplied, say by 3, as 6 : 18 : : 15 : 45
Or divided, say by 2, as 1 : 3 : : 5 : : 15, and it can make no difference whether both pairs of terms or only one be thus divided or multiplied, for the ratio remains unaltered when both the numbers compared are similarly changed. When all the terms form a regularly increasing or decreasing series it is called a continued proportion or series, thus, 2 : 4 : : 8 : 16; but if the antecedent of the second pair, bears a ratio to the consequent of the first, different from the ratio between each antecedent and its consequent, then the proportion is said to be discontinued: thus, as 2 : 4 : : 5 : 10. Here 2 and 4, 5 and 10 bear the same ratio to each other, but the proportion is "discontinued" through want of a similar ratio between 4 and 5.

"*Harmonical Proportion*," (says Nicholas Pike,) "is that which is between those numbers which assign the lengths of musical intervals, or the lengths of strings sounding musical notes; and of three numbers it is when *the first is to the third, as the difference between the first and second is to the difference between the second and third*, as the numbers 3, 4, 6. Thus if the lengths of strings be as these numbers, they will sound an octave 3 to 6; a fifth 2 to 3, and a fourth 3 to 4.

Again, between 4 numbers, when *the first is to the fourth, as the difference between the first and second is to the difference between the third and fourth*, as in the numbers 5, 6, 8, 10, for strings of such lengths, will sound an octave 5 to 10; a sixth greater, 6 to 10; a third greater, 8 to 10; a third less, 5 to 6; a sixth less, 5 to 8; a fourth, 6 to 8.

A series of numbers in harmonical proportion, is reciprocally, as another series in arithmetical proportion.

As { Harm. 10 .. 12 .. 15 .. 20 .. 30 .. 60
 Arith. 6 .. 5 .. 4 .. 3 .. 2 .. 1 } for here 10 : 12 : : 5 : 6

and 12 : 15 : : 4 : 5, and so of all the rest. Whence those series have an obvious relation to, and dependence on each other."

When numbers are compared according to their squares, it is called their duplicate ratio; and if according to their square roots, it is called their subduplicate ratio.

The sum of any arithmetical series may be found by adding together the extremes and taking half the sum, which is the mean of the series; this mean being multiplied by the number of terms will give the sum of the series. Thus;

$$2+4+6+8+10+12+14=56$$

To find the sum, 56, without the trouble of adding all the numbers, add together 2 and 14=16 and half that sum will be the mean or average of the series=8, which multiplied by 7, the number of terms, will give 56, the sum of the series. The mean may be taken without taking half the extremes, for by counting the terms the middle or mean one is found to be 8, and all on the left of 8 will be found to fall just as far short of 8 as the corresponding numbers on the right exceed 8; so that 8 is an average of the series. Six is 2 less, 10 is two more, 4 is 4 less, 12 is 4 more, &c., or add up the series and divide the sum by 7, the number of terms, and the quotient is 8. Or the idea may be made still plainer by supposing you have a floor 2 feet wide at one end and 14 at the other, the width at each foot being as stated in the series, and the length 7 feet. To find the area you may, if the sides and ends are true, take the mean or average width by a single measurement taken across the middle, and multiply this by the length; or you may take half the width of the ends; or you may take the width in any number of places at equal distances from the middle, and divide the sum of the widths, by the number of them for the mean or average width. It is plain that no advantage can result from taking many widths, unless the surface is irregular in width, as valuable boards sometimes are; for if straight and tapering regularly, one width is as good as a hundred. In that case if you draw a line lengthwise at the mean width it will be found that what is cut off towards the wide end, will if laid along the narrow end make the board, floor, or whatever it may be, of the same width from end to end. The redundancy of the wider end will just make up the deficiency of the narrower one.

The geometrical mean is less frequently needed in calculations, though it is sometimes necessary, and as in an arithmetical series, the mean doubled is equal to the sum of the extremes, and hence we find it by taking half the sum of the extremes; so in a geometrical series, the square of the mean is equal to the product of the extremes; and hence to find it we multiply the extremes together and take the square root of the product.

Rules may be given for finding any number of means that may be wanted, thus constructing a series to any ratio, or any common difference, but their elucidation is not important to our purpose.

If we have a geometrical series as 2, 4, 8, 16, 32, 64, 128, 256, &c., and place over them an arithmetical series to mark their places, thus:

1	2	3	4	5	6	7	8	9	10
2	4	8	16	32	64	128	256	512	1024

the terms of the arithmetical series are called the indices (indexes) of the geometrical series, and have this remarkable property, that the addition of the arithmetical terms corresponds to the multiplication of the geometrical. If I wish the 10th term, I multiply together any two terms the sum of whose indices is 10 and their product will be the tenth term. Take for instance the 4th and 6th, and their product 16 by 64 is 1024; so would 9th and 1st, 8th and 2nd, 7th and 3rd, 6th and 4th or, 5th squared. So suppose I wish the 15th term, I multiply the 10th, 1024, by the 5th, 32, and the product, 32768 will be the 15th. It is this principle carried out that constitutes the basis of the logarithmic system; an invention that has more effectually immortalized the name of Napier, than would the conquest of a kingdom.

Having taken a general view of this subject, we shall now present such propositions as are connected with it.

Proposition 38.

If four numbers are arithmetically proportional, the sum of the extremes is equal to the sum of the means.

2, 4, 6, 8. 2+8=4+6. Here the common difference is embraced once in the 2nd term, and twice in the 3rd, hence their sum consists of twice the first term +3 times the common difference; and the 4th is composed of 3 times the common difference+1st term, hence the sum of 1st and 4th consists of twice the 1st+3 times the common difference; just as the sum of the 2nd and 3rd was. Therefore the sum of the extremes is equal to the sum of the means; and this is true of any two means equally distant from the extremes: and if the number of the terms in the series be odd, double the middle term will be to equal the sum of the extremes.

Proposition 39.

If four numbers are geometrically proportional the product of the extremes is equal to the product of the means.

As 3 : 6 : : 4 : 8. Here the product of 3 times 8, the extremes, =24; and the product of 6 times 4 the means, =24:

and it cannot be otherwise if the numbers are proportional. Let us say, As 3 : 3 : : 4 : 4; and then multiplying the 2nd and 4th terms by the ratio 2 we have the proportion, As 3 : 6 : : 4 : 8. Now it is obvious that in the former case the products of extremes and means are equal, for they are each three times four; and when the 2nd and 4th are multiplied by the ratio, the products are still equal, for one factor of each product is multiplied by 2.

This is the principle on which the mode of solution in the Rule of Three is based. In statements in that rule, one extreme and both means are given to find the other extreme; and as the product of extremes and means are equal, if such product be divided by one extreme it must give the other. We multiply together the means to ascertain what the product of the extremes would be, and this divided by one extreme gives the other.

Proposition 40.

In any series of numbers in Arithmetical Progression, the sum of the extremes multiplied by half the number of terms, will be the sum of the series.

This may be illustrated by annexing a descending arithmetical series to a similar ascending one, thus:

1	3	5	7	9	11	13
13	11	9	7	5	3	1
14	14	14	14	14	14	14

Here we multiply the sum of the extremes 1 and 13, by 7, the number of terms, and it gives $98=7\times14$; and it is plain that this is double the sum of one series. Hence to find the sum of an arithmetical series "Multiply the sum of the extremes by half the number of terms" or what amounts to the same, half the sum of extremes by the number of terms. The latter is equivalent to multiplying the length of a board by its average width for its surface. In the series just given, 7 is the average, and if you had a board one inch wide at one end and 13 at the other its average width would be 7 inches, which it would measure across the middle.

It may be as well here to examine the five parts of an arithmetical series, already explained, and to observe how any two may be found from knowing the other three. The parts are:

1. The First Term, } called the two extremes.
2. The Last Term, }
3. The Number of Terms.
4 The Common Difference, sometimes improperly called the arithmetical ratio.
5. The Sum of the Series.

As already explained, if we have the extremes, half their sum will be the mean proportional, and this multiplied by the number of terms will give the sum of the series.

If we have the sum of the series, and the extremes: divide the sum of the series by the mean proportional, and you have the number of terms. But we may save space and be perhaps sufficiently explicit by examining the parts of the following series:

2 5 8 11 14 17 20

Here the sum of 20 and 2, the extremes, is 22; which divided by 2 gives 11 the mean proportional; and 11 multiplied by 7, the number of terms, gives 77, the sum of the series.

From 20 take 2, and we have 18, difference of extremes, which divided by 6 (one less than the number of terms) will give the common difference, 3. A little attention will make it plain that the difference between the extremes, is made up of the common difference repeated as often as there are terms after the first; or one less time than there are terms; hence the operation we have performed. And if we had the first term, 2, the common difference, 3, and the number of terms, 7, to find the last term, we need only multiply the common difference, 3, by 6 (which is one less than the number of terms) and adding 2 to the product we have 20, the last term. For 3, the common difference, enters into every term except the first, and the last term is made of the first term added to these repetitions of the common difference.

It is useless to vary the question through its other forms, as the formation of the series is sufficient to indicate their solution. We offer as an inference from the foregoing—

Proposition 41.

In any continued Arithmetical series, the greatest term is equal to the common difference, multiplied by the number of terms less 1, and having the first term added to the product.

Let us now take a Geometrical series, and see whether we can find its sum in the same way:

2 6 18 54 162 486 1458

By simple addition we find the sum of this series to be 2186. The sum of the extremes is 1460, the half of which is 730, and this multiplied by 7, the number of terms, gives 5110, which we see is incorrect. Neither will it do to multiply the general mean, 54, by 7, for this would give only 378. Place an inverted series under the above as we did with the equi-different series:

2	6	18	54	162	486	1458
1458	486	162	54	18	6	2
1460	492	180	108	180	492	1460

We find their several sums do not come out alike as they then did, and hence the reason why the half sum multiplied by the number of terms will not serve our purpose, the *products* however of the several pairs would be all alike. We might have represented the equi-different series by a diagram, 2 measures wide at one end and 20 at the other, then would the several means represent the length of lines drawn at equal distances across the board; and we might in this way construct any such series, and by measurement find the length of the parts required. Such a diagram being placed by the side of a similar one reversed, would form a parallelogram. But we could not so conveniently illustrate a geometrical series in this way. We will take a different mode. Call the sum of the above series s, then if

$$s=2+6+18+54+162+486+1458$$
$$3s=\quad 6+18+54+162+486+1458+4374$$

The latter numbers are found by multiply the former by 3, and placing their products to the right, that similar ones may fall under each other. Deducting the former from the latter, six of the terms destroy each other, leaving $2s=4374-2$, or 4372, the 2 of the subtrahend having of course to be taken from the number standing in the minuend. Then if $2s=$ 4372, once $s=2186$, the sum of the series, the same as found by addition. We hence infer—

Proposition 42.

The sum of a Geometrical series is equal to the ratio raised to a power whose exponent is the number of terms in the series, and being multiplied by the first term, and afterwards diminished by the first term, is then divided by the ratio less one.

In order to explain this proposition more fully, we may examine how the numbers of such a series are produced. They are obviously the ratio multiplied into itself as often as there are terms, less one (the ratio does not enter into the first term,) and these powers multiplied severally by the first term. In the above example, 3 is the ratio, and 3, 9, 27, 81, 243, 729, the first six powers of 3, being multiplied by 2, the first term, will give the proper terms of the series. We then assume s=*the sum of the series*, and multiply by the ratio, which makes all the terms except the first of the old equation and last of the new, identical, and in doing so, we raise the power of the ratio one term higher, making its exponent equal to the number of terms; in other words we obtain the 7th power of 3, each power being at the same time multiplied by the first term. We then deduct the first equation or single sum of series, and we have left this last product, subject to a deduction of the first term of the series. This then is as many times the true sum, as the ratio less one, and being accordingly divided by the ratio less one, will give the sum of the series.

Having therefore the first term, ratio, and number of terms, we can readily find the last term by involving the ratio to a power whose index is one less than the number of terms, and multiplying this result by the first term. So from understanding the composition of the series, we may devise rules for finding any two of the parts, from the other three being given.

Proposition 43.

The sum of any geometrical series is equal to the difference of the rectangle of the second and last terms, and the square of the first divided by the difference of the first and second term.

To understand this, take the foregoing equation $2s=4374-2$, which is as many times the sum of the first series as is expressed by the ratio less 1, hence the sum of the series may be expressed $\frac{4374-2}{3-1}$ and both terms multiplied by the first term, 2, the equality is not affected, hence $\frac{8748-4}{6-2}$ expresses the sum, and this "is the difference of the rectangle of the second and last terms, and the square of the first, divided by the difference of the first and second terms." For 4374 is the last term multiplied by the ratio, and this by 2, the first term, gives the product of "the last term by the second."

PROPOSITION 44.

The products of the corresponding terms of two Geometrical series are proportional.

2 : 6 : : 4 : 12
3 : 6 : : 4 : 8

Therefore 6 : 36 : : 16 : 96

As already explained, $\frac{2}{6}=\frac{4}{12}$; and $\frac{3}{6}=\frac{4}{8}$; hence on the principle of multiplying both sides of the equation by the same number, $\frac{2}{6}\times\frac{3}{6}=\frac{4}{12}\times\frac{4}{8}$ and multiplying gives us $\frac{6}{36}=\frac{16}{96}$ or 6 : 36 : : 16 : 96. On the same principle we may prove the truth of the next proposition, for by making the terms similar we obtain squares, cubes, &c.

PROPOSITION 45.

If numbers are proportional, so will their similar powers be.

2 : 6 : : 4 : 12, hence $2^2 : 6^2 :: 4^2 : 12^2$.

PROPOSITION 46.

If three numbers are in continued proportion, the square of the first will be to the square of the second, as the first term is to the third.

2 : 4 : 4 : 8, hence $2\times8=4\times4$. Multiplying both sides by first term, $2\times2\times8=2\times4\times4$, or $2^2\times8=2\times4^2$, hence $2^2 : 4^2 :: 2 : 8$. In the same way it may be shown to be true of any power.

PROPOSITION 47.

If three numbers form a geometrical series, the product of the extremes is equal to the square of the mean.

2 6 18. Here $2\times18=6^2$. We might show the truth of this by resolving the numbers into their constituent factors, (the 1st term and the ratio) thus 2 : 6 : 18 is equivalent to $2 : 2\times3 : 2\times3\times3$, and multiplying the extremes and squaring the mean we have $2\times3\times2\times3=2\times3\times3\times2$. In the same way we may illustrate how any number of means are to be found by

extracting the higher roots; and how the product of the extremes is equal to the product of the means, in any geometric series. The first term and ratio form by their products the terms of every geometrical series; and it is important to ascertain how they enter into every term. Indeed we may explain most propositions on the subject by considering how these factors are involved.

These are the fundamental principles of the Progressions, but to extend them through all the ten cases of which they are susceptible, would occupy space intended for matters of greater importance.

In both the Equi-different Series and Geometrical Progression, the first and last terms are called the extremes; and the intermediate the means. Addition and Subtraction in the former correspond with Multiplication and Division in the latter. The mean proportional in the former is found by taking half the sum of the extremes; in the latter by taking the square root of their product. In the former you add as much to the less extreme to make the mean proportional, as you add to the mean proportional to make the greater extreme; in the latter you multiply the less extreme by the same number to make the mean proportional, that you multiply the mean proportional by to produce the greater extreme. In the former the sum of any two terms equally distant from the extremes are equal; in the latter their products are equal.

The following proportions may enable the student to solve some apparently intricate questions and by carefully studying the principles we have laid down, he may understand them; but we have not space for their elucidation. In "Tillett's Key to the Exact Sciences" they will be found elucidated with appropriate problems.

1. The continued product of any three numbers in Arithmetical Progression, is equal to the cube of the middle number, less the square of their common difference multiplied by the middle number.

2. The continued product of four numbers in Arithmetical progression, is equal to the product of the two means, multiplied by the product of the two extremes.

3. The sum of the squares of three numbers in Arithmetical progression, is equal to three times the square of the middle number, more twice the square of their common difference.

4. The sum of the squares of four numbers in Arithmetical progression, is equal to four times the square of their mean proportional, more five times the square of their common difference.

5. The sum of the squares of five numbers in Arithmetical progression, is equal to five times the square of the middle number, added to ten times the square of their common difference.

6. The sum of the cubes of any two numbers, is equal to twice the cube of their mean proportional, more six times the mean proportional multiplied by the square of one half their difference.

7. The sum of the cubes of any three numbers in Arithmetical progression, is equal to three times the cube of the middle number, more six times the middle number multiplied by the square of their common difference.

8. The sum of the cubes of any four numbers in Arithmetical progression, is equal to four times the cube of their mean proportional, more fifteen times their mean proportional multiplied by the square of their common difference.

9. If the difference of the cubes of two numbers be divided by their difference, the quotient will be equal to the sum of the squares of the two numbers together with their product.

10. The sum of the cubes of any five numbers in Arithmetical progression, is equal to five times the cube of their mean proportional, more thirty times their mean proportional multiplied by the square of their common difference.

11. The difference of the cubes of any two numbers, is equal to the square of their mean proportional multiplied by three times their difference, more twice the cube of one half their difference.

12. The continual product of any three numbers in Geometrical progression, is equal to the cube of the middle term.

13. The continued product of four numbers in Geometrical progression, is equal to the square of the first term, multiplied by the square of the last; or also equal the square of the product of the two means or of the two extremes.

14. The sum of the squares of three numbers in Geometrical progression, is equal to three times the square of a mean proportional between the two extremes, more the square of their difference; and the product of the two extremes is equal to the square of the middle term.

15. The sum of the squares of four numbers in Geometrical progression, is equal to the square of the sum of the two extremes, more the square of the second term, multiplied by the square of the ratio less one.

16. The continued product of any three numbers in Harmonical proportion, is equal to the cube of the second term, more the cube of the difference between the second and third.

17. The sum of the squares of any three numbers in Harmonical proportion, is equal to three times the square of the middle number more three times the square of the difference between the second and third, more the square of the difference between the first and second numbers.

LECTURE VIII.

INVOLUTION, OR THE RAISING OF POWERS.

Numbers are by some writers divided into Natural or Lineal, Triangular, Quadrangular or Square, Pentagonal and other Polygonal numbers. A series increasing by 1 as a common difference, is called a lineal or natural series; as

1, 2, 3, 4, 5, 6, 7, 8, 9, &c.

And if we form another series of the sums of the numbers in the lineal series, the numbers are called Triangular numbers; thus,

1 3 6 10 15 21 &c.

They are called triangular because the numbers may be represented by points, arranged in a triangular form, thus:

```
1     3      6       10        15
.    . .   . . .   . . . .   . . . . .
      .     . .     . . .     . . . .  &c.
             .       . .       . . .
                      .         . .
                                 .
```

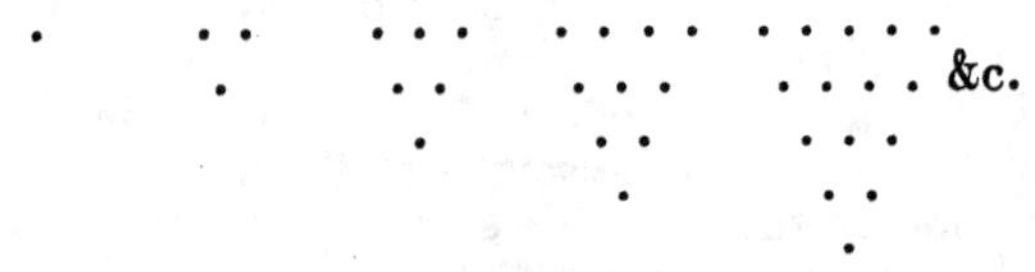

If to an arithmetical series having 2 as a common difference, we attach another series formed as above, the latter are squares, or as sometimes called Quadrangular Numbers;

and these may be formed by points disposed in a square form. Thus,

1	3	5	7	9	11 &c.
1	4	9	16	25	36 &c.

This amounts to the proposition that the numbers of every series formed by the regular addition of all the odd numbers are squares. On these distinctions some mathematicians have built propositions useful in speculative mathematics; but the only ones requiring our attention are Quadrangular or Square numbers. It would be obvious from the formation of the foregoing series that "Every number, whether prime or composite, is either a triangular number or the sum of two or three triangular numbers; a square, or the sum of two, three or four squares," and we might extend this principle to other classes but it is unnecessary. Powers, Rational Numbers, Surds, &c. have been explained, we shall therefore offer a few propositions and pass on to a general discussion of Powers and Roots, so far as they are necessary to our purpose.

Proposition 48.

Any power of a number may be found by repeated multiplication; the number itself being the first multiplicand and also the constant multiplier.

Thus, 3 the given number, may be expressed 3^1

3	the given number, may be expressed		3^1
3			
9	Square or second power,	" "	3^2
3			
27	Cube, or third power,	" "	3^3
3			
81	Biquadrate, or fourth power,	" "	3^4
3			
243	Fifth power,	" "	5^5

We might so proceed to any extent, but raising numbers to

higher powers is greatly facilitated by availing ourselves of the principle contained in

Proposition 49.

If two or more powers of any number be multiplied together, the product will be a power whose index will equal the sum of the indices of the factors.

Note.—The index or exponent of a number is a character designed to show the power to which it is to be raised; as ^{2} in the expression 3^2.

In the above we may find the 4th power by multiplying the 2nd by the 2nd, *i. e.* $9 \times 9 = 81$; and so we may find the fifth power by multiplying together the 2nd and 3rd. This principle is exceedingly important in raising numbers to any high power; and in extracting roots a somewhat similar doctrine applies, only that the indices are divided instead of being added. The above proposition is applied in raising powers by the use of logarithms.

Proposition 50.

No two numbers can have the same ratio to their respective roots or powers.

This is obvious, for no two numbers are multiplied by the same multiplier to raise them to any given power; and hence they cannot have the same ratio to their powers; neither can they to their roots. We cannot institute a proportion, and say as $2 : 5 : : 2^2 : 5^2$, for 2 is multiplied by 2 to square it, and 5 by 5. This is the reason why questions involving roots or powers of unknown quantities cannot be wrought by Position; there is no equality of ratios to constitute a proportion.

Proposition 51.

No root of a Surd can be expressed by an integer, nor by any rational fraction.

Every one who has had occasion to extract the roots of numbers has felt the inconvenience of not being able to express remainders in the shape of vulgar fractions, as you would remainders in common division: for though the value may be approximated by extending the extraction, perfect accuracy can never be reached: hence if expressed decimally, the decimal is *always* infinite. This difference between division and evolution arises from the fact that in the former the divisor is a fixed quantity, but in the latter the divisor enlarges at every step, and a remainder is neither a fractional part of one divisor, or another, but of some intermediate unknown quantity. Hence to use the divisor producing the remainder as a denominator would give a result too great; and to use the next divisor in course would give a result too small.

Let us extract the square root of 21050 by way of illustration.

```
      2̇10̇50(145
      1
      ----
 24   110
       96
      ----
285   1450
      1425
      ----
        25 remainder.
```

If we place 285 as the denominator, the fraction will be too large, since that is the proper divisor of a previous number; and if we double the ascertained root, 145, as some do, the result will still be too large. 145×2=290, and $\frac{25}{290}$=.0862+, making the root 145.0862+. This is obviously the same as adding ciphers at once to the remainder, and dividing by the divisor as a constant quantity; while by the proper mode of extraction, the divisor increases at every division. Pursuing the regular course of extraction, the root would be 145.0861+, which is only about a ten thousandth part of a unit less than the other mode.

The following rule may be used in the cube root, and the result will vary but little from the truth. It will however be rather less than the true root. "Square the ascertained root and multiply it by 3, and to that add three times the root, the sum will be the true divisor, nearly."

"It may not be amiss," says N. Pike, "to remark that the denominators both of the square and cube, show how many numbers they are denominators to; that is, what numbers are contained between any square or cube number, and the next square or cube number, exclusive of both numbers, for a complete number of either leaves no fraction when the root is extracted, and consequently has no use for a denominator; but all intervening numbers must leave remainders."

In the above example, the last figure of the root is 5, and the product is 1425; had it been 6, the product would have been 1716, for then the divisor would have been 286, and the difference between these is 291, from which 1 must be deducted, since by subtraction the difference between the numbers exclusive of both is not shown. The entire minuend is included. If I take 3 from 7, 4 remains; but if I exclude both 3d and 7th, and take only what is between them, there will be but 3, viz., 4th, 5th, and 6th.

As to abridging the extraction of the cube root, it is of small importance, since a man might be actively engaged for a long life-time and never have occasion to extract the cube root once; and if his calculations be speculative, he will find it necessary to be accurate, even though it cost a few extra figures. In thirty-two years of business I have never once, that I recollect, had occasion to extract the cube root: but the extraction of the square root is often necessary.

Proposition 52.

The number of figures in the square of any given number will never be more than twice as many as are in the root; and never be less than twice as many, less 1.

The square of 10 is 100, which is the smallest possible number that can be produced by squaring a number of two places, for the number used is the smallest possible, and the square of 99 is 9801; which is the largest possible, since there is no larger number expressed by two digits. In the latter case there are just twice the number of places in the square, that are in the root; and in the former twice the number, less 1; and this will remain true of numbers of any size whatever.

In the cube, the number of places is never more than three times the number of places in the root, nor less than three times the number, less two. $10^3=1000$, which is three times the number, less 2; and $99^3=970299$, which is just three times the places that are in the root. The same doctrine is true indefinitely onward, the 4th power having four times as many places as a maximum; the 5th five, and so on; and here you have a key to the direction to point the given number into periods of two places each, in extracting the square root; three in extracting the cube root, and so increasing the size of the periods as the powers increase; for the number of such periods determines the number of digits in the root, in consequence of the law set forth in the proposition. We might generalize our proposition still further, and say that the *product* of any two numbers can have at most but as many figures as are in the factors; and at least but one less. In squaring, the factors are alike, and hence we say twice the number of places in one factor.

The rule for extracting the square root may be demonstrated Algebraically or Geometrically; but the latter is most easily understood, perhaps, by one who has not studied the subject closely. In either case the operation is based on the theorem, that "the square of the sum of two numbers is equal to the sum of their squares, added to twice the product of one by the other."

Let us extract the square root of 576, and examine the process.

```
   5̇76(24 root.
   4
   ---
44 176
   176
   ---
```

The first figure to the left, viz., 5, in the hundreds' place, is equivalent to 500, the greatest square, in which is 400, whose root is 20, and is here represented by 2 placed in the quotient, and its square by 4 placed in the hundreds' place. The remainder is 100, and to this we bring down 76, making 176. Let the annexed square, A B C D, represent the 400, each side of which will be 20, the root found. The remaining 176 must be added to the square without destroying its shape, and hence must be added to either two or all its sides. Two is preferable. To find the length of the addition we "double the ascertained root" by which we have the length of the oblong to be added, as F G H, (except the little corner at G, which is just as large square as the addition is wide) and this divided into the area to be added gives 4 as the width of the addition that can be made; and to complete the addition we must add the 4 to the length F G, G H, by which the whole addition, including the corner, appears to be 44. This divided into 176 gives 4 as the width, without any remainder. The perfect figure may then be represented by this diagram, in which A B C D represent the square of the tens' place of the root; F G B H the square of the units' place; and the oblong E F A B, D B H I, "twice the product of one part by the other;" the whole E G C I representing the square of the sum of C D, D I=24.

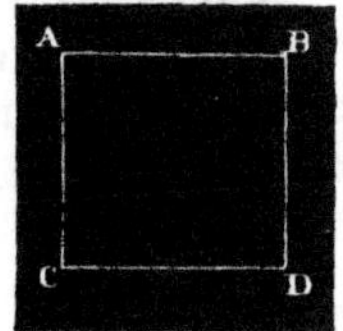

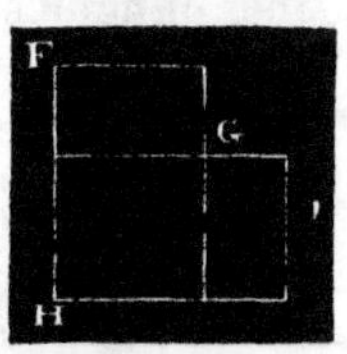

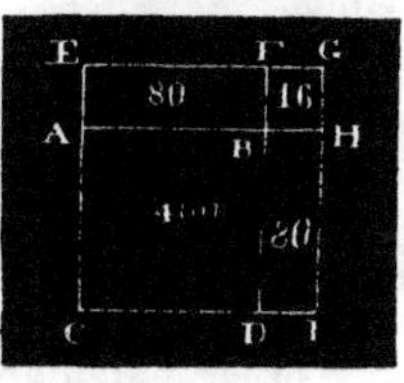

If there are more than two periods, proceed as above, and

having found two regard them jointly as one, and thus find another, and so proceed to the end of the problem.

By examining the subject attentively we can see how the root enters into the square. If there are two digits, tens and units, in the root, the square will contain the square of the tens, +the square of the units, +twice the product of the units by the tens. In the above we have 2 tens 4 units in the root: the square of 2 tens (20) is 400; the square of 4 is 16; and the product of 2 tens by 4 units is 8 tens=80, and twice 80=160. Then 400+16+160=576, the square. If the root has three digits, or places of figures, the power contains the square of the hundreds, +twice the product of the hundreds by the tens; +square of the tens; +twice the product of the hundreds by the units; +twice the product of the tens by the units; +the square of the units. Let us consider 576 as a root, then will its square consist of

"The square of the hundreds," 500^2 =	250000
"Twice the product of the hundreds by the tens" $500 \times 70 \times 2$ =	70000
"Square of tens" 70^2 =	4900
"Twice the product of the hundreds by the units" $500 \times 6 \times 2$ =	6000
"Twice the product of the tens by the units" $70 \times 6 \times 2$=	840
"Square of the units" 6^2	36
Hence 576^2 =	331776

So we might proceed to show how the parts of the root enter into the square, when there are four, five or any greater number of figures; and we might show also how they enter into the cube and other higher powers, but it is unnecessary. Or we may demonstrate the rule by a very simple algebraic process, thus—

Any square number may be represented by $a^2 + 2ab + b^2$. The root of the first term is a and we double this for a divisor, because the second term is made up of *twice* the product of the first and last terms of the root. Hence the reason is obvious.

The extraction of the square root is needed in determining the side of a square, the area being given; in comparing surfaces, they being in a duplicate ratio to each other; in determining a geometrical mean between two numbers; in determining the parts of certain geometrical figures, as of the right angled triangle, from having other parts given; and in many other calculations involving duplicate ratios, sub-duplicate ratios, &c.

The cube root is needed to find the length of one side of a cube if you have the solidity given; in finding two means between two given numbers; and in comparing the solid contents of bodies. The higher roots are also necessary when you desire to find three or more geometrical means in a series. That the roots are necessary in determining the means or terms of a geometric series, is obvious from the manner in which the powers of the ratio multiplied into the 1st term produce the remaining terms of the series.

The following positions are given as facts, which may sometimes be useful; but some of which cannot be demonstrated without algebra.

The sum of all the odd numbers commencing with unity is a square.

Every square number is a multiple of 4, or a multiple plus 1.

A square number cannot terminate with an odd number of ciphers.

If a square number terminate with 4, the figure in the 10's place will be even.

If a square number terminate with 5, it will terminate with 25.

No square number can terminate with two equal digits, except two ciphers or two fours.

No square number has 2, 3, 7 or 8 in the units' place.

If a series of numbers to any extent be arranged as in the common Multiplication table, the sum of all the numbers is a square number, the root of which is the sum of the numbers in the left hand column.

The difference between any number except 1, and its cube, is always divisible by 6.

If a cubic number be divisible by 7 it is divisible by the cube of 7.

Neither the sum or difference of two cubes, can be a cube.

If a number have 5 or 6 in the units' place, all its powers will have 5 or 6 in the units' place.

If a series of numbers beginning with 1 be extended in geometrical progression, the 3rd, 5th and 7th will be squares; the 4th, 7th and 10th cubes; the 7th being both a square and a cube.

No number which is a power of another number is divisible by a number which is not a measure or multiple of the root indicated by such power; and if the number be a measure of the root, the given number is divisible by the measure raised to the proposed power.

The difference of any two equal powers is divisible by the

difference of their roots; and also by the sum of the roots if the powers be even.

The root of a root will be such root of the original number as is indicated by the product of the indices of the roots.

Towards the close of our next lecture will be found sundry relations of different square numbers, cube numbers, &c., which naturally fall under that classification.

LECTURE IX.

RELATIONS OF NUMBERS.

There are certain relations amongst abstract numbers, their parts, sums, products, &c., that may be rendered obvious, either by algebraic, or by geometrical illustration; and as the latter is more obvious to the eye, it is preferable for such as are not familiar with Algebra and Geometry. It is true, however, that the algebraic illustration is more scientific, and all positions of the kind may be demonstrated strictly by Algebra; while the explanation by diagrams is but an *illustration*, rather than a *demonstration*. It is well adapted however to the purpose of the mere arithmetician, and for the use of such we will take up some of the common cases, and will afterwards subjoin the principles involved in the Rules given, in the shape of distinct propositions. The application of such principles is often very convenient in the solution of problems.

It seems perfectly clear that numbers may be represented by straight lines, any portion of a right line being adopted as the measuring unit. It seems equally obvious that *products* may be represented by increasing such lines in length as many times as there are units in the multiplier; and *quotients* by dividing the line representing the dividend, into as many parts as there are units in the divisor. A better mode, however, is to represent products (including squares) by rectangular surfaces; *e. g.* if we wished to represent the product of 6 by 9, we would draw an oblong figure 6 measures of any kind in width, and 9 in length; then the area, 54, will correctly express the product of these factors. So if we desire to represent the quotient of 54 by 6, the desired quantity is at once expressed, either by dividing a straight line 54, into 6 equal parts, or by drawing a diagram, 6 in width, and extending it until the area is 54; the length, 9, will express the number sought. To make the representation more obvious the diagram may be subdivided, thus:—

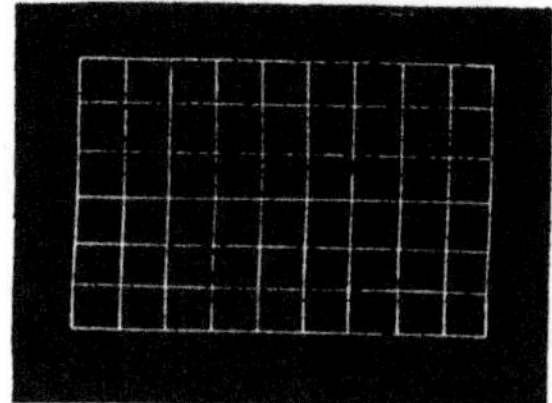

With these prefatory remarks, we shall proceed to consider the subject as divided into sundry distinct cases.

Case 1st.

The *sum* and *difference* of two numbers given, to find the numbers.

Rule. To half the sum add half the difference, for the greater of the two numbers; and from half the sum take half the difference for the less.

Or, To the sum add the difference and take half for the greater; or from the sum take half the difference and half the remainder will be the less.

Illustration. Let A B represent the less number and A C the greater, and A D the sum. Then B C will represent the difference, and the sum A D being equally divided, the point E will represent the point of division. It is then obvious that if to the half sum A E, we add E C, the half difference, the sum will be A C, the greater; and if from such half sum we deduct B E, the half difference, the remainder will be A B, the less number. The second form of the rule is equally susceptible of illustration.

A e C D
B

This is a proposition often useful in calculations, and with which it is necessary to be perfectly familiar.

Example. John and William have \$100, and John has \$30 more than William; how much has each?

$100 \div 2 = 50$, half sum; and $30 \div 2 = 15$, half difference. Then $50 + 15 = 65$ John's share; and $50 - 15 = 35$ William's share.

Case 2.

The *sum* and *quotient* of two numbers given, to find the numbers.

Rule. Divide the sum by the quotient $+1$, the result will be the less number sought; from which by subtraction the greater number is readily found.

Illustration. Let the line A B represent the sum, which may be considered as representing two numbers, one of which is three times as great as the other; hence it is composed of such smaller part and three other parts, each equal to the less, or into 4 equal parts. And thus *always* the *sum* will be composed of the quotient, and as many more equal parts as there are units in the quotient, or altogether of as many parts $+1$ as there are units in the quotient.

A ——'——'——'—— B

Example. A and B played at marbles. After A had won several they found that each had 70. A wishes to know how many he has won, but only knows that B had, at the commencement of the play, four times as many as he. How many did each begin with?

Ans. A 28, B 112.

Case 3.

The sum and product of two numbers given, to find the numbers.

Rule. From the square of their sum take four times their product, and the square root of the remainder will be the difference of the numbers.

Illustration. The sum of two numbers is 10, and their product 24. What are the numbers?

$10^2 = 100$, and $100 - (24 \times 4) = 4$ and $\sqrt{4} = 2$, difference of numbers. Then $10 + 2 \div 2 = 6$ the greater and $10 - 2 \div 2 = 4$, the less.

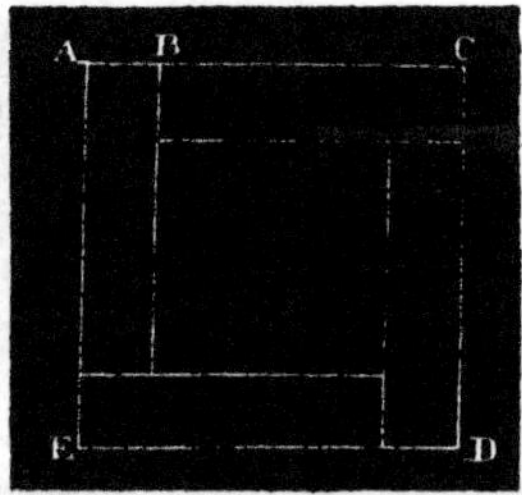

Let AB represent the less of two numbers=3, and BC the greater=9, then AC being the sum will =12. Construct the diagram A C D E, making the four parallelograms adjoining the sides of the figure as wide as the shorter line A B, and as long as the longer B C, then will they severally represent the products of A B by B C, =27, and the central remaining square will obviously be formed by sides equal 6, the difference between the numbers; and the entire diagram A C D E will represent the square of the sum of the lines A B, B C=12; the square of which is 144, the area of the figure.

Now if from the whole figure A C D E=144, we take 4 times the product of one part by the other; which will be the four oblong spaces =27×4=108, there will remain 36, which is the area of the central space, and the square root of which =6 is one side, or is the difference between the numbers A B, B C. And having the sum and difference we proceed by Case 1 to find the numbers.

Example. Five hundred rods of fencing are necessary to enclose a certain rectangular farm, which at $22 per acre will cost $1980. Required the length and breadth of the farm.

Ans. Length 160 rods. Breadth 90.

CASE 4.

The sum of two numbers and the sum of their squares given to find the numbers.

Rule. From the square of the sum take the sum of the squares, and half the remainder will be the product. Then proceed by Cases 3 and 1.

Illustration. The sum of two numbers is 10 and the sum of their squares is 52. Required the numbers.

$10^2=100$ and $100-52=48$, and $48\div2=24$ the product. Then by Case 3rd.

$10^2-(4\times24=96)=4$; and $\sqrt{4}=2=$difference.

And by Case 1st. $10+2\div2=6$, the greater, and $10-2\div2=4$ the less.

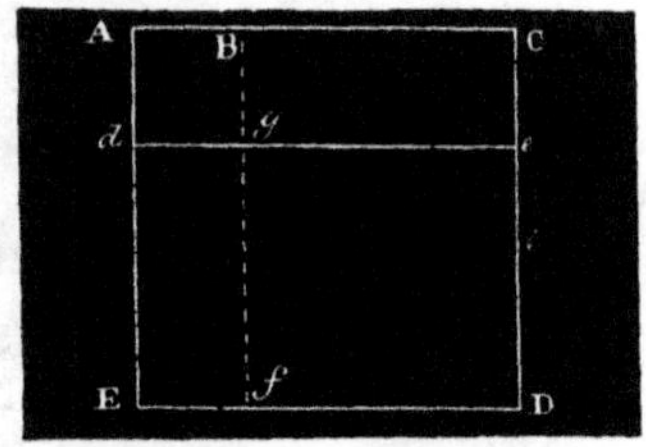

Let the line A B represent the less number and B C the greater, then A C will be the sum, and A C D E the square of the sum. Then also will A B g d=square of less, and g e D f the square of the greater. It is obvious also, from construction, that the parallelograms, B C e g and g f E d, each equal the product of one number by the other, and hence taking the two squares or "the sum of the squares" from the square of the sum, we have left the sum of the two products, half of which equals the product of the numbers.

Example. A and B have 50 guineas between them, which are to be so divided that the sum of the squares shall be 1300. How many had each supposing A to have the greatest number? *Ans.* A 30, B 20.

Case 5.

Given the sum of two numbers and the difference of their squares, to find the numbers.

Rule. Divide the difference of their squares by their sum, and you will have the difference of the numbers.

Illustration. The sum of two numbers is 13, and the difference of their squares 39. What are the numbers?

39÷13=3 their difference. Hence 8 and 5 are the numbers required.

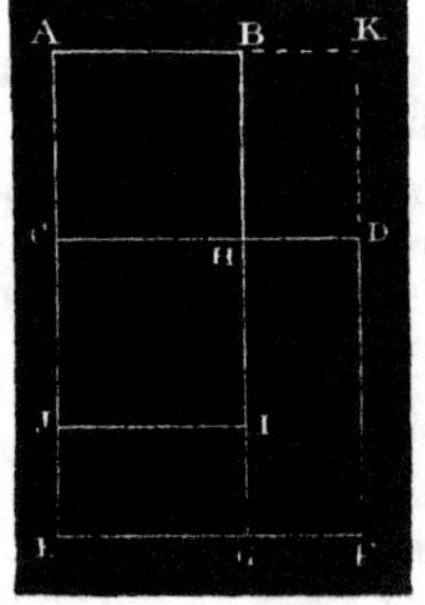

Draw C H=5; and H D=3, then will C H= less number and C D the greater. Form the square C H I J on C H, and C D F E on C D, then will the space H D F E J I represent the difference of the squares. Consider the space J I G E as placed above H D, along the line H B, as represented by the dotted space; then will B G= sum of the numbers =13, and the area B K F G (which is the difference of squares) being divided by 13 gives 3, the difference of the numbers. From which the numbers are readily found.

Example. A man has two square farms, one containing 1000 acres more than the other; and to enclose both with a fence 10 rails high and 2 panels to the rod, required 64000 rails. How many acres in each? *Ans.* 1562½ and 562½.

Case 6.

The sum of the squares, and the difference of the squares of two numbers given, to find the numbers.

Rule. From the sum take the difference, and half the remainder is the square of the less, which taken from the sum of the squares will, of course, give the square of the greater. Then extract for the numbers.

Illustration. The sum of the squares of two numbers is 89, and the difference 39; required the numbers. 89—39=50, and 50÷2=25 and $\sqrt{25}=5$, the less number; and 89—25 =64, and $\sqrt{64}=8$, the greater.

In the preceding diagram, A B H D F E expresses the sum of the squares, A B and C D, and taking away H D F I J E, we have A B I J twice the sum of the less left; or we might add the difference to the sum, and half the result would be the square of the greater.

Example. Two companies of boys went in search of nuts, and each boy got as many nuts as there were boys in his company. Moreover all the boys in the larger company got 225 nuts more than all the boys in the smaller company; and the whole number collected was 1025. How many boys were there? *Ans.* 20 in the less company and 25 in the greater.

Case 7.

Given the difference of two numbers, and their product, to find the numbers.

Rule. To the square of the difference add four times the product, and the square root of the sum will be the sum of the numbers. Then proceed by Case 1.

Illustration. The difference of two numbers is 3 and their product 40; required the numbers.

$\sqrt{(3^2+(40\times4))}=13=$ sum of numbers. Hence 8 and 5 are the numbers.

By turning to the diagram under Case 3, we perceive that it represents the square of the sum of the numbers, and that it is composed of the square of the difference at the centre, and four times their product around it. Hence the reason of the rule is obvious.

Example. Two travelers, A and B, were asked how far they

had travelled. A said he had travelled 50 miles farther than B, and B said that if the number of miles that he had travelled were repeated once for every mile A had travelled, the distance would be no less than 75000 miles, which would reach three times round the earth. How far had each travelled?

Ans. A 300 miles, B 250 miles.

Case 8.

The difference of two numbers, and their quotient given, to find the numbers.

Rule. Divide the difference by the quotient, *less one*, for the smaller number, and add it to the difference for the larger.

Illustration. Let A C and A B represent the numbers, then C B will represent the difference; and to divide the difference into portions each equal to the quotient, we must make the number of such parts one less than the number in the quotient, for the quotient itself is one part.

C
A—|—|—|—|—B

Example. Says a father to his son, I am 10 times as old as you, and I was 45 years old when you were born. What is the age of each? *Ans.* Son 5 years, father 50 years.

Case 9.

Given the difference of two numbers and the sum of their squares, to find the numbers.

Rule. From twice the sum of the squares take the square of the difference, the square root of the remainder will be the sum of the numbers. Then proceed by Case 1.

Illustration. The difference between two numbers is 3, and the sum of their squares 89, what are the numbers?

$\sqrt{(89\times2-3^2)}=13$ the sum of the numbers, whence they are easily found.

Let A B= one number and B C the other, then will A B G F= the square of the former and B C L J of the latter. I L D K is also the square of the smaller number, and E K H F the square of the greater; the only portion twice embraced being the clouded space G H I J= the square of the difference between the numbers. The whole figure A C D E being the

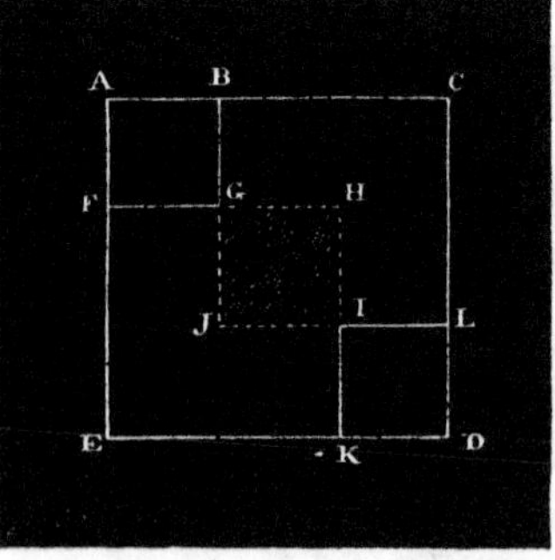

square of the sum of the numbers, obviously contains twice the square of the less and twice the square of the greater, except that the clouded part is repeated. Hence taking twice the sum of the squares, and deducting once the square of the difference will leave just the area of the figure, A C D E, which is the square of the sum of the two numbers.

Example. The perpendicular of a right-angled triangle measures 10 rods more than the base, and the hypotenuse is 50; required the area. *Ans.* $3\frac{3}{4}$ acres.

Case 10.

The difference of two numbers, and the difference of their squares given, to find the numbers.

Rule. Divide the difference of the squares by the difference of the numbers, the quotient will be the sum of the numbers. Then proceed as in Case 1.

The reason of this will be obvious from Case 5.

Example. A field of corn is 10 rods longer than it is wide, and the square of the width is 700 less than the square of the length. What is the area of the field. *Ans.* $7\frac{1}{2}$ acres.

Case 11.

The product of two numbers, and the sum of the squares given, to find the numbers.

Rule. Add twice the product to the sum of the squares; the square root of the sum will be the sum of the numbers. Then proceed by Case 3.

Illustration. By reference to the diagram at Case 4, it is obvious that the square of the sum of two numbers, is equal to twice the product, added to the sum of the squares. Hence the reason of the rule is obvious.

Example. The diagonal of a rectangular 30 acre lot is 100, and it is 10 rods longer than wide. What are the length and breadth of the lot? *Ans.* 60 rods by 80.

Case 12.

Given the product of two numbers, and the difference of their squares, to find the numbers.

Rule. Add one fourth of the square of the difference of squares to the square of the product; take the square root of the sum; from this root take half the difference of squares, and the square root of the remainder will be the less number.

This is deduced from an algebraic process, and as it involves the fourth power of the unknown quantity, it cannot be illustrated by a diagram.

Example. A bought a piece of cloth, giving as many dollars per yard as there were yards; while B bought a finer piece, giving likewise as many dollars per yard as there were yards, and it was found that B's cost $28 more than A's; and had B paid the same rate per yard that A paid, his cloth would have cost him $48. What did each pay, and what quantity of cloth did he get? *Ans.* A got 6 yards = $36, B got 8 yards = $64.

CASE 13.

The product of two numbers, and their quotient given, to find the numbers.

Rule. Divide the product by the quotient, and the square root of the result will be the less number.

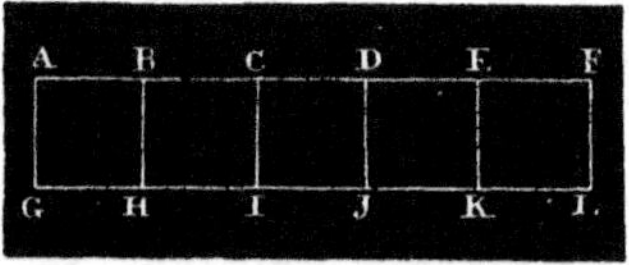

Let A F represent the larger number and A G the less. Complete the parallelogram A F G L. Divide the line A F in the points B C D &c., making each of the lines A B, B C &c., equal to A G, and draw B H, C I &c., parallel to A G. It is manifest that the parallelogram will represent the product of the numbers; and the number of the squares A H, B I &c., will be the quotient of the product by the quotient. Hence the reason of the rule is clear.

Example. A bought a quantity of calico for $3.60, and the number of yards was $2\frac{1}{2}$ times the number of cents he gave per yard. What did he give and how much did he get? *Ans.* 30 yards, at 12 cents.

CASE 14.

Given the product of two numbers, and the square of the difference, to find the numbers.

Rule. Add four times the product to the square of the difference, and the square root of the sum will be the sum of the numbers. Then having the sum and product, proceed by Case 3.

Using the diagram by which Case 3 is illustrated, it is clear that having the central square (the "square of the difference") and one of the rectangles or products, we have but to complete the square of the sum by adding three other rectangles; or in other words by "adding 4 times the product to the square of the difference." The reason of the rule is entirely obvious.

Example. Being engaged in laying out an oblong garden, which contained just an acre and a half, I found that having cut off from one end a square area, the greatest square that I could form in the remainder contained just 64 square rods. What were the length and breadth of my garden?

Ans. 12 and 20 rods.

Case 15.

The quotient of two numbers, and the sum of their squares given, to find the numbers.

Rule. Divide the sum of the squares by the square of the quotient, plus one; and the square root of the result will be the less number.

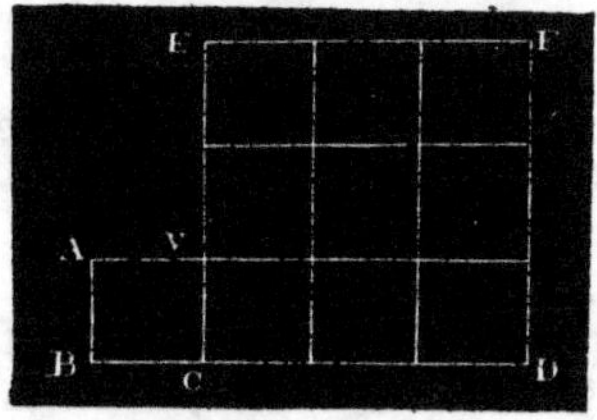

Let B C represent the less number, and C D the greater, then A C will be the square of the less, and D E of the larger; the whole figure A D E representing the sum of the squares. The square of the ratio or quotient of C D by B C will obviously express the number of small squares, each equal to A C, contained in D E. Adding one to this for the square A C, and dividing A D E by the sum gives one of the smaller squares —the square root of which will be the less number.

Example. A and B spent $100 for cloth, each paying as many dollars per yard as he got yards, and A got $\frac{3}{4}$ as many yards as B. How much did each get?

Ans. A got 6 yards, and B 8 yards.

Case 16.

Given the quotient of two numbers, and the difference of their squares, to find the numbers.

Rule. Divide the difference of squares by the square of the quotient, *less one;* the square root of the resulting quotient will be the less number.

A reference to the preceding figure will make this plain.

Example. A and B bought cloth, each paying as many dollars a yard as he got yards, and B got $1\frac{1}{3}$ times as many yards as A, and paid for it $28 more than A paid. How much did each buy? *Ans.* A got 6 yards, and B 8 yards.

Case 17.

The quotient of two numbers, and the square of their sum given, to find the numbers.

Rule. Divide the square of the sum by the square of the quotient plus one; the square root of the resulting quotient will be the less number.

Illustration. Let A E= the less number, and E D= the greater; then A C will represent the square of the sum. A glance at the figure will show that the number of small squares in A C will be equal to the square of the number of times E A is contained in D E *plus* one. Hence the reason of the rule.

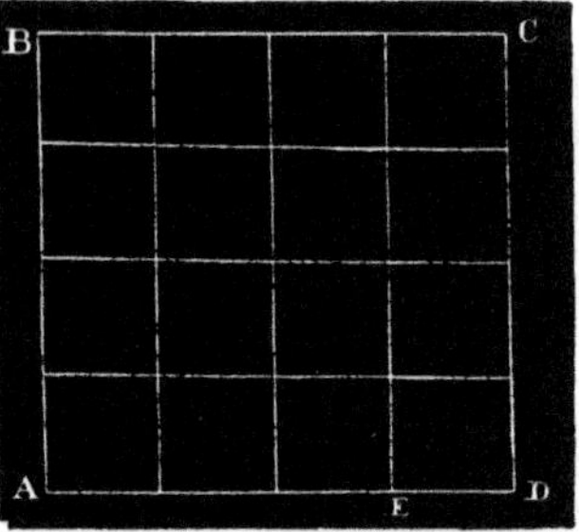

Example. Says John to Henry, I have three times as many coppers as you have, and if both were added together, and expended in apples, at as many apples for a copper as there are coppers, they would purchase 400. How many coppers had each boy? *Ans.* John 15, Henry 5.

Case 18.

The quotient of two numbers, and the square of their difference given; to find the numbers.

Rule. Divide the square of their difference by the square of the quotient *less* one. The square root of the resulting quotient will be the less number.

Illustration. Let A B= less number, and B C the greater; and D C the difference. Then the number of small squares in D E F C will manifestly be equal to the square of the ratio of A B to B C (*i. e.* the quotient of B C by A B) *less* one. Hence the rule.

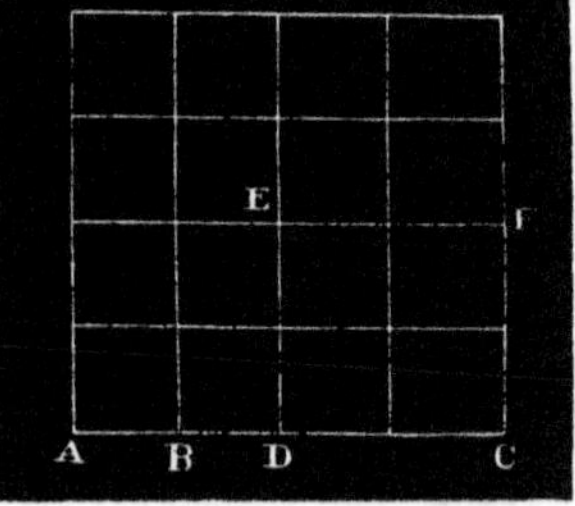

Example. The ratio of two numbers is 7 and the square of their difference 324. What are the numbers?

Ans. 3 and 21.

Case 19.

Given the square of the sum, and the sum of the squares of two numbers, to find the numbers.

Rule. From the square of the sum, take the sum of the squares, and twice the product of the numbers will remain. Then subtract four times the product from the square of the sum, and the remainder will be the square of the difference. Then having the square of the sum and difference, extract the roots and proceed by Case 1.

Illustration. Let A D H C = square of A D the less, and D E F I the square of the greater, then will these two squares represent the sum of the squares, and A E G B the square of their sum, A D and D E. In addition to the sum of the squares, the square of the sum embraces the two rectangles C H J B and I F G J, each equal to the product of the numbers. Hence the reason of the rule is quite obvious.

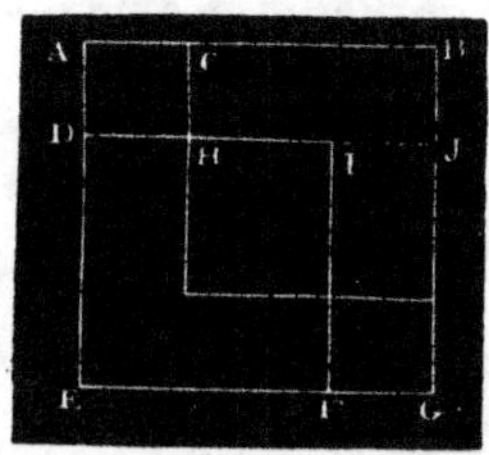

Example. The square of the sum of two numbers is 144, the sum of their squares is 80. What are the numbers?

Ans. 4 and 8.

We might extend problems of this kind almost indefinitely, but it is thought that the foregoing will make the subject reasonably familiar, and show how the subject may be illustrated. To show how far such cases might be varied, we may take the case preceding the last of the above, and make four distinct cases, viz:

1st. The quotient and square of difference.
2nd. " " " difference of squares.
3rd. " " " square of sum.
4th. " " " sum of squares.

And thus we might vary and extend the combinations of two numbers only, and if we increase the powers, or the number of numbers, the combinations would become too great for discussion in a merely incidental chapter. Several of the foregoing are convenient in the solution of problems, but a multiplicity of such rules would burden the memory of most calculators, and lead to perplexity rather than promote facility. A large book might be written on this branch alone of the subject, but it would be infinitely better for the student to strike at once at the root of the matter by making himself familiar with the principles of Algebra.

We will now take up in the shape of propositions the principles on which the foregoing problems are based, adding such additional illustrations as may be thought proper; and as those already given are based on Geometry, we will now use Algebra.

Proposition 53.

The greater of two numbers is equal to half their sum, *plus* half their difference; and the less is equal to half their sum, *minus* half their difference.

Proposition 54.

The less of two numbers is equal to their sum divided by their ratio *plus* one.

Let a represent the sum, b the ratio and x the less number, then bx=greater.

And $bx+x=a$

Separating factors $(b+1)x=a$

Dividing $x=\frac{a}{b+1}$

Proposition 55.

The square of the sum of two numbers, less four times their product equals the square of their difference. Let x and y be the numbers.

Then $(x+y)^2=x^2+2xy+y^2$

Taking away $4xy$ or 4 times product.

Leaves $x^2-2xy+y^2$, or the square of their difference.

When quantities are included in a parenthesis, or placed under a vinculum they are to be taken together. Thus $(a-b)\div2$ denotes that b is to be taken away from a and the remainder divided by 2. So $\overline{14+2}|^2=16^2=256$. Without the vinculum its value would be $14+2^2=18$.

Proposition 56.

The sum of the squares of two numbers, *more* twice their product, equals the square of their sum.

$$x^2+2xy+y^2=(x+y)^2.$$

Proposition 57.

The difference of the squares of two numbers divided by the sum of the numbers, equals their difference.

$$\overline{x^2-y^2}\div\overline{x+y}=x-y.$$

Proposition 58.

The sum of the squares of two numbers, *less* the difference of their squares is equal twice the square of the less, and the sum of the squares, *more* the difference of the squares, is equal to twice the square of the greater.

$$\overline{x^2+y^2}-\overline{x^2-y^2}=2y^2;\ \text{and}\ \overline{x^2+y^2}+\overline{x^2-y^2}=2x^2.$$

Proposition 59.

Four times the product of two numbers, plus the square of their difference equals the square of their sum.

$$4xy+x^2-2xy+y^2=x^2+2xy+y^2=(x+y)^2.$$

Proposition 60.

The difference of two numbers divided by the ratio, *less* one, equals the less number.

Let a be the difference, r the ratio and x the less number.

$$rx-x=a$$

Separating factors, $(r-1)x=a$

Dividing $x=\dfrac{a}{r-1}.$

Proposition 61.

Twice the sum of the squares of two numbers, *less* the square of the difference equals the square of the sum.

$$2x^2+2y^2-(x^2-2xy+y^2)=x^2+2xy+y^2=(x+y)^2.$$

Proposition 62.

The difference of the squares of two numbers divided by the difference of the numbers, equals the sum of the numbers.

This is just the same in principle as Proposition 57.

Proposition 63.

The square of the sum of two numbers, *less* the sum of their squares, equals twice their product.

This is obvious; $(x^2+2xy+y^2)-(x^2+y^2)=2xy$.

Proposition 64.

The square root of one-fourth of the square of the difference of the squares of two numbers, plus the square of the product; diminished by half the difference of the squares will be the square of the less number.

Let x and y be the numbers, a the product, and b the difference of squares.

Then $xy=a$ and $x^2-y^2=b$

By division $x=\frac{a}{y}$

Substituting $\frac{a^2}{y^2}-y^2=b$

Multiplying $a^2-y^4=by^2$

Transposing $y^4+by^2=a^2$

Completing square $y^4+by^2+\frac{b^2}{4}=a^2+\frac{b^2}{4}$

Extracting $y^2+\frac{b}{2}=\surd(a^2+\frac{1}{4}b^2)$

Transposing $y^2=\surd(a^2+\frac{1}{4}b^2)-\frac{1}{2}b$

And $y=\surd(\surd(a^2+\frac{1}{4}b^2)-\frac{1}{2}b.)$

Proposition 65.

The product of two numbers divided by their ratio, equals the square of the less.

Let x=less, r=ratio and a=product. Then rx=greater, and $rx^2=a$.

Hence $x^2=\frac{a}{r}.$

Proposition 66.

The sum of the squares of two numbers divided by the square of the ratio, plus one, equals the square of the less number.

Let $x=$ less number, $r=$ratio and $b=$ sum of squares. Then $rx=$greater number.

And $r^2x^2+x^2=b$

Or $(r^2+1)x^2=b$

And $x^2=\frac{b}{r^2+1}$.

Proposition 67.

The difference of the squares of two numbers divided by the square of the quotient *less* one, equals the square of the less number.

Let $x=$less number, $r=$ratio and $a=$difference of squares.

Then $r^2x^2-x^2=a$

Or $(r^2-1)x^2=a$

And $x^2=\frac{a}{r^2-1}$.

Proposition 68.

The square of the sum of two numbers divided by the square of the ratio *plus* one, equals the square of the less number.

Representing the less number, the ratio, and the square of the sum, by x, r, and a respectively, we have

$$(rx+x)^2=a$$

Or $r^2x^2+2rx^2+x^2=a$

And $(r^2+2r+1)x^2=a$

Hence $x^2=\frac{a}{(r+1)^2}$.

Proposition 69.

The square of the difference of two numbers divided by the square of the quotient *less* one, equals the square of the less number.

Let r=ratio, x=less number and a=difference of squares.

Then $(rx-x)^2=a$

Or $r^2x^2-2rx^2+x^2=a$

And $(r^2-2r+1)x^2=a$

Hence $$x^2=\frac{a}{(r-1)^2}$$

To the foregoing, which include all the principles embraced in the former part of the present lecture, the following may be added.

Proposition 70.

The product of any two numbers is equal to the square of their mean proportional, *less* the square of half their difference.

Let x and y be the numbers, then will $\frac{x+y}{2}$ be their mean proportional and $\frac{x-y}{2}$ their half difference.

And $\left(\frac{x+y}{2}\right)^2=\frac{x^2+2xy+y^2}{4}$; and $\left(\frac{x-y}{2}\right)^2=\frac{x^2-2xy+y^2}{4}$;

and $\frac{x^2+2xy+y^2}{4}-\frac{x^2-2xy+y^2}{4}=\frac{4xy}{4}=xy$.

Proposition 71.

The sum of the squares of any two numbers is equal to twice the square of their mean proportional, added to twice the square of half their difference.

For $\left(\frac{x+y}{2}\right)^2\times2=\frac{x^2+2xy+y^2}{2}$; and $\left(\frac{x-y}{2}\right)^2\times2=$

$\frac{x^2-2xy+y^2}{2}$; and $\frac{x^2+2xy+y^2}{2}+\frac{x^2-2xy+y^2}{2}=\frac{2x^2+2y^2}{2}$

$=x^2+y^2$.

Proposition 72.

The square root of the product of any two numbers, is equal to the product of the roots extracted separately.

This is manifestly true, and may be thus illustrated. Take the numbers 16 and 25, and their product 400. Obtaining the square root is dividing the numbers into two equal factors, and 16 and 25 the factors of 400 are thus divided, in effect, by 4 and 5, and consequently must divide 400 by the product of 4 and 5, in order to obtain a number that shall bear the same *relation* (not *ratio*) to 400 that 4 does to 16, or 5 to 25. An algebraic demonstration might be given, but it is tedious.

The principle contained in this Proposition is true whatever may be the number of *numbers*, or the *roots* involved.

Proposition 73.

The square of the arithmetical mean between two numbers, is a mean between the product, and half the sum of their squares.

For $\overline{xy+\frac{x^2+y^2}{2}}\div 2=\frac{x^2+2xy+y^2}{4}=\left(\frac{x+y}{2}\right)^2.$

The following propositions might be demonstrated like the preceding, but we will leave the proof for the reader's amusement.

1. If a number be divided into three equal parts, the product of one part by the sum of the other two, taken from the square of half the whole number, is equal to the square of one sixth of the number.
2. The product of the sum of two squares, by the sum of two squares is also the sum of two squares.
3. The sum of two squares prime to each other, can only be divided by numbers which are also the sums of two such squares.
4. A number which is the sum of a square, and double a square, can be divided only by numbers which are equal to the sum of a square and double a square.
5. If a unit, or any number considered as a unit, be divided into two parts, the sum produced by adding the first part to the square of the second, is equal to the sum produced by adding the second to the square of the first.

The list might be continued indefinitely, but we forbear.

LECTURE X.

REVIEW OF PRECEDING LECTURES, COMPARATIVE VIEW OF THE RULES OF ARITHMETIC, &c.

Before passing to the application of the preceding principles in the elucidation and solution of Arithmetical Problems, we will take a rapid review of the matters already discussed, and then proceed to compare the rules of arithmetic, and show the principles on which they are based.

In our first Lecture we made some remarks on the Study of Arithmetic, and traced the History of the subject from its earliest period down to the present age. In our second, we discussed pretty fully the Principles of Numbering; and especially according to our scale of notation. In our third, we took up the Properties of Numbers; and more particularly certain properties incident to our scale. In our fourth, we considered pretty fully the important doctrine of Prime and Composite Numbers, Measures, Multiples, &c., and showed how they may be applied to useful purposes. In our fifth and sixth, we entered very fully into the investigation of Fractional Quantities, whether in the Vulgar or Decimal form. The student may there find some things not to be met with elsewhere. In our seventh, the doctrine of Proportion is fully considered, and he that carefully studies what is there laid down, cannot fail to understand the numerous arithmetical operations that are based on this doctrine. Our eighth lecture was devoted to the consideration of the Involution and Evolution of quantities, so far as we thought it necessary to our purpose, and consistent with our main design of making our course useful rather than speculative. Our ninth lecture was devoted to the consideration of certain Relations of Numbers, often found exceedingly useful in the solution of problems. What is there said, will, if well understood, explain rules and operations frequently met with, and generally without note or explanation.

We will now devote a brief space to a general consideration of the subject, and a comparative view of the different rules or divisions usually found in treatises on arithmetic.

From the number and variety of these, we might well suppose the subject to be very complicated, or at least that the principles are exceedingly various. In order to show how far this opinion is correct, we will enumerate the rules of the subject as generally classified, and show the range and object of each, and how far they serve to elucidate each other.

Notation is the first step in written Arithmetic, for until numbers are written they are not visible to the eye, and it is to this operation that the name Notation is given; and then *Numeration* teaches to read the numbers written. The word Numeration has, however, a much more extensive meaning, as expressive of numbering in its widest sense, without reference to written numbers. Most writers class both these under the general head of Numeration; but the operations are distinct, and the names express the processes of writing or noting down and numbering. As generally understood, both terms have reference to the elementary operations of writing down and reading whole numbers, but they are properly applied to the operations of expressing and reading numbers in every part of the science; and it is very common to speak of the Decimal Notation, Fractional Notation, &c.

Having learned to express numbers in writing and to read them, our next step is to learn to find the sum of two or more numbers; this we call *Addition;* and as numbers are divided into Integers, or whole numbers, and Fractions, including properly Compound Numbers and Decimals, Addition is found in our books under each of these heads; but it is the same operation still, and though varied in its mode to suit these different modifications of numbers, the same principle is involved in all. In whole numbers we add our column of units, and set down the unit figure of the result under the units' column, carrying the tens to the tens' column. In the compound rules we add the column of the lowest denomination, and divide the result by the number of that name which makes one of the next greater, setting down the overplus under the column which produced it, and carrying the quotient to the next column. Now these two operations are precisely the same in principle, and only differ in form from the fact that in whole numbers it is not necessary to divide, since if we were to do so, the units' figure would always remain to be set down, and the quotient would be just the figures that we now carry; for dividing by ten never alters a figure. In fractions, if we would add into one sum, we bring all to the same denomination, by bringing all to a common denominator, and then having added up the numerators, for they express the quantity or value of each fraction, we divide by the common denominator, for that is the number of such parts in a unit, the overplus is set down as a fraction, and the units carried to the units. The same principles apply, therefore, in adding fractions, whether common or decimal, so that though there are several rules bearing the name of *Addition*, it is one and the same thing.

The word "*Rule*" may require a passing notice. A rule, in its ordinary acceptation, is something to direct, a law or instruction by which we are to be guided and controlled; and hence the rules with which books of instruction abound. But in arithmetical works, the chapters or various classifications are called rules, as well as the instructions given for the different operations, and it is in this sense that Addition, Subtraction, &c., are called rules.

When we desire to add together several similar numbers, or to find the sum of a number repeated a number of times, as 1354, 1354, 1354, 1354, the operation is shortened by changing the mode, and instead of setting down the numbers four times and adding them together, we throw it into a different form by which both figures and labor are saved. But let us first work it in the old way:

$$\begin{array}{r} 1354 \\ 1354 \\ 1354 \\ 1354 \\ \hline 5416 \\ \hline \end{array}$$

It seems highly probable that, instead of adding the several fours successively to each other, the operator would soon learn to say 4 fours are 16; the 6 units he would set down in the units' place, and carry the 1 ten to the tens' place. Here again he would be disposed to make short work of the fives, by saying 4 fives are twenty, and 1 to carry makes 21; and thus he would proceed to the end of the operation. But the task of setting down a number repeatedly would after a time become burdensome, and the operator would look out for some easier plan. Perhaps none would strike him sooner than the expedient of setting down the number once, and setting underneath it the number of times it should be taken. Following out this idea he would probably adopt a process something like this:

$$\begin{array}{r} 1354 \\ 4 \\ \hline 5416 \\ \hline \end{array}$$

and thus he would fall at once upon the process we call *Multiplication*, which, *so far as whole numbers are concerned*, is but a brief mode of adding numbers, and would naturally be invented from *Addition*, by any accountant of ordinary ingenuity.

"*Multiplication*," most of our elementary treatises say, "teaches to find what a number amounts to when repeated a given number of times." This definition applies well enough to whole numbers, and so far as multiplication is but a compendious form of addition, it is correct. Perhaps, too, if we look to the etymology of the word as derived from the Latin *multiplico*, meaning to increase or make many, the definition is correct; but it is not comprehensive enough to embrace multiplication of fractions, for there the number multiplied is often diminished, instead of being repeated any number of times. But still, though there is this difference between the multiplication of integers and fractions, the analogy is complete—the principles are the same; and equally so in compound numbers and decimals. If a number be multiplied by 2 it will be repeated "a given number of times," but if we multiply it by the fourth of 2 ($\frac{1}{2}$) the product will be one fourth as much; but here our definition fails, since instead of being repeated "times," it is only taken half a time.

The second operation upon numbers is taking one number from another, which is the reverse of adding one to another. This is called *Subtraction*, (from *subtraho*) a name signifying a taking from. It is not only necessary to take one number from another, but sometimes to find how often one number may be taken from another before the larger will be exhausted: or as it is usually termed, to find how often the smaller number is contained in the larger. This latter operation is called *Division*, and though it is not so easy to fall into the mode of operation from the process of subtraction, as to glide into multiplication from addition, it is not less evident that subtraction is the most simple and natural mode of effecting division.

Let 45 nuts be distributed among 9 boys, or rather given them to distribute amongst themselves, and if ignorant of numbers their practical operation will probably be to take 9 nuts from the pile, at a time, and distribute them around, giving one to each, and repeating the operation until all are exhausted; when each may count what he has received. This is a kind of mechanical division. But another idea may strike them, if they cannot readily make the distribution. They may set down the number 45 and subtract 9 at a time until all are gone, and then count the number of their subtractions. This I have witnessed; and I may add that to some extent it is the mode we all pursue in our ordinary process of division. Let us for illustration divide 1860 by 15.

```
15)1860(124
   15
   --
    36
    30
    --
     60
     60
     --
```

Here from 1860 we first take away 1500, (for 15 in the hundreds' place is 1500,) and 360 remains; from this we take 300, and 60 remains, and lastly we deduct this 60 also. The numbers 36 and 30 in the tens' and hundreds' place are equivalent to 360 and 300. From which it appears that even in our ordinary mode of division, subtraction is its basis. Instead, however, of taking in the above instance 15 only at a time, we took first 100 times 15, then we took 20 times 15, and lastly 4 times 15, making in all 124 times 15. From all which we infer that 15 is contained 124 times in 1860; or that 1860 would make 124 groups of 15 each.

From this it appears that instead of four elementary operations upon numbers, as seems indicated by the four elementary rules, there are but *two*, by one of which a number is increased by having another added to it; and by the other it is diminished by having a number taken from it; the rules called Multiplication and Division involving no new principle, but only different modes of operation. It is true that Division gives a number bearing a relation to those that produced it, different from either *sum* or *difference*, but this is found by a compound operation of adding and subtracting, and the finding of it involves no new principle.

Multiplication and Division, as well as Addition and Subtraction, are found in the compound rules, vulgar fractions, and decimals, but they are the same operations there as in whole numbers, differing in no wise except so far as the character of the numbers renders necessary; and this may be a favorable moment to urge upon the student's attention, that the first principles of this science, and indeed of all mathematical science, are few in number. All the sublime calculations of the Astronomer, by which he measures the celestial orbs and their revolutions, and all the secrets of the Analytic art, consist in properly applying the two operations that we have attempted to explain, viz: making numbers *greater* or *less*.

There are certain operations or relations amongst numbers that it is necessary to understand, and for this purpose the *Rule of Three* is introduced, which considered as a rule of pure science, aims at nothing but to find a fourth proportional to three given numbers. It is one of the few rules that would be retained if we limited our arithmetical researches to the principles of science only, for the purpose of becoming skilled in the higher branches, without applying the subject to common business purposes; but then the questions would be totally different, for we should hear nothing in them of the value of pork and potatoes, nor how much work A, B or C could do. The following would be the kind of questions asked:

As 3 : 4 : : 5 to what number?

As 8 : 2 : : 12 to what number?

What number bears the same ratio to 20, that 20 does to 40?

What number bears the same ratio to 15, that $\frac{1}{2}$ of 10 does to $\frac{1}{5}$ of 60?

This would be the description of questions asked, and in order that they might be perfectly intelligible, it would be necessary to embrace the doctrine of Geometrical Progression, for without some knowledge of this, the nature of ratio and proportion could not be well understood; since the mode of operation is derived from that series. As a kindred subject, adapted to throw light on Geometrical Progression, we must introduce also Arithmetical Progression, or the doctrine of Equi-different Series; and as fractional quantities in every variety would be involved, that subject would require attention, for there would be no running of remainders into inferior denominations, they being entirely an invention of practical Arithmetic.

It may be as well here to explain the difference between Arithmetic considered exclusively as a *science*, and Arithmetic considered as an *art*, or as it is generally termed *Practical Arithmetic*. The elementary rules are the foundation of both, but as soon as we pass these, a distinction arises. Arithmetic *as a science* then aims to unfold the properties and relations of numbers, while commerce seeks to apply the principles thus developed, *as an art*, to aid its purposes in business transactions.

Having disposed of the elementary rules in *whole numbers*, science would forthwith take up the same operations upon fractional quantities in the common and decimal form; but this does not suit commerce, and hence our earliest lessons in broken numbers are taken in what are called compound rules, because the fractions, and fractions of fractions, that there present themselves, are called by various names. A pound Troy is the unit of the system of weights called Troy; an ounce is $\frac{1}{12}$ of a lb.; a dwt. is $\frac{1}{20}$ of an oz.; and is hence $\frac{1}{20}$ of $\frac{1}{12}$ of a lb.; a grain is $\frac{1}{24}$ of a dwt. and hence is $\frac{1}{24}$ of $\frac{1}{20}$ of $\frac{1}{12}$ of 1 lb.

If the wants of science alone were consulted we would have no compound quantities, and with these the Reduction of such quantities would disappear also; and all parts of integers would assume a regular fractional form. Next would come the doctrine of *Proportion*, of which the Rule of Three is a branch, and this subject abounds in relations and principles adapted to throw light upon the science, and to enable the operator to unravel the mysteries of numbers. *Permutation* and *Combination* are scientific, the former teaching the number of changes of which any given number is susceptible, and the latter what combinations or given numbers may be formed of some other given number. The *Raising of Powers* and the *Extraction of Roots*, are scientific operations, and *Position* may be considered scientific also, being based entirely on the doctrine of Proportion; and of no practical use in commerce or the mechanic arts.

These are about all that science needs, and now let us see how commerce and mechanism extend the catalogue.

The great variety of compound quantities has been already alluded to, and every boy remembers how he has toiled in the labyrinth. These result from weights, measures, and money being grouped in sets to suit purposes of trade; and from these the rule called *Practice* results. It is a mode of multiplying these compound quantities or multiplying by them, by converting the inferior denominations into equivalent vulgar fractions of the higher denominations. Interest and Discount are purely business rules, and are nearly allied; indeed the latter is embraced in the former and is identically the same with the case in interest in which the rate per cent., the time, and the amount are given to find the principal; for the present worth is the principal. It is true that in general, the calculation in the case in Interest, as given in the books, refers to past time, to a debt now due, the amount of which we know, and the original principal of which we seek to find; while in Discount the debt is supposed to be due at a distant day, and we seek to find a present sum, a principal, that at an agreed rate will at the proper time equal the debt. It is to find a present sum equal in value to a given larger debt due at a future day. *Equation of Payments* might also be classed under the head of Interest.

Barter and *Loss and Gain* are also purely commercial Rules, designed to train the mind in the mysteries of traffic; and the same may be said of the *Millers' Rule*, *Tare and Trett*, *Alligation*, *Fellowship*, *and Exchange*. These rules advance no new principle, and are only retained in our books for convenience of classification; the problems appropriate to them might be blended in a general mass of "Promiscuous Questions," if by so doing, the system would be in any wise simplified.

In most trains of causes and effects, the effect is proportionate to the cause, and if we can by any means ascertain the ratio between them, we can from having one find the other. A purchases twenty pounds of sugar, for which he pays $2; B purchases 30 pounds, and it is easy to see that if the price per pound is the same as A paid, the whole cost must be in proportion to the cost of A's as 20 lbs. are to 30 lbs. This relation or ratio between cause and effect in one case, and cause and effect in another, gives rise to the application of Proportion, (or that modification of it called "The Rule of Three,") to commercial purposes. The Rule of Three is therefore both a scientific and a commercial rule; and from its extensive applicability to both purposes, has been aptly called the *Golden Rule.*

But though it has long been extensively applied in Arithmetic, the rule may be dispensed with altogether, and some exult in discarding it from their systems; it is doubtful, however, whether they simplify the subject by rejecting it; since they only change their mode of operation, and do in another way what may be as well or better done, in many cases, by this.

In many instances the value of commodities may be more briefly found by multiplication only; and in others by Practice, though since the general adoption of Federal money, Practice is much less used than under the old currency of pounds, shillings, and pence.

All the rules we have named as scientific, are also applied to practical purposes, either by tradesmen or mechanics. Extraction of the Square Root is seldom needed in commercial Arithmetic, but by mechanics it is often found necessary; and in some of their calculations even the Cube Root is necessary, though not frequently. Mechanics and practical men have need of many applications of this science peculiar to themselves, though they generally blend it with Geometry as in measuring, calculating forces, &c. Men of science apply it to their own peculiar purposes, the Mariner to Navigation, the Astronomer to calculating the motions of heavenly bodies, the Surveyor to finding the area of his surveys, and every one to his own purpose. We shall endeavor to show, hereafter, how the subject may be applied to some of the purposes of each.

The principles, as we have seen, are few in number, and if these be well understood, their manifold applications are easily comprehended. But one thing seems very clear, the rejection of the usual classification will not simplify the subject, any more than the Scriptures would be simplified by blending all the chapters and verses of each book into one; or the traveller be benefitted by taking away the milestones from the highway. It is neither the number nor the scarcity of divisions that makes a subject clear or obscure, but it is the manner in which it is treated. If arbitrary rules are dogmatically given and blindly followed, there may be scarcely the difference of a shade between two modes and yet they may appear entirely distinct. It is necessary, therefore, to look behind the words of the rule in all cases, and see the principle on which it is based; for if possessed of this, even complex operations become simple.

We shall now proceed to consider in the course of a few succeeding lectures, somewhat in detail, the subjects to which a knowledge of Arithmetic is applied; and then having examined carefully the relative merits of the Synthetic and Analytic systems, we shall devote a few lectures to the solution of problems, with such annotations as may be thought profitable.

LECTURE XI.

INTEREST, DISCOUNT, INSURANCE, ANNUITIES, &c.

Before entering upon an investigation of the different modes of calculating interest, it may be interesting to bestow some attention upon the history of the subject, that we may be better prepared to understand it.

Amongst the Jews a law existed that they should not take interest of their brethren, though they were permitted to take it of foreigners. "Thou shalt not lend upon usury to thy brother; usury of money, usury of victuals, usury of any thing that is lent upon usury; unto a stranger thou mayest lend upon usury; but unto thy brother thou shalt not lend upon usury." (Deuteronomy xxiii, 19, 20.) After the dispersion of the Jews they wandered through the earth, but they yet remain a distinct people, mixing, but not becoming assimilated with the people amongst whom they reside. Still looking to the period when they shall return to the promised land, they seldom engage in permanent business, but pursue traffic, and especially dealing in money; and if their national policy forbids their taking interest of each other, they show no backwardness in taking it unsparingly of the rest of mankind. For ages they have been the money lenders of Europe, and we may safely attribute to this circumstance the prejudice, in some measure, that still exists even in our own country against such as pursue this business as a profession. The prejudice of the Christian against the Jew has been transferred to his occupation, and from the days of Shakspeare, who painted the inexorable Shylock contending for his pound of flesh, down to the present time, the grasping money lender, no less than the grinding dealer in other matters, has been sneeringly called a Jew.

For ages the taking of any compensation whatever for the use of money was called usury, and was denounced as unchristian; and we find Aristotle, the heathen philosopher, gravely contending that as money could not beget money it was barren, and usury should not be charged for its use.—The philosopher forgot that with money the borrower could add to his flocks and his fields, and profit by the produce of both.

As the commercial transactions of the civilized world increased in extent, and men found the advantage of using the capital of others, the prejudice against compensating the owner for its use gradually abated, and while a reasonable compensation received the name of interest, the opprobrious epithet *usury* was reserved for an unreasonable demand.

But still the progress of this change in popular opinion, seems in most countries to have been watched with jealousy, and we find accordingly that legislators interfered early to establish the rate of compensation. The Roman law allowed 12 per cent. per annum, but Justinian reduced it to 4 per cent. In the 13th year of Queen Elizabeth, it was first tolerated by law in England, and the rate restricted to 10 per cent.; but a statute of James I. reduced it to 8; a subsequent one of Charles II. to 6; and a still later of Queen Anne to 5, at which rate it still remains established; the penalty for receiving a higher rate being a forfeiture of treble the money lent. In Ireland it is 6 per cent.; in the West Indies 8; in Hindostan 10 or 12, and at Constantinople it is said that 30 per cent. is a common rate. In the United States the rate varies from 5 to 8 per cent.; the law in some of the States inflicting a severe penalty for receiving a higher rate than is allowed by law, while in others it is left very much to the contracting parties, the law only establishing a rate in the absence of special bargain. The exceptions to 6 per cent. in our own country are New Hampshire and Louisiana, 5; New York, South Carolina, Michigan and Wisconsin, 7; Georgia, Alabama, Mississippi, and Florida, 8. All debts due the United States are charged only 6 per cent., even in those States where the law allows a higher rate.

The question whether this restriction upon the freedom of contracting parties is necessary and expedient is one of grave import, and its discussion would require far more space than we can appropriate to it here. Persons are apt to suppose that there must be a reason for any opinion in which we find mankind concur generally, and this is correct; but we are inclined to think that in this case there is more of feeling than of reason in the opinions and views of most of us; and it is worthy of remark, that the views of men are constantly becoming more liberal on this subject; and our laws, instead of imposing new restrictions, are inclined to remove old ones.

There are, without doubt, instances in which the recklessness of individuals should be checked by legal restraint, but it is believed that in most instances men may be safely trusted to decide for themselves, and that where competition is left free to act, the price of money, like that of every other commodity, will find its proper level; for men will decide best for themselves, according to the exigency of their wants, or their prospects of gain, what they can afford to pay for the use of money. It is neccessary, however, that a rate be established for the many cases in which no contract is made.

If we define usury to be the taking of *illegal* interest, then the law becomes the measure of the creditor's conscience, and what is usury to-day, may by a change of the law, cease to be usury to-morrow. The deed is not evil in itself, but right or wrong as the law declares it so to be.—But if we define it to be the taking of exorbitant or unreasonable interest, when the debtor is in the creditor's power, it becomes a question of Morals, and is more naturally to be decided by a consideration of all the circumstances, than from the application of any single arbitrary rule. The dealer in money may oppress, and with perhaps greater facility than most others, yet it appears difficult to show that he who "grinds the face of the poor" by oppression in any other shape, is in any wise a better man. It is true, however, that money being the representative of value, the medium of exchange, the "*open sesame*," by which all terrestrial wealth may be attained, seems to have a fascination about it, that makes some men reckless in their engagements to obtain it; and legal restriction may be necessary on the same principle that lotteries and gambling in general require it. Restriction, too, serves to check recklessness and give a correct tone to public sentiment; from which no doubt benefit results.

Though it is now agreed that Simple Interest is just and right, it is not agreed that the debtor who fails to pay his debts shall be compelled always to pay interest on the interest due, and which he unjustly withholds; though if he pays his debt with honorable punctuality the creditor may place the proceeds at interest or use them in his ordinary expenditure, and the money received for interest will be found just as valuable as that received for principal. Instead of punishing the delinquent for withholding the property of his creditor, he is now rewarded by being permitted to use it free of charge. Even Paley, with all his ethical ingenuity, can find no moral reason for such a policy; but he thinks it may be well enough as an obstacle thrown in the way of procuring money without labor. But before men obtain money labor must be bestowed, and we do not perceive the justice of striking the staff from the hand of trembling age, or depriving the widow and fatherless of the support for which the husband and father toiled. It would be as just to encumber the farm, as the cash capital, which a man toils to procure for his family, as a means of support when he is gone.

We have not room, however, to enter at large into a discussion of questions connected with the propriety of legal restrictions on the subject of interest, and shall pass on to a consideration of the calculations occurring in business.

There are four elements in every calculation of this kind, any three of which being given, the other may be found, viz: Principal, Rate, Time, Amount.

The *Principal* is the sum placed at *interest.*

The *Rate* is the proportionate compensation, and is usually reckoned at so much for a hundred of the denomination of the debt for a year; familiarly designated at so much *per cent. per annum.* The words "*per cent.*" or centum "*per annum,*" being a Latin expression meaning *for a hundred*, *for a year.* *Cent.* being an abbreviation of *centum*, is followed by a period.

The *Time* is the length of time the principal is permitted to bear interest.

The *Amount* of the debt is the sum of the interest and principal added together.

The subject of Simple Interest in our school Arithmetics is generally divided into four cases, viz:

CASE 1. In which the Principal, Rate and Time are given to find the Amount.

CASE 2. In which the Principal, Amount and Rate are given to find the Time.

CASE 3. In which the Amount, Rate and Time are given to find the Principal.

CASE 4. In which the Principal, Interest and Time are given to find the Rate.

In business the first of these operations is much more liable to occur than any of the others. Case 3 is the same in form and principle precisely with the rule or calculation called Discount. We will give an example in interest and vary it through the several cases.

1. If a note for $450 be permitted to run at legal interest in Ohio for 7 years, what will be the amount?

$450	
6	
27.00	Interest for 1 year.
7	
$189.00	Interest for 7 years.
$450	Principal.
$639	Amount.

Here we have Principal, Rate and Time given to find the Amount. If we knew the interest of one dollar for 7 years, it is very plain that the interest of $450 would be 450 times as much; or even if we had the interest on one dollar for a year, we could find the whole interest by multiplying by the time, and the number of dollars. But instead of having the interest of one dollar given us, he have the interest for $100, and we may proceed to find the interest on $1 by dividing the $6 by 100, by which we should obtain 6 cents; or we may first multiply by 6 and afterwards divide by 100; for if we multiply by the interest of $100, we shall obtain 100 times too much interest. Having found $27, the interest on $450 for one year, we multiply it by 7 to find the interest for 7 years, to which the principal being added, we have $639, the amount.

If we consider the 6 as 6 cents, the interest of one dollar, the work will be precisely the same, the two figures cut off for cents corresponding with those now cut off in dividing by 100; and as you have only multiplied by the interest of $1 you will not then divide by 100. Some calculators pursue this mode, setting the 6 down as .06 of a dollar; which is equivalent to 6 cents.

2. By varying the question we make it correspond with Case 2; thus—A note of $450 has been at interest at 6 per cent. per annum until it amounts to $639 dollars: how long has it been running?

We proceed first to find what amount of interest has accrued, which we do by subtracting the principal, $450, from the amount, $639, which leaves $189, the whole interest. Then we find that the principal will produce $27 in one year, thus $450×.06=$27.00, and, As $27 : $189 : : 1 year : 7 years, the time the note bore interest.

3. A note has been bearing interest for 7 years at 6 per cent. and now amounts to $639; what was the original principal?

If any other sum were placed at interest for 7 years at 6 per cent. it is evident that their amounts would be proportionate to their principals, or their principals to their amounts: *i.e.* as the amount of one is to the amount of the other, so is the principal of one, to the principal of the other.

In solving this case it is usual to find the amount of $100 at the rate and for the time given, and then to say, as this amount is to the given amount, so is $100, to the principal required. It is evident, however, that the $100 is only used to obtain a proportion, and that any other sum would serve just as well, some prefer finding the amount of $1 for the time and at the rate given, and this divided into the given amount will give the principal from which it was derived.—Let us try this question by both modes—

$100 at 6 per cent. for 7 years, amounts to $142; then

As $142 : $639 : : $100 : $450, the principal required.

Or,—$1 at 6 per cent. for 7 years amounts to $1.42; and

639÷1.42=450, the result as before.

The latter mode is less usual, but is based on precisely the same principle as the other.

4. A note for $450 was put at interest 7 years ago, and now amounts to $639; required the rate per cent.?

Here we find by subtraction that $450 has produced $189 interest in 7 years; and dividing by 7 we find that it produced $27 in one year; and if $450 produced $27 in 1 year, $100 must produce $6, the rate *per cent. per annum* required.

It is evident that though we usually fix the rate or ratio of interest to the principal, by giving the interest on $100 for 1 year, it is not necessarily so established; for we may use any other amount, and for any other time, but custom and convenience sanction the former mode. Instead of saying, "What will $450 amount to in 7 years at 6 per cent. per annum?"

we may say, "What will \$450 amount to in 7 years, if \$20 bring \$1.80 interest in 18 months?" These two questions produce the same result, but the former is more convenient of solution.

But though the principle of solution is thus readily established, modifications less simple frequently occur in practice. The length of time during which interest has been accruing, is seldom limited to even months, and even if there be a given number of months, they may or may not average so many twelfths of a year; and in a large sum it may make a difference of several dollars. Let for instance a sum of \$10,000 be at interest for 5 calendar months, ending with June 30, and it will be but 150 days; but the next five months would contain 153 days, and the interest on \$10,000 for 3 days would at 6 per cent. be \$4.93+. The former would not be $\frac{5}{12}$ of a year, and the latter would be rather more.

Where entire accuracy is desired, calculation by days is indispensable; but in small amounts the difference would not compensate for the trouble.

In finding the interest for any number of months at 6 per cent. a very common method is to multiply the principal by half the number of months, for the interest in cents, if the principal be dollars. That this must produce the proper result is very apparent when we consider that the interest of a dollar is 6 cents for 12 months, or half a cent a month, at 6 per cent. The rule applies to no other rate.

For days, some calculators multiply the principal by $\frac{1}{6}$ the number of days, and if the principal be expressed in dollars, the product will be the answer in mills. But this is on the supposition that 360 days make a year, which is not true.

Required the interest of \$960 for 63 days at 6 per cent?

```
      $960
       10½=⅙ of 63
      ----
       480
      9600
      ------
Ans. $10.08.0
      ------
```

The interest of any sum at 6 per cent. for a year is equal to $\frac{6}{100}$ the principal, and if a year consisted of just 360 days, the interest for the whole being $\frac{6}{100}$ of the principal, the interest for the sixth part of 360 days=60 days, will be $\frac{1}{100}$ of the principal, or as many cents as the principal contains dollars. The interest of a dollar at this rate would be one cent for 60 days, or $\frac{1}{60}$ of a cent for 1 day, and if $\frac{1}{60}$ for one day, it will be $\frac{1}{10}$ of a cent, or 1 mill for 6 days; and hence multiplying by $\frac{1}{6}$ of the days must give the number of mills contained in the *interest.*

But the year consists of 5 more days than 360, and 5 days are $\frac{1}{73}$ of a year, so that this mode will give $\frac{1}{73}$ too much interest. In the question given above, the excess would be rather less than 14 cents, but in a year it would have been nearly a dollar.

This is the mode adopted in banks generally, and in addition to charging thus they take the interest in advance, by which their profits are still farther advanced. For instance, if you borrow $600 for 60 days they charge $6 interest which they keep back, and pay you $594. They loan also for short periods, by which they receive all the advantage of compound interest.

What is the difference between the interest of $1000 for one year at 6 per cent., and the interest of the same sum loaned according to bank rules?

$1000
6

$60.00, Interest by ordinary mode.

By Bank mode—

$1000 for 60 days=$10, and this being paid in advance, may be loaned and will produce 10 cents, and this again being loaned would produce 1 mill, and we might thus descend forever in theory. But descending no farther, we have for the first 60 days $10.10.1 interest, and adding this to the principal we have $1010.10.1 to loan at the commencement of the next 60 days.

The interest on this sum for the next 60 days will be $10.-10.1, which being paid as before will produce 10 cents 1 mill, and this again will produce 1 mill, making in all $10.20.3; and making the whole amount $1020.30.4.

The interest for the 3d 60 days will be $10.20.3; and this will produce 10 cents, 2 mills; and this 1 mill, or $10.30.6, and the amount will be $1030.61.

The interest for the 4th 60 days will be $10.30.6+10 cents 3 m.+1 m.=$10.41, and the amount will be $1041.02.

The interest for the 5th 60 days will be $10.41+10 cents 4 m.+1 m.=$10.51.5, and the amount will be $1051.53.5.

For the 6th 60 days the interest will be $10.51.5+10 cents 5 mills+1 m.=$10.62.1, and the amount will be $1062.15.6.

There are still five days remaining of the year, and for that five days the additional interest will be 88 cents 5+mills, and this added to the last amount will make $1063.04.1, being

$3.04 more than by an ordinary loan. This is equivalent to 6.304 per cent.; so that loaning at Bank rates on loans of 60 days, is equal to loaning by the year at 6.304 per cent. *compound* interest. This concerns only the mode of loaning, and does not consider the privilege banks have of loaning in the shape of their own notes that do not bear interest, for men's notes that do, to the extent of two or three times the amount of capital they have. Loaning money by banks is usually called bank discount; and it may be proper to remark that 3 days, called *Days of Grace*, are added to the time notes have to run before they are considered due, and liable to protest for non-payment.

Though as already remarked, the mode of calculating by days, allowing 365 to the year, is more exact than by months, owing to the inequality of length among the latter, it would make ordinary business calculations tedious, and unless the amounts were very large, the difference would not be worth the trouble. It is very common, therefore, to use half the number of months, where that will answer the purpose, and make the time months and halves, thirds, fourths, &c., without being particular as to a day or two, one way or the other. Thus from May 1 to August 15 is $3\frac{1}{2}$ months; or to September 20th is $4\frac{2}{3}$ months; or to October 25th is $5\frac{5}{6}$ months.—Or we may find the interest for a year, and take parts for months and days according to the rule of Practice, if we prefer that mode to the Rule of Three. For short periods, 30 days to the month is accurate enough.

In banks and offices where heavy calculations are frequently made, Interest Tables are generally resorted to.—These give the amount of $1 from 1 day to several hundred, so that the operator has only to find the amount of one dollar for the desired number of days, and this multiplied by the given principal will give the amount required. Other tables contain the interest of different sums from $1 to 50 or 100, but they are all based on the same simple principle, and are equally easy of application. One of the most popular tables of this kind is Rowlett's.

If when a sum of money is due, instead of receiving the interest, the creditor adds it to the principal and charges interest on the amount as a new principal, and so adds the interest successively to the principal as it falls due, the process is called charging *Compound Interest*, in contradistinction to *Simple Interest*, which is charged on the original principal only.

A sum of money loaned at 6 per cent. simple interest, will double itself in 16 years 8 months, but at compound interest it will double itself in 11 years 10 months and between 22 and 23 days, if the payments be annual: if at shorter intervals it will sooner double itself.

Ethical writers and law makers seem to have set their faces against allowing compound interest, and yet they can give no satisfactory reason why it should not be allowed.—There is no pretence but that a man has a perfect right to settle with his debtor and take a new note for both principal and interest, whenever his debt is due, or he may compel him to adopt the generally more unwelcome alternative of paying up; and yet it would require a great degree of metaphysical acumen to show the difference in a moral point of view, between taking a new note and charging interest, and charging interest without the formality of a note.

The question might be asked, and with reason, whether there is no proper limit to the frequency of the operation of adding the interest to the principal. Shall it be annually, semi annually, quarterly or oftener? for it is evident that the shorter this period is, the more rapidly will the interest increase. We think there is a limit on both hands. To add it daily, or monthly, or even quarterly, would scarcely allow the borrower to make the profits on his loan pay the interest; and on the other hand to extend the additions beyond a year would do injustice to the lender. The principle of taking interest for money is certainly based on the supposition that though not of itself prolific, the money may be expended in that which will yield an increase, and the profit of this increase is to be divided between him who owns and him who wields the capital. Some regard may therefore be properly paid to the time in which a return of profit is ordinarily received. If it were a proper partnership, the division should be made at each return of profit; but in ordinary loans the capitalist's share is neither contingent as to time or amount; but being fixed at something like an average of both, is made a certain sum.

Calculating compound interest by the ordinary rule of adding each year's interest for a new principal is a slow process when the number of additions is great, either from their frequency or the length of time the sum bears interest. But this difficulty is greatly lessened: indeed it is almost removed by the use of tables, by logarithms, &c. As it is, however, little more with us than a theoretical calculation, it is not necessary to pursue the subject farther.

Connected with the subject of interest is the doctrine of Discount, Equation of Payments, Annuities, Perpetuities, &c., and the mode of charging interest on notes on which partial payments have been made. The common mode of calculating

Equation of Payments, is not strictly correct, though usual in business. The true mode would be to find the present worth of all the payments at some assumed rate of interest, and then find in what time such present worth would amount to the given sum.

We have already remarked that the rule called Discount is identical with the 3d Case of Interest; it being but an operation to find the principal or present worth from knowing its amount after bearing interest for a given time at a given rate. The word Discount, or Rebate as it used to be called in the old Arithmetics, means something to be taken off or abated; and in its simplest form means a given per centage on a gross sum, without reference to time. Thus if a dealer offers to make a discount of 20 per cent. on a claim of $1000 of which he is perhaps doubtful, he generally designs to deduct 20 per cent. or one fifth of the whole sum, being $200. In this case the purchaser, if he collects the debt, makes 25 per cent. profit on his investment.

But when this word is used in reference to time, as when I say, "I will allow 10 per cent. *per annum* on a note of $1000, having 12 months to run," I do not mean that I will allow 10 per cent. on $1000, but that I will allow such a deduction as will enable the purchaser to make 10 per cent on what money he pays for the note. In other words I mean to deduct so much that if the net proceeds, which are usually called the *Present Worth*, be put at interest at 10 per cent. for 12 months, the result will be $1000. It is evident then that the operation is neither more nor less than to find the principal, when we have the *rate*, *time* and *amount* given. To pursue a different mode, would be to allow the shaver interest both on what he keeps back, and on what he pays.

This being the case the following somewhat paradoxical effect follows: If a sum of money be placed at interest, the amount of interest accruing will always be in proportion to the time in which it accrues. If in one year it produce a given sum, in two years it will produce twice as much, in three years three times as much, and so on. *But this is not true of Discount.* If I have a series of ten bonds for $100 each, due in ten successive years, and I agree to sell them at a rate of discount that will take off $12 from the first bond, it will not take 24 from the second, 36 from the third, and so on, for at this rate the discount on the last bond would be $120, being $20 more than the amount of the bond; a position evidently absurd, for however distant may be the day of payment, the bond must be worth something; and however great may be the per centage of discount allowed, there must be a *present value*, that placed at interest for the time and at the rate proposed will amount to the face of the debt.

If the discount on $106 for one year at 6 per cent. be $6, what will it be for two years?

If the discount on $106 for one year at 6 per cent. be $6, what will it be at 12 per cent.?

That discount is not in proportion to time is evident, therefore, from the absurdity to which such a conclusion would lead. But the reason of this is easily seen. The per centage of discount is always estimated on the present value, and this sum constantly diminishes as the remoteness of the time of payment increases. In the question proposed, $100 will be the present worth if the note has one year to run, but if it has two years to run, the present value will be 94.64\frac{2}{7}$, which being put at interest for two years at 6 per cent. will amount to $106. Had we deducted $6 for the second year, the balance would have been $94, which being placed at interest for 2 years at 6 per cent. would amount to only $105.28. As the time increases the difference between discount and interest increases more and more rapidly, and nearly in proportion to the square of the time. If in 6 months the difference be $1, in 12 months it will be very nearly $4, in 2 years $16, as may be shown by calculation.

Inasmuch then as Discount is the interest on the present value, it will always be less than the interest on the given sum, at the same rate, and for the same time; and it is just as much less as the interest on the true discount. Take the above example; the interest on $106 for 2 years is $12.72, the discount is 11.35\frac{5}{7}$, to which add the interest on 11.35\frac{5}{7}$ for 2 years, viz: 1.36\frac{2}{7}$, and the sum will be $12.72.

There is a striking analogy between Discount and Compound Interest. In both the principal is constantly changing. In the former it is becoming less, in the latter greater. In neither, therefore, can the result be proportionate to the time or rate, but will always be to the principal. The compound interest of a sum of money forms a constantly increasing series, during successive years; but the discount is a constantly decreasing series. At Simple Interest the annual accretion is a constant quantity.

The same reasoning will show that discount does not increase in the same ratio as its per centage increases. The discount on $100 for 4 years, at 2 per cent., would be 7.40\frac{20}{27}$; while at 4 per cent it would amount to only 13.79\frac{9}{29}$ instead of being double what it was at 2 per cent.

The increase of money at Compound Interest is neither proportionate to time or per centage. If the time is doubled, the interest is more than doubled, for there is interest on the accruing interest; and if the rate is doubled, the interest is more than doubled, for there is interest on the increased rate.

The word *par*, a Latin word signifying equality of value, is frequently used in connection with discount. Thus we speak of bank notes, &c., being above or below *par*, accordingly as they are worth more or less than specie, which is considered the standard of par value. The deficiency of such value is called the discount to which they are subject.

What is the difference between the interest of $800 for 5 years at 6 per cent. per annum, and the discount of the same sum for the same time, at the same rate?

The present worth may also be found by approximation, but not generally to much advantage in practice; the direct mode being quite as short, and more satisfactory. To find it by approximation, take the per centage on the gross sum, as you would in interest, and subtract the result from the gross sum for the *imperfect present worth;* then find the interest of the interest first found and *add* it to the imperfect present worth, for an approximate present worth; then find the interest of the last interest and *subtract* from the approximate present worth; and thus proceed adding and subtracting alternately until you reach the required degree of accuracy.

Required the present worth of $150 due in one year, allowing discount at 6 per cent. per annum?

	$150	$150	
	6	— 9	
Interest	9.00	141	imperfect present worth.
	6	+ .54	
Int. of Int.	.54	141.54	approx. present worth.
	6	— .03.24	
do	.03.24	$141.50.76	do do
	6	+ .00.1944	
do	.00.1944	$141.50.9544	do do

By the common mode—

$100
6
—

As 106 : 100 : : $150 : $141.50.9; present worth as found above, and by continuing both processes, they may be made to coincide to any extent.

It is evident that discount may be allowed at compound interest, as readily as at simple interest, though the calculation would be rather more troublesome. In this way a man who desires to make compound interest on his money can determine the value of a note he may wish to purchase.—To effect this, let the compound interest, instead of the simple, of one dollar for the time and at the proposed rate be taken for a divisor, and the quotient will be the present worth allowing compound interest.

Another business operation of a kindred character, already alluded to, is *Equation of Payments*. Where two or more payments are due at different times, and it is desired to pay all at once by averaging the time, the operation is called *Equation*, or *Equalizing Payments*. If I owe $100 due in 6 months, and another $100 in 12 months, I may average the payments by paying both sums at the end of 9 months. I thus keep $100 for 3 months after it is due and pay the other hundred 3 months before it is due; and this, were discount and interest the same in amount, would just equalize the profit and loss. But we have shown that this is not true. I obtain the use of $100 for 3 months by paying the other 100 three months before it is due; and as I would only be entitled to discount, which is less than interest for such advance payment, I am a gainer by the amount of difference. The discount on $100 for 3 months is $1.47.7+ the interest is $1.50, so that I make rather less than 2 cents 3 mills by the equation.

ANNUITIES, which are sums of money payable annually or at other stated periods, are very common in some communities, and especially in old countries, where retired wealth is abundant. PERPETUITIES are perpetual annuities, payable at stated periods forever. The rent of real estate, held in fee simple, may be considered a perpetuity. Calculations for the purchase and sale of both these, abound in English and Irish systems of Arithmetic, for in those countries investments of that kind are very common, the living of thousands being derived from such sources; but in this new country they are little known.

The purchase of a farm or other freehold is upon the same principle as the purchase of a perpetual annuity. If we seek to buy property merely to rent out, and without regard to rise in value, or other such consideration, the calculation is very simple; but there are generally various considerations of situation, capacity of improvement, rise in value, &c., &c., that influence the actual purchaser. The general principle however, of the calculation, holds good, and if closely scanned would probably prevent many an improvident purchase.

15

The difference between the value of a long lease and of a freehold estate is less than most persons suppose. If an estate that yields \$60 per annum, be leased out for a hundred years, the *Reversion* or ownership after the expiration of the lease, will be worth very little. If the purchaser of the reversion be allowed 6 per cent. compound interest for his money, the reversion will be worth but \$2.94.6$\frac{7}{8}$. If we calculate the value of such an estate in perpetuity, and then deduct the value of the reversion, we shall know the value of the lease.

If a note for \$100 due 60 years hence be sold at a discount of 6 per cent. simple interest, it will fetch but \$21.73$\frac{21}{23}$; and if at 6 per cent. compound interest, it will fetch but \$3.03+, and had it been a hundred years, the value at simple interest would have been but \$14.28$\frac{4}{7}$, and at compound interest it would be 29 cents 4+mills.

This difference may seem incredible, but it must be recollected that calculations at compound interest partake of the nature of the man's purchase, when he agreed to give a farthing for the first nail in the horse's shoe, a penny for the second, and so on in a quadruple ratio to the last.

To make this calculation we find that \$1 at 6 per cent. per annum will in 100 years amount to \$7 at simple interest; then,

\$ \$ \$ \$
As 7 am't. : 100 am't. : : 1 pr. : 14.28$\frac{4}{7}$, principal.

And at compound interest \$1.06, the amount of \$1 for 1 year, being raised to its hundredth power, will be \$339.3020 73383716, which is the amount of \$1 at compound interest for 100 years. This divided into \$100 will give .29.4+ the present value; or it may be thus stated, As 339.302073383716 : 1 : : 100 : .29.4+, which amounts to division at last.

That the amount of \$1 raised to a power indicated by the number of years will be the amount for that number of years may be thus shown,—As \$1 principal is to \$1.06, its amount, so is any other principal to its amount. But as the amount of each year is the principal for the next, we have the following proportions—

\$ \$ \$ \$
As 1 : 1.06 : : 1.06 : 1.06^2 am't 2d y'r and pr. for 3d; then,
As 1 : 1.06 : : 1.06^2 : 1.06^3, am't 3d y'r and pr. for 4th; then,
As 1 : 1.06 : : 1.06^3 : 1.06^4, am't 4th y'r and pr. for 5th.

And this may be carried to any extent, either by regular succession, or multiplying the amounts whose indices make the number of years desired, as the amount for the 3d year multiplied by the amount for the 4th year will give the amount for the 7th year, according to the established laws of *Involution.* See Prop. 49, page 124.

Having found the amount of $1 for the required number of years, it is obvious that multiplying it by any number will give the amount of such number of dollars. I cannot multiply the amount of $1 by $150 or any other number of dollars, but I can multiply by 150 because the amount of $150 will be 150 times as much as the amount of one dollar. It is on this principle that Interest tables are constructed.

But we are wandering from the subject of annuities. Not only are annuities and estates granted for years and forever, but they are often granted for lives; and in Europe determining the value of an annuity for life is a very common calculation. To determine the probable duration of such an annuity, tables are constructed giving the average value of annuities for every age, and these are founded on long and close observation of the duration of life; or rather perhaps the ages of such as die. In infancy the uncertainty of life is very great, but it diminishes as we approach manhood, increasing again in old age to a greater and greater extent. In a calculation of this kind we have two points to consider: the uncertainty of life, and its probable duration. A young infant has a chance of a longer life than a man of 30, but there is so much greater probability of its dying prematurely that the uncertainty more than counterbalances the possibility of a long life, and it would cost less to buy an annuity for life for such infant than for the grown person. At 8 years of age an annuity is more valuable than at any other period, as the dangers of infancy are passed and there is a fairer prospect of long life than before or afterwards.

It has been estimated in England, that "Of 6 or 7 children born in the same year, only 1, on an average, attains to 70 years; of 10 or 11, one may arrive at 75; of 17, one may reach 78; of 25 or 26, one lingers on to 80; of 73, one advances to 85; of 205, one realizes 90; of 730, one prolongs his existence to 95; and of 8179, one may complete a century."

"The average life of a child of one year of age, and that of a young man of 21 years, have been estimated at 33 years. A man of 66 years of age has an equal chance of life with a new born infant. An individual of 10 years of age has a probable expectancy of 40 years more of life; at 20, he may reckon on nearly 33½; at 30, on 28; at 40, on 22; at 50, on 16½; at 60, on 11 years and 1 month; at 70, on 6 years 2 months; at 75, on 4½ years; at 80, on 3 years 7 months; and at 85, on 3 years." These estimates are based on European observations, and should no doubt be modified in some degree to suit change of climate and situation; but the general features may serve to give an idea of the mode of estimating annuities, &c., on lives.

Sometimes estates, annuities, &c., are granted for a single life, sometimes for two or more, both living at the same time, sometimes to terminate at the death of either, sometimes of all. These uncertain or contingent estates as they are called, assume various forms according to the fancy or wish of grantees or grantors, but though necessary and desirable where lands are dear, they are seldom resorted to in this country, where land is cheap, and every one of ordinary industry and enterprize may possess his land in fee. A man in England wishing to provide for his children, one, two, or three, or more in number, may be able to purchase a life estate for their lives, that would not be able to purchase an estate in fee for them. Where such estates are common, rules for determining their value are necessary, but here they would be a useless encumbrance in a book.

It remains to close this subject by an investigation of the several modes of calculating balances, where partial payments have been made upon notes or other claims.

To calculate such balances, different modes have been adopted by accountants and sanctioned by courts, but the subject is still matter of controversy, and will be until the ridiculous prejudice against compound interest is exploded; for to escape this bugbear we are driven constantly into injustice towards either the debtor or the creditor. If it were understood that interest being withheld more than one year, should be added to the principal, then we would only have to charge the interest on the debt to the end of each year, and do the same upon the payments, and thus make an annual balance or settlement; and this course pursued annually to the time of general settlement would show the just and equitable balance. But this is not permitted.

There are two modes in general use. One which is called the *commercial* or *mercantile* mode, (because often used by merchants) charges interest on the whole debt for the whole time, and then deducts the several payments with interest from the time of payment to the time of settlement; the remainder is accounted the proper balance due.

The other mode, which is substantially the course pursued in our courts of justice, estimates the interest up to the time of the first payment, and then deducts the payment; the interest is thus added and the payment deducted at each successive payment down to the time of settlement.

In this case it is usual if the payments are made within less than a year of each other, to estimate the interest to the end of the year, before deducting; and should a payment be made at any time of a sum less than the interest due, the interest is not added and the payment deducted, since the addition being greater than the deduction, the interest-bearing principal would be increased; and as it would be so increased by an addition of interest, it would cause interest to bear interest, which is the monstrosity to be especially guarded against.

At a hasty glance the above modes would seem to promise the same result, but let us examine them a little more closely. By the mercantile mode every dollar that is paid becomes an interest-bearing principal, and offsets so much of the creditor's principal, let the amount of interest due be ever so great. If I owe a friend $1000, that has been bearing interest for 5 years, it amounts to $1300; suppose I pay him $1200, and let it rest for 5 years longer, what will I owe him? I certainly owe him $100 balance and 5 years interest added will make it $130. But let us calculate. He charges me with $1000 and 10 years' interest, making $1600, and I charge him with $1200 and interest for 5 years, amounting to $1560, deducting this from his claim leaves me only $40 in debt, and had settlement been postponed 4 years longer my friend would have been $8 in my debt; for though I certainly owed him $100 more than I paid him, yet I can charge him interest on $1200 per annum, and he can only charge me interest on $1000.

By this mode the payment of interest is postponed to the last, and the debtor is not required to pay the interest due even on that portion of the principal which he liquidates. Suppose I owe $400 on which there is one year's interest=$24, and I pay $200, I by this mode of calculation pay half the principal, and though I owed $12 interest on that half, I am neither required to pay this interest, nor to pay interest upon it. In the case of *The Miami Exporting Company* vs. *The Bank of the United States*, 5 Hammond, p. 261, this principle was presented, and the Supreme Court of Ohio, decided, (but disclaiming the establishment of a general rule,) "That where a sum of money is paid on a debt which is due 'on or before' a particular time not yet arrived, the payment should be applied to the payment of principal and such proportion of interest as has accrued on the principal thus extinguished." But this concerns the payment of money not due, and in such case the rule, even as we have given it, works no injustice; for if I borrow a sum of money, to be paid 10 years hence, bearing interest, but with no stipulation to pay the interest before the principal is due, then I can with no propriety be called upon to pay any part before that time, and if I do pay I should be entitled to draw interest from the time of payment. This is perfectly consistent, and differs entirely from the case of money wrongfully withheld after it becomes due; for whether it be the inability or the dishonesty of a debtor, that prevents payment, it is equally a wrong upon the creditor.

But to place the injustice of the commercial mode of calculation in a still stronger light, we will suppose A to borrow of B $1000 and to pay him annually $60, which is just the interest, and matters run on for 25 years, B neither demanding nor receiving any part of the principal: the parties then wish to settle. By this mode of calculation B would be found in A's debt. Can such a mode be just? If it be asked how this can be, we have but to refer to the principles already laid down. B could charge interest on $1000 only, but every dollar that A paid would by this mode draw interest; and from the time he made his 17th payment he was drawing a larger interest than B, and in $24\frac{8}{33}$ years, his debt would pay itself. The interest has completely eaten up the principal, as Pharaoh's lean kine of old destroyed their fellows.

Were the modification we have alluded to adopted, this could never be the case, for the interest would be constantly paid up on the portions of principal liquidated, and though the creditor might be compelled to wait a long time for part of his principal and the interest due upon it, he could not be compelled to wait for interest yet due on principal long ago paid. And should he be compelled to wait a few years he would be in no danger of being brought in debt by his own claim.

Let us now examine the other mode, and see whether it does even-handed justice. If the payments are made at intervals of just one year, and in sums exceeding the interest, it would seem plain that the interest should be first paid and the remainder of the payment be applied to diminish the principal, for where there is no contract to interfere, a man's capital certainly should yield him a return once a year. This effect will be produced by adding the interest to the principal and deducting the payment.

But payments are sometimes made at intervals less than a year. This is true, but we can allow interest on them to the end of the year, though in justice this can seldom be asked, for payments are much more frequently withheld more than a year, and what is justice to one should be also to the other; both should therefore be brought to an annual settlement, or neither. Sometimes payments are made of less than the interest due, and here we must not strike a balance lest the principal be increased. To avoid this it is usual to carry the payment to the next one, so that it may be considered merely a credit on the interest to that time. But this is a useless nicety, growing out of the old prejudice against taking interest on interest; for suppose I owe $1000 and aim to pay the interest annually, but this year it suits me to pay $70, which is $10 more than the interest due, and the principal is reduced to $990; next year I fall $10 short of paying the interest, but this cannot restore the $10 deducted from the principal the year before. Again I am not able to pay the first year's interest until six months after it is due, but the second time I pay just at the end of the second year; this, however, is less than a year since I paid before, and it would be wrong to strike the balance, then, though it was perfectly right to keep my friend waiting 6 months the year before!

It has always appeared to me strange that persons who have such a dread of compound interest, &c., do not contrive some mode to obtain discount on interest that is paid on periods less than a year. I borrow $1000 for 3 months, or if you choose, borrow it for a year and pay in 3 months, the interest will then be $15, but if I pay it, the owner may loan it and make 75 cents on it during the remainder of the year, and thus obtain more than six per cent. for his money! It is true that I might keep it 3 months after it is due, but as the owner of the money would be the loser in that case, it is quite a different matter.

There is another reason why annual settlements would be better than even this mode. It is by this mode the interest of the debtor to pay as seldom as possible, and as little as possible at a time, unless he can pay enough to reduce the principal; and thus an inducement to dishonesty is held out. By the former mode the interest is considered as never due, while there is any principal to be paid; by the latter, as always due, unless paid up within less than a year preceding.

Interest may be calculated so as to be added to the principal momentarily as it falls due; but we have already extended our discussion to greater length than we designed. If the interest be thus compounded, a sum will double itself at 6 per cent. in 11.552 years.

We have seen it stated that a suit was some years ago carried through the courts of Connecticut, in which the amount claimed arose entirely from the different modes of calculating interest. The plaintiff calculated by the latter mode we have given, and claimed several hundred dollars, the defendant by the commercial, and pleaded full payment. The latter succeeded in proving the custom at the time and place of contract, and gained the suit; but so far as we know the mercantile is not recognized as a general rule in the courts of any State in the Union.

In the courts of our own State it has been decided that the proper mode of computing interest is "Where more is paid than the interest due, to compute the interest up to the time of payment and apply the sum paid to pay the interest, and the balance to the principal. If less is paid than will pay the interest, the payment is applied *pro tanto* to the interest as far as it goes."—*Hammer* vs. *Neville et al.*, *Wright's Reports*, p. 169, A. D. 1832.

In Virginia also a similar decision was made as long ago as the time of Judge Wythe, that distinguished jurist adding a proviso "That the fœnerating operation should not be too frequently repeated."

Our Supreme Court has likewise decided that where a sum is due in several annual payments, the interest on all the payments to be paid annually, interest may be charged on the several sums of interest as they fall due. But it is not decided that interest on the accruing interest may be charged. *Watkinson* vs. *Root*, 4 Hammond, p. 373, A. D. 1830. In *Redish's Exrs.* vs. *Watson, Holcomb et al.*, 6 Hammond, p. 510, it is decided that the present law fixing the rate of interest has reference alone to money *due*, and that for money until due, the contract price, however great, may be enforced. This opinion was given by Judges Wright and Wood, Judge Lane dissenting, and the remaining judge sick.

In *LaFayette Benefit Society* vs. *Lewis*, 7 Hammond, p. 80, the court decided that a contract for the payment of more than 6 per cent. cannot be enforced in the courts of Ohio, under the law of 1824. This of course overruled the decision in 6th Hammond. In *Spaulding* vs. *Bank of Muskingum*, December Term, Court in Bank, 1841, it was held that "Where illegal interest has been voluntarily paid, it cannot be recovered back." But by a law passed February 18th, 1848, it is provided that all payments of money or property, made by way of usurious interest, whether made in advance or not, shall be deemed and taken, as to the excess of interest above the legal rate, as payments made on account of principal. But this does not affect the interests of *bona fide* holders of notes not due, and the holders having, prior to purchase, had no notice of the usury.

That our true position in reference to anti-usury penalties, established rates of interest, compound interest, &c., may be understood, we would remark that we consider these questions as entirely distinct. It may be necessary for the law to regulate the rate of interest; that the inexperienced and indiscreet may be in some measure protected. So long, however, as all things in nature, including all kinds of capital, increase as well by the increase of the increase as from the original stock, we expect to believe that money capital should not form an exception; and whether the law establishes the rate high or low, or leaves it to be agreed upon by the contracting parties, this principle is not affected.

INSURANCE, like Interest, is calculated as so much per cent.; and perhaps it may be well enough, before proceeding farther, to give a few explanatory remarks on *percentage.*

By common consent amongst mathematicians and men of business, this expression has been adopted to signify ratio; and we hear it used figuratively to express emotions and sensations as well as number and quantity. The valetudinarian says, "I feel 50 per cent. better than I felt yesterday;" the swain loves his mistress "a hundred per cent." better than ever. "It is 20 per cent. colder to-day than yesterday." In social intercourse we hear it said, "The population has increased 40 per cent. in the last five years."—"The deaths have diminished 10 per cent." "The fly has shortened the wheat crop 50 per cent.; but this rain will add to the corn a hundred per cent." The introduction of this expression into familiar intercourse shows that it is appropriate to its purpose.

In calculating numbers there is quite a convenience in being able to divide by 100 by merely cutting off two figures.

What will 8 per cent. on $150 be?

$ 150 8	Or,	As	$ 100 : $ 150 : $ 8 8
Ans. $12.00			1,00)12,00
		Ans.	12.

What will 3 out of every 37½ on $150 be?

Here is precisely the same problem in effect; but instead of being able to solve it in two lines, we must proceed by an operation requiring several lines; part of which we omit.

As 37½ : 150 : : 3 : 12, *Ans.*

In consequence of the convenience of this mode of expression, it is used in expressing the rate of Interest, Discount, Insurance, Commission, &c. To find the amount at any given percentage on any sum, we have only to multiply the sum by the rate and cut off two figures from the right as is done above; and the reason of which is shown by the statement.

Insurance is of various kinds. The most common insurance is upon houses, stores, and other property liable to be destroyed by fire; and these are called Fire Insurances.—Another class is called Marine Insurance, and is upon ships, boats, &c. The person desirous to have his property insured pays to the insurer, or underwriter as he is frequently called, who is generally agent of an insurance company, a small sum, which is regulated by a percentage on the amount insured, and receives in return a written or printed instrument called a *Policy*, which sets forth that if the insured property which it describes, be destroyed within the stipulated time, or if a marine insurance, upon a stipulated voyage, then the insurer shall pay the amount insured, if the loss amounts to an agreed sum. The rate of insurance depends on circumstances. A brick or stone tenement, standing apart from other buildings, and occupied by a private family, would be insured at a very low rate, perhaps at a half per cent. per annum (50 cents on the hundred dollars); for in the first place there would not be great danger of the building taking fire, and in the next place the substantial character of the building would afford a better opportunity for extinguishing fire without an entire destruction of the building. But on the other hand if the building be in close contact with others, or be of wood, or if occupied for some purpose that would subject it to great risk of taking fire, the rate of insurance would be perhaps double the former; and in many instances insurances will not be granted. In insuring, a specific amount is agreed upon, and this should be less than the property is really worth, otherwise one great inducement to care is taken away; and if property be insured too high, it might be destroyed by the owner, for the purpose of obtaining its value of the insurers. But though a definite amount is fixed, it becomes necessary for the loser to show that he lost that amount; for if I have an insurance of a thousand dollars on a building, or on my furniture, I may have it partially destroyed, or even entirely destroyed, and yet not lose $500. In that case I can only recover what I lose; but if my loss equal or exceed the amount insured, I receive the amount of insurance.

Some kinds of property are subjected to a higher rate of insurance from the difficulty of removing it in safety, in case of fire; although there may be no greater original liability of taking fire. A drug establishment, where chemical preparations are made, is very liable to take fire and it is difficult to save such stock from a burning building.

In view of these circumstances insurers divide risks into

Not Hazardous, Hazardous, and Extra Hazardous, reserving some kinds for special contract. The rate on these several classes is different according to the degree of risk.

It may be asked how this classification has been made up, and how the rate of insuring property is established. They are both the result of long observation of instances of fires, their origin, and the facility of rescuing property in danger. But though it could be ascertained that of all the property in a given territory, a given percentage is annually destroyed, it would still be very uncertain on what particular pieces of property the loss would fall, and hence the business of insurers is always liable to great uncertainty. The profits for a time may be very great and then may be swept away; or in one place they may be fortunate and in another part of the country unfortunate, but balancing their accounts they may do a fair average business. With a heavy capital and business scattered far and wide, they do not feel ordinary losses, though such as would ruin the business of individuals, and their risks being scattered, no single calamity is likely to involve them in ruin. To avoid this, insurers are careful not to take large risks on a single piece of property, or on property so connected that the loss of part will probably produce the loss of the remainder. In extensive conflagrations, however, such as that which laid waste so large a portion of New York in 1836, even wealthy insurance companies are liable to be involved to bankruptcy.

It is sometimes said that if insurance companies can insure at a profit, individuals may afford to run their own risk. But it must be borne in mind that an individual may be ruined by a single fire, and having his capital destroyed may never be able to resume business; but as we have already remarked, the risks of insurance companies are, or should be scattered, that a loss in one place may he retrieved by gains in others. In small matters we may safely run our own risks, since a loss would not be felt; and all ordinary losses are small to a strong company.

The industrious trader makes up his annual cargo for New Orleans and embarks his entire capital on the bosom of the Muskingum or Ohio, and if in his anxiety to make profit he declines to insure, some fatal snag in the stream, or some leak in his vessel may blast his hopes by the destruction of his entire cargo. But if he is content to gain rather less, and to be safe from ruinous loss, he pays his insurance, and if he makes less he is safe from the destruction of his capital; and should loss overtake him, he has his insured amount with which to recommence business. But if his capital be large he may lose a cargo, or several of them perhaps, without deranging his business.

Where the law imposes no restriction on the amount which a company may insure with a given capital, a question arises as to the amount which it would be morally right to insure; for it is plain that it would be dishonest for a company to insure an amount beyond its ability to pay, should loss take place.

Suppose a company has a capital of $100,000, what risk would it be morally right for such company to assume?

It is here evident that if the company takes risk to this amount only, it will be able to pay the full amount of damages, though the property should all be destroyed. But probably no company ever restricted its liabilities to so narrow a limit. On the other hand, it is clear that if with this capital only, the company should insure to the amount of several millions, the security of the insured would be inadequate, for the losses on so large an amount might be beyond the ability of the company to pay: as the premiums received would or might be divided and would form no part of the paying fund. And a company, knowing that it could only lose its capital, and not caring whether the insured were made safe or not, might be induced by the hope of large dividends to extend its nominal insurance to a most improper extent. It would be gambling at other men's risk; though a company would have a fair right to presume that all its insurances, scattered in various places, would not be lost; and that it would be safe in taking an amount of risks beyond the amount of its capital.

But how much beyond? Suppose it be found that the average annual loss by fire, for twenty years, in the State of Ohio, upon the whole value of buildings in the State, be one dollar's worth in every five hundred, or 20 cents on the $100 worth; then the average risk would be as 1 to 500.—So long, therefore, as the capital of the company is not less than as 1 to 500 on the amount of risk, it would afford some security to the insured, that security diminishing as the ratio would approach that limit; but we cannot think that an insurance company would be justifiable in approaching to the neighborhood of that ratio. After passing the ratio of average loss, the insured would part with his insurance fees and receive in return a greater risk than he before was liable to from fire, since the company would have such an amount of risks on hand, that its capital would be in greater danger than other men's property; and extending the risks, the capital of the company would be more and more exposed until its destruction would become inevitable. If therefore a company wishes to continue its corporate existence, it must pay some regard to the amount of insurance it undertakes.

Of course in our estimates we regard the average loss, taking a large region of country or a great length of time, or both these together; for though there may be unfortunate cities, or neighborhoods, or unfortunate times for fires, these blend in the general mass, and we are able to reduce even the freaks of fortune and the sport of accident to something like system.

There is another species of insurance that sometimes meets with bitter denunciation from such as do not look into the nature of the operation. It is *Insurance upon Lives.*—The ignorant hear persons speak of Life Insurance, and think the parties engaged are impiously striving to thwart the will of the Almighty and to avoid death; instead of being engaged in a very harmless and rational business transaction.

If an individual has a salary from which he is able to save annually a small amount, and he is anxious to provide for his family at his death, he goes to a Life Insurance office and agrees according to his age, health, &c., to pay annually while he lives, a stipulated sum, in consideration that at his death his wife or family shall receive a sum agreed upon, say $1000, $2000, or whatever sum may be stipulated. The individual may die the next day, but his family receives the amount agreed upon; and on the other hand he may live to pay more than his family will receive, and the probability is in favor of that supposition, for were it not so, the insurers would do a poor business. But the advantage to the insured is that his family has a certain provision made for them.

On the other hand, an individual wishing to be released from the cares of business, and having a sum of money on hand, gives his ready money to the insurer on condition of receiving a stipulated periodical sum for life; and this may be paid annually, semi-annually, quarterly, or otherwise, as agreed upon.

In this case the man may give a thousand or ten thousand dollars and die the next day, and on the other hand he may live to receive many times what he paid. He has the advantage, however, of knowing that by this arrangement his living is as sure as the solvency of his underwriter; and though had he lived upon his capital, it might have been more than sufficient to support him for life, and thus have left a surplus for some survivor, it might on the other hand have proved insufficient, and have left him to perish of want or become a public charge. It is a neat way of being a man's own executor, and is often an excellent one. None wish him dead; and

16

no disappointed heirs wrangle like hungry hyenas over his pecuniary remains.

Here as before, the insurer may lose largely in some instances and gain in others, but the probability is that he will gain upon an average; and thus be enabled amidst gains and losses to do a fair business. His estimates and hopes are based on calculations founded on observations made upon bills of mortality, as we have already suggested when speaking of Life Annuities.

We have extended our remarks beyond their designed limits; for which the importance of the subjects discussed must be our apology; but we are aware that much might yet be said without exhausting the materials.

LECTURE XII.

OTHER USES TO WHICH A KNOWLEDGE OF ARITHMETIC IS APPLIED.

It was remarked on a former occasion that the science of Arithmetic is "alike indispensable to the scholar and the man of business, and must remain of primary importance through all the vicissitudes of time." Our present object is to show with some degree of minuteness the purposes to which it is daily applied. A good knowledge of Arithmetic, as a science, is to its possessor what familiarity with the use of tools is to the mechanic; the former can with a little care, apply his powers of calculation to any purpose required, and so may the latter apply his familiarity with the use of tools, to any operation requiring their use. But in each case some special knowledge will be found necessary. Every branch of business has its peculiar calculations, and every trade has its peculiar operations.

The speculative mathematician studies Arithmetic that he may understand the philosophy of numbers: that he may perceive their properties and relations, and use them in prosecuting his investigations in the higher branches of mathematical science. The Astronomer calculates the orbits and cycles of the planetary world, announcing phenomena that will happen ages hence ; as well as such as happened long before mankind had sufficiently emerged from barbarism to understand the reason of what they saw. By calculating the times of eclipses in former ages, the Chronologist fixes the dates of events that occurred while the modes of dividing time were yet imperfect. This division of time appears to us a very simple matter, for we know just how many days are in a year, and how long the period is from one change of the moon to another, and a thousand other facts that are familiar to us as household words. But let us refer to man in a state of nature, as we did in pointing out the difficulty of learning the use of numbers; and how would he measure time? He would see the sun rise and set, and the succession of day and night would soon become familiar to him; while he would in due time learn that the seasons of cold and heat follow each other in succession; but the precise period of time necessary to embrace all the changes of the seasons would be difficult to learn; for we find the seasons blend together, so that none could decide when the one is gone or the other come. The changes of the moon would arrest his attention, but it would be difficult for him to decide how long a period elapses from one full moon to another: and he might well doubt whether the periods of changing seasons, and of the moon's aspects, were always the same. The Indians of our own country measure their time by days, moons and snows; the last term being expressive of years: but they do not know how many days are included in a year.

To determine the precise length of a year was a problem that long puzzled men of science. The Egyptians were probably amongst the first to solve it, for in addition to their early discoveries in other branches of science, the yearly overflowing of the Nile was to them a circumstance of deepest moment, and it was annually preceded by the heliacal rising of Sirius, the Dog Star; hence it was natural that they should look for this harbinger of the overflowing waters with much anxiety. They counted the days that intervened, between one appearance and another, and thus measured their year, for they made their year to commence with this appearance of the star. But it was found that the star constantly rose a little later, and in four years it rose a day later. Corrections were applied and leap year was introduced; but still in a few hundred years it was found that the calendar and the seasons were not keeping pace together; and Pope Gregory XIII, called together the learned, once more to adjust the calendar, lest winter should ultimately fall in July and summer in January. After ten years of investigation they reported the correction necessary and it was made by dropping ten days in October, 1582.

Most catholic countries at once adopted the change, but the protestants were less prompt, for Great Britain did not adopt it until September, 1752, by which time another day had been lost, and it was necessary to call the 3d of September the 14th, in order to bring the calendar and seasons once more together. This was done by act of Parliament, and leads to what is called *Old Style and New Style.* Prior to that time the year had commenced on the 25th of March, but it was at the passage of the above law enacted, that from and after the last day of December, 1751, the year should commence on the first day of January. This gives rise to such dates as 1764-5, &c. And so long as both styles were used this was necessary, to prevent a misunderstanding of a whole year. The prejudices of the ignorant were strong in favor of the old style, because they had been accustomed to it, while the intelligent saw the necessity of a change in order to preserve uniformity with other nations, and to keep the months to their places in the seasons. It is said that though great pains had been taken to prepare the public mind for the change, the superstition and ignorance of the populace of England were so great, that when a son of Lord Macclesfield was a few years afterwards a candidate for a seat in the House of Commons, the mob pursued him calling out "Give us back the eleven days we have been robbed of." Lord Macclesfield and Dr. Bradley had been active in effecting the change, hence their prejudice against the son; and several years afterwards when the venerable Doctor was broken down with affliction, it was thought a judgment upon him for having engaged in so impious an undertaking. They seemed to think that the omitted days had been stricken from the lives of men as well as from the imperfect Julian Calendar.

Pleasant as it might be to pursue the divisions of time through all their changes, and to show why some months have 30 and others 31 days, and why February has but 28; as well as how different nations reckon time and adjust dates, it would detain us too long from the goal at which we are anxious to arrive. We can only hint at points that may arouse the student's curiosity, and show him that without the aid of Arithmetic we would know but little of the divisions of time; the distances of the planets from us; the times of their revolutions, and all the wonders of Astronomy.

The Navigator calls Arithmetic to his aid, and traverses the trackless ocean. Both he and the Astronomer use it in connexion with Geometry, and hence their calculations belong to the class called Mixed Mathematics. The Astronomer has to do principally with circles, and the laws which govern bodies moving in them, while the navigator is more concerned with triangles, plane and spherical, and the effect of winds and tides. Many of their calculations are however nearly allied; for the way-marks of the mariner are not upon the waters, but in the heavens above.

The calculations of the machinist are various, and involve many principles for which he must look into the laws of physical science. If he seeks to propel his machinery by water power, he must study the laws of Hydraulics and Hydrostatics. He must learn how fluids move and how they tend to act when not in motion. Here he will meet with the Hydrostatic Paradox, and other phenomena no less incomprehensible to him who has not closely investigated the subject. If he calls to his aid the strength of steam, he must study its laws; and whatever motive power he may use, he must study the principles of the Mechanical Powers in constructing his machinery; that he may not be found warring against the ever constant laws of nature. To undertake even a sketch of the calculations which the machinist will find necessary in his business, would be to write a book on this subject alone. By Machinists we mean to include Millwrights and all others employed in the construction of machinery; and they have many books especially adapted to their use.

To express the effect of a steam engine it is usual to compare it with the power of a horse, probably from steam engines being often used in their first introduction, to perform the work otherwise done by horses: hence we speak of engines of 10, 20, or 50 horse power. The power of a horse is however variously estimated, and consequently this expression is very indefinite. SMEATON considers a horse capable of raising 22916 lbs. one foot high in a minute. DESAGULIERS makes it 26500: WATT 33000, while GRIER, in his *Mechanic's Calculator*, makes it 44000. In speaking of Animal Strength as an effective agent, GRIER says "There is a certain load which an animal can just bear but cannot move with it, and there is a certain velocity with which an animal can move but cannot carry any load. In these two circumstances it is clear, that the exertion of the animal can be of no avail as a mover of machinery. These are, as it were, the extremes of the animal's exertion, where its effect is nothing; but between these two extremes, there must be weights and velocities with which the animal can move, and be more or less efficient.

If one man travel at the rate of three miles an hour, and carry a load of 56 lbs., and another move at the rate of 4 miles an hour and carry a load of 42 lbs., the speed of the first is 3, and the load 56, the useful effect may therefore be estimated as the momentum=168. The other carries only 42 lbs., but travels at the rate of 4 miles an hour; therefore, in the same way, his useful effect will be $4\times42=168$, the same as before: hence the effects of these two men are the same. It will not be difficult to show, that in the same time they perform the same quantity of work. For the first will in six hours carry 56 lbs. $3\times6=18$ miles, as he travels at the rate of 3 miles an hour; and if he be supposed to carry a different load, but of the same weight every mile, he will in the six hours have carried altogether $18\times56=1008$ lbs.; but the other carries in the same way, 4 times 42 lbs. every hour, that is 168 lbs. in one hour—therefore in 6 hours he will have carried $168\times6=1008$ lbs., the same as the other.

It will now be seen what is meant by the phrase useful effect, and from what has been observed above, we will be led to conclude, that when the load is the greatest which the animal can possibly bear, the useful effect is nothing, because the animal cannot move; and when the animal moves with its greatest possible speed, the useful effect will also be nothing, for then the animal can carry no load; and it becomes a very useful problem to determine where between these two limits, the load and speed are so related that the useful effect of the animal will be the greatest. By investigation it has been found that the maximum effect of an animal will be when it moves with $\frac{1}{3}$ of its greatest speed, and carries $\frac{4}{9}$ths of the greatest load it can bear."

The Carpenter, the Mason, the Farmer, in short every one, has his appropriate calculations; but we shall not enter upon a detail of them. We remarked on a former occasion that the blacksmith cannot even iron a double tree without calling to his aid the principles of this science, if he would do the work correctly. Many questions in this book are framed with a view to suggesting hints of useful applications of the science. A little investigation and simple calculation would often save men a deal of trouble. I recollect once seeing a man endeavoring with all his ingenuity to gain power to his ox mill by lengthening his beam for the cattle to draw at, to a great extent, and then adding an extra wheel to recover the motion he had lost. Another was busy at work inventing ways and means to throw the water that had turned his wheel, back into his forebay, that it might do the same again. Redheffer spent his life in the idle hope that he could disturb nature's balance and invent *Perpetual Motion;* but his labors ended in placing an old man in a garret to turn his crank. Hundreds have sought the same invention, and with similar want of success. The same certainty may be reached in estimating the power of mechanical agents and the operation of physical causes, as in any other business calculation, if men will only acquaint themselves thoroughly with the laws of action by which these causes are influenced and apply to them the close reasoning which they do in mathematical calculations generally.

The calculation of interest is deemed of such importance in business matters, that we have devoted an entire lecture to its consideration; and we shall do the same with the doctrine of *Wheel Carriages*, because it is a subject of practical importance, and we know no treatise to which the student could be referred for accurate information.

In applying Arithmetic to the purposes of life, the doctrine of *Weights and Measures*, and of *Coins and Currency* demand attention. To secure uniformity in weights and measures has always been an object of importance, but at the same time one difficult to accomplish. Nature does not furnish any standard of length or capacity with which all other terrestrial bodies may be compared, for her productions are ever varying, and we perhaps could scarcely find any two shells or other natural productions of precisely the same capacity, or of the same length, which we might use as standards; hence when great precision is required, we must adopt or rather construct, a standard with which to compare. In the infancy of science nature is however resorted to for standards of measurement, and the names of both ancient and modern measures show that the *hand*, the *span*, the *arm*, and the *foot* have been the first tests of length, amongst all nations. The savage in adjusting his bowstring or his angling line would probably resort to the *finger's width;* or its *length*, to the ancient *cubit*, or distance from the tip of the elbows to the end of the middle finger; the *ell*, or distance from the tip of the longest finger of the extended arm to the middle of the breast; while double this, or the entire distance from one hand to the other when the arms are extended would be the *fathom*. To this day the fathom is used to express the length of cordage, and hence to express the depth of water, which is measured with the sounding line. How natural to suppose that such a mode would be adopted to measure the length of a line, and having sounded the water, thus to determine its depth! but for measures upon the surface of the ground, the *foot*, or the *step* would be preferred.

In our artificial system, the measuring unit is a *Barleycorn*, the length of three of these being called an inch, and twelve inches a foot; but we must know that barleycorns are of very unequal length; and if the standard measures were lost, it would be a difficult task to replace them from such originals. On the other hand we are told that the foot was originally the unit, and that its length was determined in the year 1101 by measuring the foot of his majesty, HENRY the First, king of England, and some say that the yard was established by measuring his arm. Be this matter as it may, the standard has long been fixed, and every precautionary measure is adopted to preserve it from being lost; though it might be restored, as we shall presently show, if it were lost. With the divisions and multiples of the measuring unit such as inches, yards, rods, miles, &c., every one is familiar; and it is entirely obvious that measures of length being established, measures of superficial extent would be drawn from them, and equally so that measures of capacity would also be. The extent of a foot being fixed upon, the length of a line would be determined by the number of times it would contain a foot. Then the extent of a given surface would be determined by the number of times it would contain a square surface a foot in length on each side; and the size of a solid body by the number of solid blocks, a foot in length on each side, that could be made from it. Determining a measure of length, determines therefore all other measures. Liquid and Dry Measure are as completely solid or cubic measure, as measuring stone or earth would be; though from the nature of substances thus measured, the quantity could not be so readily estimated by taking their several dimensions in the usual way; hence a vessel whose capacity in cubic inches is known, is filled with the fluid or the loose substance, sought to be measured, and the quantity determined by the number of times it will fill the vessel.

The original English standard yard measure is kept with great care in the Tower at London, and certified copies are furnished to officers of the customs, and all others who are willing to pay the fee. The Standard measures of modern times, not only of length but of capacity, are made with great accuracy, and of materials least liable to be affected by changes of temperature; or of such a combination of materials as may counteract such changes. For the purpose of accurate adjustment they are placed on frames and their length and divisions adjusted by miscroscopic observation, which may render the twenty thousandth part of an inch measurable. To touch them with the warm hand, or even to blow the breath upon them, would cause expansion, invisible to the naked eye, but the accurate micrometer would mark the difference.

Dr. Locke, of Cincinnati, who visited Europe in 1837 for the purpose of obtaining philosophical apparatus, and to obtain copies of the English standards, gives rather a ludicrous account of some of their *ancient* standards, which like their ancient customs are cherished because they are old. "I was permitted" says he, "to examine the ancient standards in the exchequer: some of them are of extremely rude construction. The standard yard of queen Elizabeth, which was in use until within eleven years past, is a rude bar of brass, neither straight nor square, but such an approximation as might be expected from the hammer and anvil of a bungling blacksmith. The ends are neither square nor smooth; the divisions of halves, quarters, &c., are marked by notches, deeply and broadly scored as if cut with a jack knife in a rod of wood, some of them straggling obliquely at least the twentieth of an inch. It had been broken asunder at some unknown period, and re-united by a dovetail, but so badly that the joint is nearly as loose as that of a pair of tongs." He adds the remark of another gentleman, that "a common kitchen poker would make as good a standard."

Various standard yards have been at different times constructed, and though differing very slightly from each other, say two or three thousandths of an inch, they are still so nearly alike that any one could be restored if destroyed, for they have been carefully compared and their differences recorded. But should the original and all its copies be destroyed, there are means by which the measure could be reconstructed.

The space through which a body falls in a second of time, is known, but there would certainly be much difficulty in making observations of a falling body with the necessary accuracy, if indeed the precise velocity of the falling body is known. The length of a degree upon the surface of the globe, would furnish a second mode of restoring the lost measure; but this would require a very laborious and careful geodesic survey. Still it would be a means of restoring that which had been lost. But a third, and much superior mode would be to adjust a pendulum to beat seconds, or any other given rate, at a certain latitude, and from its length to establish the lost measure. Any required time, an hour for instance, may be measured by the motion of the heavenly bodies, without the use of clocks or watches, or it may be measured by means of a dial. All then that would be necessary would be to adjust the pendulum by trial until it would beat at the rate of 3600 times in one hour, and having verified this by proper observation, its length in inches would be known, from past discoveries.

The British Parliament, in 1824, enacted that if the standard yard should at any time be lost, a new standard should be made, bearing the ratio of 36 to 39.1393 to a pendulum that would beat seconds at the level of the sea, in a vacuum, in the latitude of London. For it is well known by experiment that such a pendulum must be 39.1393 inches in length, 36 of which would of course be 3 feet, or one yard. The laws of Ohio make a provision very similar. Pretty general uniformity exists in the measures of the United States, except perhaps in Louisiana, but in England and its dependencies, local customs have always triumphed to a very great extent over Parliamentary enactments. In some counties the perch, for instance, is 6 yards, in some 7, and in others 8. In Cunningham measure it is 6¼ yards, in Forest Measure 8 yards, in Woodland or Burleigh Measure 6 yards, and yet not one of these is the legal length; that being 5½ yards as in the United States. Others of their measures and weights are almost as various.

In Ohio, by an act of 5th March, 1835, it is declared that "the unit or standard measure of length and surface, from whence all other measures of extension, whether they be lineal, superficial, or solid, shall be derived and ascertained, shall be the yard, as used in the State of New York, on the 4th day of July, 1776." The yard alluded to is the same as that of the English exchequer, as appears from an act of the New York legislature of April 10, 1784.

Our several measures of length and capacity bear to each other no uniform ratio, and hence much difficulty arises to learners in the study of arithmetic, and to business men in applying it. To obviate this the French adopted a decimal scale of measures; and in 1818, the Congress of the United States referred the subject of decimalizing our Weights and Measures to John Quincy Adams, who was then Secretary of State. He did not report until 1821, when he reported adversely to the project. His report enters very minutely into the history of the subject, and shows what the French nation had done; but it has not satisfied the American people that his reasoning is correct, though it caused the project to be abandoned at that time. The great advantage of our system of Federal money over the old currency, shows us how convenient it would be to have the ratios of all our compound quantities decimalized.

Few things are more uniform than the weight of pure water

at any given temperature, hence the system of *Weights* is readily derived from the system of Measures. A cubic foot of pure water at its maximum density, which is at the temperature of 40° of Fahrenheit, weighs precisely 1000 ounces Avoirdupois, or 62½ lbs. This is the standard provided by law in Ohio, as the unit of weight from which all others shall be determined. The Avoirdupois pound being divided into 7000 equal parts furnishes the Troy grain, from which the several weights in that system are readily determined. The pound and ounce of Apothecaries' weight are similar to those of Troy. In Ohio, the law declares that 25 lbs. (not 28) shall be a quarter of cwt.; that 100 lbs. shall be a cwt. and that 2000 lbs. shall be a ton.

The English system, on which ours is founded, does not seem to have been based on the weight of a cubic foot of water; but had reference to the articles *wine* and *wheat*, and the weights used in the balance were the coins of circulation; for then pounds, shillings and pence, were not only money but weights also. The penny was equal in weight to "32 wheat corns, in the midst of the ear," and 20 pennies made an ounce, or as we yet have it "20 pennyweights make an ounce." The weight of the penny was afterwards reduced, so that it would counterpoise no more than 24 grains of wheat. EDWARD III, and other English sovereigns, reduced the weight of their coins, by making a larger number from a given quantity of metal, but retaining their nominal value, until now a £ money and a lb. weight no longer balance each other. The *Easterling* pound was long used at the English mint, and hence the coinage of England is called *Sterling* money to this day.

The unit of *Liquid Measure*, as recognized by the laws of Ohio, is a gallon, of such capacity as to contain at the mean pressure of the atmosphere, at the level of the sea, eight pounds of distilled water, at its maximum density. This is less than 231 cubic inches, being only $221\frac{23}{125}$ cubic inches. The *Dry Measure* gallon contains just one fourth more. From these all other measures are derived. We have thus hastily sketched the connexion between measures of capacity and measures of dimension and of weight. The pendulum which is fixed by a law of nature establishes the length of the standard *yard*, from this by multiplication and division, all other measures of length are established, and from these measures of surface and solidity follow of course. Having established these we form a cubic box of one solid foot, and filling it with pure water of proper temperature, &c., we establish the weight of 62½ pounds avoirdupois, and from this we establish the weight of one pound, and hence by multiplication and division all avoirdupois weights are obtained. We know the pound avoirdupois contains 7000 grains Troy, and thus the means of forming that series become known; or we may effect it from knowing that 144 pounds avoirdupois will make 175 pounds Troy. Having weights and measures of dimension, we fix our standard of measures of capacity. So that were all our standards of weights and measures stricken from existence, nature furnishes us the means of restoring them in the precise form in which they now exist.

The English measures of capacity, as the ground work of our own, remain to be noticed. The English wine gallon of 1266 is thus defined: "32 wheat corns from the midst of the ear make the weight of a penny, 20 pence an ounce, 12 ounces a pound, *eight pounds a gallon of wine*, and eight gallons a London bushel, which is the eights part of a quarter;" probably so called from its being $\frac{1}{4}$ of a ton, which is 32 cubic feet of water and would be nearly so of wheat, even with the present bushel. Thus wheat and silver money were made the tests and standards of each other; and the weight and measure of wheat and wine were proportioned to each other. The *measure* of wheat was at the same time a *weight* of wheat, and the measure of wine expressed a weight of wine, and whether these important commodities were bought and sold by weight or measure the result was the same. And in reference to wheat, the means were furnished to find from the weight the number of kernels also, a fact not indeed very important; but tending to give system to the scheme. Several laws were enacted at various times, by which standards of different sizes were established, some apparently through ignorance of existing regulations, and others to effect what it was thought would be salutary changes, by producing a nearer approach to uniformity in the measures used for different purposes and in different places. The ale gallon was established as the eighth part of the measure of a bushel which should contain of wheat the weight of eight gallons of wine, or as we may ascertain from the specific gravity of wine it must contain 64 pounds of wheat of their standard, and be in capacity equal to 2256 cubic inches, the eighth part of which is 282 the cubic inches in the ale gallon. The wine gallon had been before changed from 224 to 231 cubic inches, so that the wine and ale gallons are in the same proportion to each other as the Troy and Avoirdupois pounds. The bushel, finally, however, was established at $2150\frac{4}{10}$ cubic inches, the eighth part of which is still called a gallon, and must contain of course $268\frac{4}{5}$ cubic inches; but to distinguish it from the others it is called the Dry Measure gallon. Under the present English system as established in

1826, some of these measures are slightly modified, but the minutiæ of detail would fill a volume.

In 1790, Prince TALLEYRAND moved the subject, and in 1793 measures were adopted by the National Convention of France for the establishment of an improved system of Metrology, and in 1795 the system was given to the world. The twofold object sought to be attained in this system is uniformity and a decimal ratio in the denominations. Instead of various weights and measures adapted to different purposes as the English have them, they sought to establish one set of measures of extension and connected with this a set of measures of capacity and weights. They wished to found their system on something natural, and for this purpose they determined to measure the length of a quadrant of the meridian, *i. e.* a line drawn from the equator to the pole, passing through Paris, and to take the ten millionth part of this extent as their unit of measurement. It is not easy, however, to see what great advantage is to result from this standard being founded on a natural basis, unless by so founding it other nations would be induced to adopt the same standard. After some years employed in the survey by some of the best mathematicians of France, it was determined that the quadrantal arc measured 5130740 toises of 6 feet each, the ten millionth part of which is 3 feet 11.296 lines of the old French measure, which is 3.2809167 feet, English measure. This was the standard unit of the system, and was called the *Metre*, (or Measure) from which by multiplication and division, all other measures were established. It may be proper to remark, that in determining the length of such quadrantal arc, it is not necessary to measure the whole arc, but only a degree, or such part of a degree, as determined by an astronomical observation; for from this the length of a degree may be readily found.

☞ Table of the New French Measures.

1 *Metre*,= $\frac{1}{10000000}$ of the distance from the Equator to the Pole, or 3.078444 French feet=3.2809167 English feet.

1 *Deca-Metre*,=10 Metres.

1 *Hecto-Metre*,=10 Deca-metres.

1 *Kilo-Metre*,=10 Hecto-metres.

1 *Myria-Metre*,=10 Kilo-metres.

The measures were also carried *downward* by division, thus the *Deci-Metre* was the 10th part of the metre, the *Centi-Metre* was the 10th part of the deci-metre, the *Milli-Metre* was the 10th part of the centi-metre. The prefixes to express measures greater than a metre being from the Greek and those less from the Latin. These are their measures of length, and from these the measures of surface are obtained in the usual way by squaring. The unit of field or agrarian measure is the *are*, equal to 100 square metres. The *centiare* is one square metre, and the *hectare* 100 ares, or 10,000 square metres.

The unit of solid measure is the *stere* or cubic metre, and the *decistere* which is a tenth of the stere.

The liquid measures are the *litre* or cubic deci-metre, the *deca-litre* of ten times the capacity of the litre; and the *deci-litre*, which is the tenth part of the litre.

The measures for dry goods are the litre, the deca-litre, hecto-litre, and kilo-litre, which are respectively once, 10 times, 100 times, and 1000 times the deci-metre cube.

The basis of the system of weights is the *kilogramme*, which is the weight of a deci-metre cube of distilled water at the temperature of 40° of FAHRENHEIT'S thermometer: the thousandth part of this is the *gramme* or unit of the system of weights. The *decagramme*, *hectogramme*, and *kilogramme*, are respectively 10, 100 and 1000 grammes. The *quintal* is 100 kilogrammes, and the *millier* 1000 kilogrammes. The *decigramme* is the tenth part of the gramme, the *centigramme* the hundredth part of the gramme, and so on.

The French government at the same time adopted decimal divisions of money and time, but with all the devotion of the French nation to the new order of things, its introduction into general use was slow, and twelve years after its final adoption in 1799, the government was obliged to modify the law by permitting a different division to be used for certain purposes. How far it has now been adopted by the people to the exclusion of all others we are not prepared to say. The Chinese have for ages had all their subdivisions in weights, measures, &c., adjusted in a decimal ratio.

In ordinary business operations, great carelessness is usually indulged in, in measuring and weighing. The law of Ohio very justly requires that all surveys should be made with a horizontal chain, and thus that the measure of a line extending over a hill should be the same exactly as that across its base; but this precaution so necessary to accuracy, is too often but partially observed, and hence if one or more lines of a survey pass over hilly ground, they will be longer than they should be, and the survey, however accurately made in other respects, will not close. When this is the case, it is usual to add one half the difference to the shorter lines, and subtract the other half from the longer ones, thus making them balance midway and of course giving an area greater than the horizontal surface. That the surface of a hill is greater than that of its base is certainly true, but more stalks of grain cannot stand upon it, nor would it take a greater number of palings to make a fence over a hill than across its base, though it would require a greater length of horizontal rails and more posts, the rails being placed parallel with the hill side.

In cases where extreme accuracy is required, as in the French survey to determine the length of 90° of the meridian for the purpose of fixing the length of the *metre*, their unit of length, it becomes necessary to use a measure that will be as little as possible affected by moisture and by changes of temperature. Dry pine or deal rods will be less affected in length by variations of heat and cold than a metal chain, but then it will be much affected by moisture. The common impression is that timber does not shrink endwise in drying, and that it is not changed in length by moisture or dryness. It is true that the expanding and shrinking across the grain is much greater than lengthwise, but it is materially affected both ways.

In the French survey for determining the *metre*, their measures were formed of rulers of platina and copper. Some Swedish mathematicians, who were employed in a similar survey at the same time used iron bars covered near the ends with silver. General ROY in England used at first deal rods perfectly seasoned and effectually secured from bending, but they were so much affected by moisture as to take away all confidence in the result. Afterwards glass tubes 20 feet long, enclosed in wooden frames, were used, and allowance made by a certain rule for changes of temperature. With these a line of 27404.08 feet, or about 5.19 miles, was measured on Hounslow heath, and several years afterwards the same line was remeasured by General MUDGE, with a steel chain, 100 feet long, constructed like a watch chain. The chain was constantly stretched to the same tension, supported on troughs laid horizontally, and allowances made by an ascertained rule for changes of temperature. The result when compared with General ROY's measurement with glass tubes, showed a difference of only $2\frac{3}{4}$ inches, which is a small matter in a line more than 5 miles long.

We have already remarked that coins were originally weights also, and we find a similar connexion between the money, weights and measures, mentioned in the Scriptures. In our own country, the money system is made very simple by the adoption of a decimal ratio. Prior to the adoption of this, there was no uniformity of currency amongst the States. A shilling, which is an imaginary sum, not a coin, varied in value, in the several States, so as to make ordinary business calculations very perplexing to such as were not familiar with them. These different modes have given place generally to the simple one of dollars and cents; and in our own state such a currency as pounds, shillings, &c., is not recognized by the laws; hence if a note were given for any number of pounds, shillings, &c., the meaning would become a question of fact to be inquired into through witnesses.

The power to coin money and fix its value is granted by the constitution to Congress alone, with the restriction that only gold and silver shall be made a legal tender in the payment of debts. Cents are therefore not a legal tender. The present system of Federal money was adopted on the 8th of August, 1786. The eagle was then made to contain $247\frac{1}{2}$ grains of pure gold, and $22\frac{1}{2}$ grains of alloy; but Congress on the 21st of June, 1834, enacted that from and after 31st of July of that year, the eagle should contain 232 grains of pure gold, and 26 grains of alloy, making 258 grains of standard gold. By the law of 1786, the value of silver to gold was 1 to 15; by the law of 1834, it was made as one to 16. By the latter act the value of all the gold in circulation was increased $\frac{1}{15}$. Eagles of the old coinage may be known by their having a liberty cap on the head of the figure; they are worth $\$10.66\frac{2}{3}$.

The weight of the silver dollar remains as at first established, $371\frac{1}{4}$ grains of pure silver, and $44\frac{3}{4}$ grains of alloy. The copper cent weighs 208 grains.

We might pursue the subject yet farther, and show that a knowledge of Arithmetic is indispensable in every branch of business and science; but what has been said will probably be sufficient for our purpose.

LECTURE XIII.

PROPORTIONS AMONGST LINES, SURFACES AND SOLIDS.

In a former lecture the subject of proportion amongst numbers was fully discussed; in the present we propose to consider this subject very briefly in reference to Lines, Surfaces and Solids; embracing in part the comparison of similar parts of similar figures and also a comparison of dissimilar figures. Some knowledge of this subject is entirely necessary to any one who would learn to measure, or study the principles of Mechanics and the laws of Motion; hence we introduce it, though not strictly a part of the doctrine of numbers.

Lines are proportionate to each other simply as their lengths, since they have but one dimension; but similar surfaces are to each other as the squares of their like linear dimensions; and similar solids as the cubes of their like linear dimensions. If for instance there are two circles, the diameter of the first being 4, and the second 8, we might suppose the first to be just half as large as the second, since its diameter is just half as great; but their areas are in proportion as the squares of their diameters, and the square of 4 is 16, the square of 8 is 64, hence they are as 16 to 64, or as 1 to 4; the area of the larger being 4 times the area of the smaller. If we would compare a ball 4 inches in diameter with another 8 inches in diameter, the ratio in solidity, and consequently weight, will be as the cube of 4=64, to the cube of 8=512, or as 1 to 8; the larger being 8 times as large and heavy as the smaller. Hence a ball two inches in diameter, if melted and cast into one inch balls, would make eight of them.

It is desirable at the outset to obtain a clear idea of this law, for it always appears paradoxical to the young beginner, and a distinct understanding of it is important. The question is sometimes asked, "Which will vent most water in a given time, a single auger hole an inch in diameter, or two holes each of half an inch in diameter,—equal areas venting equal quantities." Almost every one who is unskilled in mathematics, when he hears this question for the first time, is ready to exclaim, *they will vent equally*. Yet a little reflection will show that this is not true, for if you describe a circle an inch in diameter, and within it draw two others, each half an inch in diameter, there is still much left not embraced in the small circles; or the same idea may be made plainer perhaps by describing an inch circle on a board, and attempting to bore it out with a half inch auger. The four holes that must be bored will run into each other, but as much will remain uncut, as is thus cut twice. (See Fig. 1.) To ascertain the exact proportion we must square the $\frac{1}{2}$ inch, the diameter of the small holes, $=\frac{1}{4}$, and take twice that square, for there are two holes, equal $\frac{1}{2}$; then the square of 1, the diameter of the larger hole, $=1$, which is twice as much as both the small ones; therefore, the inch hole will vent twice as much as both the half inch holes.

Fig. 1.

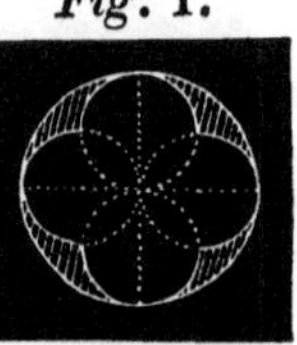

That this must be so is more easily shown if the figures be square. If I have a square plat of ground measuring 2 rods on a side, and I increase it so that a side shall measure 6 rods, then the plat is made 9 times as large; for the square of 2 is 4, and the square of 6 is 36, and 4 : 36 : : 1 : 9. The first plat contained 4 square rods, the last 36, and you might lay off 9 such as the first upon the surface of the second.

A mile square of land in our public surveys, is called a *section*, and contains 640 acres; but $\frac{1}{2}$ a mile square is only $\frac{1}{4}$ of a section, and contains only 160 acres.

When we look at an iron ball weighing one pound, and form an idea of one that shall weigh a large number of lbs., say 40 lbs., we are almost sure to fancy one larger than the reality, for we forget that every additional inch of diameter forms a coating half an inch thick over the entire surface of the ball, and that a ball 3 inches in diameter will weigh as much as 27 balls one inch in diameter; and that an inch more added to the diameter will make it weigh as much as 64 of one inch.

When two lines are compared, it is evident that both must be considered as finite, so that the smaller may be increased until it shall equal the greater, otherwise there could be no ratio between them. Time cannot be compared with eternity, since the latter is without limit. A second of time bears some assignable ratio to the duration of time from the creation of the world to the present hour; but what is the ratio between a million of years, and duration that shall never end?

PARTS OF THE SAME FIGURE.

LET us first consider lines that are parts of the same figure, as the diameter and circumference of a circle, the legs and hypotenuse of a right angled triangle, &c.

Finding the ratio between the Diameter and Circumference of a circle is frequently called Squaring the Circle, or the Quadrature of the Circle. It is a problem that long puzzled mathematicians, but its farther investigation is now generally abandoned; they being satisfied that the exact ratio cannot be found, while for all practical purposes it was long since discovered with sufficient accuracy. Some have supposed that it might be done mechanically, but any attempt of the kind would be useless.

ARCHIMEDES, who was born at Syracuse 287 years before CHRIST, discovered that if the diameter of a circle be 1, the circumference will be between $3\frac{10}{70}$ and $3\frac{10}{71}$, and hence he gave the practical ratio as 7 to 22 or 1 to $3\frac{1}{7}$. At a much later date, METIUS, a German mathematician, established the ratio for practical purposes at 113 to 355; but he carried the calculation theoretically much farther, finding the ratio accurately to 17 places of decimals. VIETA, a French mathematician, having before carried the ratio to 11 places. At a still later period another Dutch mathematician, LUDOLPH VAN CEULEN, investigated the subject with great care and carried the calculation much farther than any of his predecessors. He proved that if the diameter of a circle be 1, the circumference will be greater than

3.14159265358979323846264338327950288, but less than 3.14159265358979323846264338327950289

and he was so pleased with his discovery that he desired that the numbers might be engraved upon his tombstone, which was done, as may be seen at St. Peter's Church, at Leyden. He appears, however, to have effected his calculation by dint of labor, rather than fertility of invention, for he used only the tedious mode of calculation long before adopted by ARCHIMEDES. SNELLIUS, of the same country, adopted a much shorter process by which he fully proved the accuracy of VAN CEULEN's calculation.

There is a common notion with workmen that 3 times across a circle is equal to once round it, but this would be too wide of the mark for reasonable accuracy. It is true, however, that if you take six steps with the compasses, opened to the width of the radius, or that distance of the compasses used in describing the circle, they will just step round the circle, and hence the idea that with double the radius, or the whole diameter, it will only take 3 steps; but this is not true even in stepping, for you cannot step round a circle at all if you take the entire diameter in your dividers; you will merely step across it. Stepping round the circle in this way does not give the length of the circle, but the periphery of a polygon of a number of sides corresponding with the steps taken; the curve of the circle being constantly a little longer than the side of the polygon. If the radius be used, the circle will be laid off into a hexagon. (See Fig. 2.) As we increase the number of sides, the sum of all the sides, or the perimeter of the figure, will approximate more and more nearly to the circumference of the circle; and if we enclose the circle in a similar and corresponding polygon, the length of the circumference will be somewhere between the perimeters of the two polygons. Enclosing the circle between an inscribed and circumscribed polygon, was the mode adopted by ARCHIMEDES, and by his successors down to the time of VAN CEULEN, inclusive; later mathematicians, however, have adopted briefer and better modes: but they involve principles which we cannot here discuss.

***Fig.* 2.**

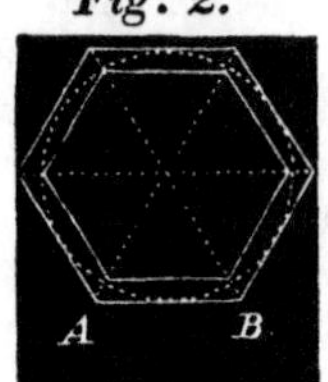

In arriving at the ratio 7 to 22, ARCHIMEDES used polygons of 96 sides, and by extending the number of sides to 32768 we should obtain the ratio true to 7 places; how immense then must have been the labor of VAN CEULEN to extend the calculation to 35 places! The operator can know when he is correct to any given number of places by the length of the inscribed and circumscribed polygon coinciding to such extent; for as the circumference of the circle is between them it must be thus far the same. In VAN CEULEN's proportion given above, the first number gives the length of the inscribed, and the second the circumscribed polygon; the circle being between them.

It might be matter of interesting amusement to ascertain by comparison how far the proportion of VAN CEULEN falls short of perfect accuracy, but no calculation that could be made would bring it within the grasp of the human imagination. If a ray of light, travelling at the rate of 12000000 miles per minute, had been despatched at the moment of creation, and had continued its unabated speed until the present hour, allowing that period to have been 6000 years, it would have travelled only 34989120000000000 miles, 17 figures being sufficient to express the number; while 35 are necessary to express the number of parts into which the diameter must be divided, that one of those parts may express the ratio of the *greatest possible* error to it; and how much less it may be we cannot tell. Now when we consider that every cipher added to a number increases it tenfold, we are ready to exclaim that the error bears to unity a less ratio than a grain of sand to the globe we inhabit. But if this be the case when the decimals are carried to only 35 places, what must it be when carried to 156 places, as has been done by the later mathematicians, but still without finding the exact ratio. This would be accurate enough to create a universe by.

For practical purposes it is usual to consider the circumference 3.1416 when the diameter is 1, this being within less than $\frac{1}{100000}$ of a unit of the truth, and being sufficiently difficult to remember and tedious to apply. Taking this as accurate, and the diameter of figure 2, as 1, $\frac{1}{6}$ of the circumference would be .5236, which is equal to the curve arc upon one side of the polygon; the side itself being equal to the radius, and hence it is .5; so that the curve exceeds the straight line by .0236.

The radius of a circle is a line drawn from the centre to the circumference, as a *ray* emanating from the centre, and if these radii be drawn from the centre to every part of the circumference, the circle will appear as a star formed by a glowing light in the centre of the circle. As these pass only from the centre to the circumference, they are but half the diameter, and they will be to the circumference as $\frac{1}{2}$ to 3.1416, or as 1 to 6.2832.

The chord of 60° of a circle, is equal to the radius, as may be seen in figure 2, where the circumference being divided into 6 equal parts, each must contain 360°÷6=60°. In that figure the triangles are equilateral, and hence equiangular.

The following is the most accurate proportion yet calculated, and as perfect accuracy seems not to be attainable, it is not likely that the subject will undergo farther investigation. If the diameter be 1, the circumference will be 3.1415926535897932384626433832795028841971693993751058209749445923078164062862089986280348253421170679821480865132823066470938446460955051822317253594081284802.

This ratio furnishes a number of the constant multipliers used in measuring. Call the above number n; then—

1. The number .7854, (or more properly .7853, &c.,) is one fourth of n.

2. .707106, &c., used to multiply the diameter of a tree by in order to find what it will square, is the square root of one half of n; and .225079, &c., used as a multiplier of the girt or circumference, for the same purpose, is the product of the

above multiplied by the quotient of n, divided into the diameter.

3. To find how large a tree is required to make a given square beam, we may multiply the side of the square by 1.4142, &c., for the diameter; or by 4.44288, &c., for the circumference. The former is the square root of twice n, and the latter the square root of twice n, multiplied by n.

4. The number .5236, &c.. used in finding the solidity of a sphere, is one sixth of n.

5. We are sometimes directed to square the circumference of a sphere and multiply by .3183, &c., for the surface. This is the diameter 1, divided by n.

6. We may square the circumference of a circle, and multiply by .07958, &c., for the area. This is the diameter 1, divided by 4 times n.

We might proceed to point out many other numbers, used in calculation, but it is unnecessary. Whenever you find a constant decimal having reference to a circle, or almost anything circular, you may rest assured that it is derived in some way from the ratio of the diameter of a circle to its circumference. These numbers might of course be carried by division to as many places as n itself, but it is not necessary. The reason of these multipliers will be obvious from the purposes to which they are applied.

One of the most interesting and valuable proportions in the parts of any known figure, exists amongst the sides of a right angled triangle. It is the often spoken of 47th Proposition of EUCLID's 1st book; and so valuable was it deemed by PYTHAGORAS, who discovered the truth involved, that in his heathen piety and superstition, he caused a hundred oxen to be sacrificed in thankfulness to the gods for enabling him to make the discovery. The proposition is, that *In every right angled triangle the square of the side subtending the right angle, is equal to the sum of the squares of the two sides containing the right angle;* or in other words, the square of the Hypotenuse is equal to the sum of the squares of the Base and Perpendicular. If the base of a right angled triangle be 3, and the perpendicular 4, then the hypotenuse will be 5; for $3^2+4^2=25$, and $5^2=25$. The Persians are said to call this proposition *The Bride*, from the large number or family of other propositions dependant upon it. It enters into the demonstration of many propositions by the mathematician, and is used by the mechanic in his workshop, though he is not always aware of the principle to which he is indebted, for the calculation he uses. Builders use it constantly in squaring the foundations of their houses, by forming a right angled triangled triangle. They measure 6 feet along one side, 8 along the adjoining, and then adjust the timbers so that the points where these measures fall, shall be 10 feet apart. That these will form a square or right angle is evident from the fact that $6^2=36$, $8^2=64$, and $36+64=100$, which is just the square of 10. Any other numbers bearing the same ratio will serve as well, only that these are about the right length for accuracy. 3, 4 and 5 make the lines rather short, since a slight inaccuracy in the measures would throw the building very much out of square by the time the measures are extended to the size of a building, it is usual besides for builders to have a 10 feet measure, and this serves to take the dimensions. 6, 8 and 10 are double these numbers; 9, 12 and 15 suit very well, being triple 3, 4 and 5. 12, 16 and 20 are quadruple, and we will find that all multiples of 3, 4 and 5 will be in proper ratio.

The surveyor's Traverse Table is constructed on this principle; the Latitude and Departure being the legs, and the distance or measured line the hypotenuse of a right angled triangle, the angles of which are determined by the bearing of the line. Architects often find use for this ratio in calculating the length of braces, rafters, &c. If, for instance, it were necessary to find a brace, that placed 12 feet from the building would support a point 16 feet from the ground, we would have the base and perpendicular of a right angled triangle, of which the brace would be the hypotenuse; and by squaring the base, 12, and the perpendicular, 16, and adding their squares together, we would have the square of the hypotenuse, the square root of which would be the hypotenuse.

$12^2=144$
$16^2=256$

4'00 (20, hypotenuse.
4

00

We could determine the height of a kite floating in the air, by ascertaining the point over which it is perpendicular, and measuring the distance thence to the place where the string of the kite is held; this would be the base, and the string would be the hypotenuse, from the square of which the square of the base being taken, the square of the perpendicular will remain. But if the base be not measured truly to the point directly under the kite, the triangle will not be precisely right angled;

and in practice the cord or line by which the kite is held would be inclined to form a curve, and would be rather longer than a straight line from the point at which it is held, to the kite. For this an allowance must be made.

Suppose we have the length, width and height of a room, and we desire to find the length of a cord that will reach from the floor at one corner, diagonally to the ceiling at the opposite. We could first find the diagonal of the room on the floor by means of the length and width of the room, and this diagonal would become the base, and the height of the room the perpendicular of another triangle, the hypotenuse of which would be the cord sought.

This principle is sometimes applied to the measurement of inaccessible lines, as where a pond or other obstacle prevents the line from being measured on the ground. In that case let two other lines be measured so as to make the unknown line one side of a triangle, and from the two sides that can be measured the third is easily found.

We might here introduce various proportions between parts of triangles, as well as some other plane figures; but the limits assigned to our present lecture will not admit of any thing like a satisfactory explanation of the subject; we shall, therefore, pass them over, and shall hereafter solve a few problems involving principles which may be sought for in the Elements of Euclid, or some other treatise on Geometry.

SIMILAR PARTS OF SIMILAR FIGURES.

Next to the consideration of proportion amongst parts of the same figure, is the ratio between similar parts of similar figures. Here we find the parts proportionate simply according to other corresponding parts. As the diameter of one circle, is to the diameter of any other circle, so is the circumference of the first to the circumference of the second; or between parts of *similar* triangles—as the base of one, is to the base of another; so is the perpendicular or hypotenuse of the first, to the perpendicular or hypotenuse of the second.

The analogy between parts of the triangle is applicable to many valuable purposes. The determination of heights and distances, both by instruments and comparison of shadows depends on this principle. If we compare the shadows of different trees or houses or the like, standing on the same plane, we shall find the shadows respectively proportionate to the heights of the objects producing them. If we set up two poles one 50 feet high and the other 100, the shadow of the latter will be twice as long as the shadow of the former. You might thus determine the height of a church steeple from the shadow of your walking stick. It would seem that if the rays of light issue from the sun as from a luminous point, that they should diverge constantly as any thing else does that issues from a centre, and hence that the above proportion would not be exactly correct. But even if this be strictly so, the distance of the sun from the earth is so great that we may safely consider the pencils of rays in the same vicinity as parallel. If in travelling 96,000,000 of miles they have separated but a few feet, any variation within a short distance may be safely disregarded.

If then the shadow of a pole increases as the length increases, we have a ready mode of ascertaining the height of any object by the shadow, for we have only to measure the length of a pole and setting it up, we have the following proportion:

As the shadow of the known object,
Is to the shadow of the unknown object,
So is the height of the known object,
To the height of the unknown.

A very convenient mode of finding heights by means of two sticks of unequal height, is explained in Parke's Arithmetic, page 143. It is based on the principle we have been discussing.

SIMILAR SURFACES.

We have already stated that similar surfaces are to each other as the squares of their like linear dimensions. One circle is to any other circle, as the square of the diameter or circumference of the former, is to the square of the diameter or the circumference of the latter. So similar triangles are to each other in area as the squares of their like parts; and the same may be said of any other shaped figure. If for instance there be two hexagons, (see Fig. 2,) a side of one measuring 3 feet and the other 6 feet, the latter will contain 4 times as much surface as the former. The hexagon, the square, and the equilateral triangle, are the only figures of equal sides that can be made, by placing a number of them together, to fill entirely any space. The bee in the construction of its honeycomb furnishes an instance of the hexagon; which is the only figure that will suit its purpose; being nearly enough round, and yet without waste of space.

The surfaces of similar solid figures bear the same relation to each other, as plane figures generally. The surface of a sphere or ball 2 inches in diameter, is to the surface of another sphere 4 inches in diameter, as 2^2 to 4^2, *i. e.* as 4 to 16, or 1 to 4; and the same may be said of any surface, whether it be a plane figure or the entire surface of a solid body.

SIMILAR SOLIDS.

As lines, with length only, are related to each other, simply according to the extension of that dimension: and as surfaces with their twofold dimensions of length and breadth are related as the square of their similar lines; so solids, with their threefold dimensions of length, breadth, and thickness, are related as the cubes of their like linear dimensions, which we have already explained.

DISSIMILAR SURFACES.

The comparisons we have made have been between similar figures, we shall now introduce a comparison between some dissimilar surfaces.

Amongst the most important and common superficial figures are the Circle and the Triangle, concerning whose lines as related to others in the same figure, and as related to similar lines in a similar figure, differing in size, we have already treated; it remains now to show the relation in area between these figures.

The area of the *circle*, like its circumference when the diameter is established, cannot be accurately ascertained, though its relation to the triangle is demonstrable with perfect accuracy, for a circle is equal in area to a triangle of which the radius of the circle is the perpendicular, and the circumference of the circle the base. But here you observe is the same difficulty in finding accurately the base of the triangle as the circumference of the circle, and hence the area partakes of the inaccuracy which cost Archimedes, Metius, Van Ceulen and others so much labor.

From this analogy is derived a very convenient rule for finding the area of a circle if the diameter and circumference are both given; viz—Multiply half the circumference by half the diameter; or take one fourth the product of the circumference by the diameter.

That the area of the circle is equal to such triangle, may be thus shown. Let any regular polygon, as a hexagon, (See Fig. 2,) be inscribed in the circle, and it is evident that the area of the polygon will be divided into a number of triangles corresponding with the number of sides which it has, and the area of the figure will be found by multiplying half the length of a side by the distance from the centre of the figure to the middle of the side or base; and that being the area of one triangle, must be multiplied by the number of triangles for the area of the entire figure; or the same result may be had from multiplying half the perimeter of the figure, which will be half the sum of the bases of all the triangles by their height. Let the number of sides of the polygon be made very great, and of course very short, and the perimeter of the polygon will ultimately coincide infinitely nearly with the circumference of the circle, and the perpendicular will be infinitely near the semi-diameter or radius of the circle; and still the position will hold true that the figure consists of a succession of triangles united at the apex, and the area of the whole will be found by multiplying half the sum of their bases by their perpendicular height.—This approximation of the perimeter of a polygon inscribed in a circle, to the circumference of a circle, is the same principle that was used by Archimedes in his attempts to find the ratio of the diameter of the circle to the circumference, and we have already shown how nearly that approximation has been carried to the truth. By a different and more scientific mode of proof, the above proposition may be shown in a light perhaps more satisfactory to some minds.

A *circle* is to its circumscribing square as .7854, &c., to 1; and an ellipsis is to its circumscribing parallelogram in the same ratio. If we multiply the two diameters of an ellipsis together, the square root of the product will be the diameter of a circle of equal area.

The *triangle* is equal to a parallelogram of equal base and half its altitude, whether the triangle be right angled or otherwise, and hence the rule for finding its area is to multiply the base by half the perpendicular height; or half the base by the whole height; or the base by the height and take half the product.

DISSIMILAR SOLIDS.

It has been already shown that the law of proportion between similar solids, is as the cube of their like linear dimensions; it remains to point out the relation between some dissimilar solids.

As the circle, in surfaces, contains the greatest surface within the same bounds, so the *sphere* in solids contains the greatest solidity within the same surface; there being no corners in the one or the other. Both in solidity and surface, the sphere is $\frac{2}{3}$ its circumscribing cylinder, if we take into consideration the surface of the ends of the cylinder; the surface of the sphere being equal to the curve surface of the cylinder, without its ends.

If the diameter of the sphere be 1 inch, its solidity will be .52359, &c., of an inch, or as we generally say, .5236; and the solidity of the circumscribing cylinder will be .78539, &c., or as we may say for convenience .7854 of a cubic inch; of which .5236 is just $\frac{2}{3}$.

By the "circumscribing cylinder" is meant a cylinder that may be just circumscribed around the sphere; or a cylinder, like a piece of stove-pipe, into which the sphere may be dropped, and it will just fit its cavity and be of the same height.

The surface of an inch globe is 3.1416 square inches; and the curve surface of an inch cylinder is 3.1416, to which .7854 being added for each end, we have the whole surface 4.7124, of which 3.1416, the surface of the globe, is just $\frac{2}{3}$; so that both in solidity and surface the globe is $\frac{2}{3}$ of its circumscribing cylinder. The surface of a sphere is equal to four times the area of a great circle of it.

But if the sphere bears to the circumscribing cylinder, the same ratio both in surface and solidity; how is it to contain greater solidity under the same surface than the cylinder does? The answer to this must be sought in the fact that solidity increases more rapidly than surface, and hence if you reduce the cylinder until it measures no more than the sphere enclosed in it, the surface will be made greater as compared with the solidity of it; but if you increase the sphere until it has 50 per cent. more matter in it, which will make it equal in solidity to the cylinder, the surface will not be equal to that of the cylinder;—it will not be 4.7124, hence it has greater solidity under a given surface than the cylinder has. Surface increases as the square only, while solidity increases as the cube of the linear dimensions of the solid body: a globe 2 inches in diameter has four times the surface of one only an inch in diameter; and it has 8 times the solidity. If the diameter of the base of a cone be equal to the perpendicular height of the cone, then the solidity will be half that of a sphere whose diameter is equal to the base of the cone, and $\frac{1}{3}$ of a cylinder of the same base and altitude as the cone.

From which it appears that the height and diameter being equal, the solidity of a *cone*, *sphere* and *cylinder* are as 1, 2, 3. If a cylinder of equal diameter and height be made to contain 3 pints of water, and a cone of equal diameter and height as the cylinder be introduced, one pint will run out; and if the cone be taken out and a sphere of the same diameter be introduced, another pint will run over. These proportions were amongst the discoveries of ARCHIMEDES, that prince of mathematicians.

It is frequently important for workmen to know how to determine the size that a given log will square, or on the other hand to know how large a log must be to make a square beam of a given size. To find what a tree will square, multiply the girt or circumference by .225, the result will be the side of the square that may be formed from it, near enough for all practical purposes. To ascertain the girt of a tree necessary to make a given sized square beam, multiply one side of the beam by 4.443.

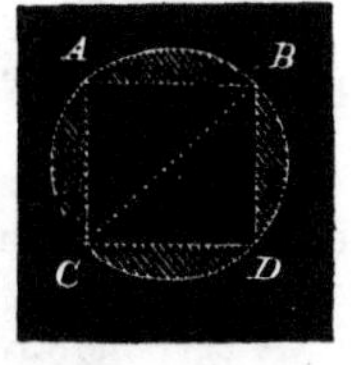

But as these arbitrary multipliers are liable to be forgotten, a little investigation will enable the operator to make his calculations from principle.

Let the circle A C D B represent the end of a log, from which the largest practicable square beam is to be formed. The end of the beam will be represented by the square A B D C, of which C B is the hypotenuse, and is at the same time the diameter of the circle. Then if we know the circumference of a tree we can readily ascertain the diameter by dividing by 3.1416, and as the square of the hypotenuse is equal to the squares of the two legs, if the piece is to be square, the square root of half the square of the hypotenuse will be one side, A B or A C. If the piece is to be larger one way than the other, the diameter is still the hypotenuse; and its square must be divided accordingly.

If on the other hand we know the size of the timber, whether it be square or of unequal depth and thickness, we have but to ascertain its diagonal C B, and we know the diameter of the tree.

18*

GENERATION OF GEOMETRICAL FIGURES.

IMAGINING how bodies may be formed or generated by motion, or built up of parts, often assists in forming a clear idea of them.

The *straight line* may be considered as generated by the motion of a point constantly moving in the same direction, and leaving a track as it proceeds.

The *circumference* of a *circle* as formed by the motion of a movable point carried round a fixed point or centre, and kept constantly at the same distance from it.

A *circle* as formed by the motion of a straight line carried round a point; and if carried less than entirely round, a *sector of a circle* is produced.

A *parallelogram* as formed by the motion of a line carried at right angles to its length.

A *triangle* as formed by the movement of a point, while expanding into a line.

A *cycloid* is described by the motion of a point in the periphery of a wheel rolling forward.

A *parallelopipedon* is produced by the motion of a parallelogram, at right angles to its plane.

A *cone* by the revolution of a right angled triangle around its perpendicular; or the motion of a point expanding into a circle: if we can imagine the expansion of that which has no dimensions.

A *cylinder* by the motion of a circle at right angles to its plane, or the revolution of a parallelogram on one of its sides as an axis.

A *prism* by the motion of a polygon.

A *sphere* by the revolution of a semi-circle on its axis.

In this way we may imagine lines, surfaces and solids to be generated, and frequently our clearest ideas are thus formed. It is the basis of the doctrine of *Fluxions*, in which figures are not supposed to be made up of a collection of distinct parts, but as formed by the *flowing* of other figures; as a line by the flowing of a point, a surface by the flowing of a line, a solid by the flowing of a surface. It is from this the science takes its name; the word Fluxion, signifying a *flowing*. The notion of generating figures in this way was however thought of long before it was built up by LEIBNITZ and NEWTON into a beautiful and efficient science.

The doctrine of *Differential and Integral calculus*, which has in a great measure superseded the old form of Fluxions, supposes the quantities to be generated, not by a uniform increase or flowing motion, but by the successive additions of infinitely small portions. The method which descends from quantities to their elements, is called the Differential Calculus; while that which ascends from the elements to the quantities is called the Integral Calculus.

After GULDIN and KEPLER had treated of figures as generated by a flowing motion, and supposed the extent of surface to be found by multiplying the line into the extent passed through in describing a surface, and the size of solids by multiplying the generating surface by its flowing space, CAVALLERIUS advanced a new view of the subject. He supposed a line to be made up of an infinite number of points; a surface of an infinite number of lines; and a solid of an infinite number of surfaces. These were considered *Indivisibles*, or the elementary principles of the figures produced; and though we know that lines could never be produced by using points which have no dimension, any more than lines could form surfaces, or a pile of surfaces having no thickness could be ultimately built into a solid body, yet it was a step in the march of science.

We may conceive a prism to be made up of an infinite number of small prisms—a pyramid of small pyramids, and a sphere of a great number of small prisms uniting their apices at the centre as grains of corn upon the cob tend towards the centre. A large number of such small prisms having spherical bases, and properly proportioned sides, if laid so that their points would all meet at the centre, would form a perfect sphere. They must, however, be hexagonal, square, or equitriangular, in order to fit with each other, leaving no vacancies, unless the sides be made unequal.

In explaining the rule for finding the surface of figures, we may often make our explanations more clear by imagining the surface to be divided into many small squares; as if we are attempting to prove that an oblong 7 inches long and 4 wide, will contain 28 square inches, we may imagine the figure laid off as follows and really divided into square inches.

So if we are seeking to find the cubic inches in a block 7 inches long, 4 inches wide, and 5 inches thick; we may imagine that the base is laid off into 28 square inches, and sawed into 28 square prisms, each containing 5 solid inches, and multiplying the number of little prisms (28) by 5, the solid inches in one, we have the solidity of the whole.

A tangent to a circle is a line drawn touching the circumference, and at right angles with a line drawn from the centre to the point of contact. If a stone be whirled in a sling, and suddenly released, it will fly off at a tangent from the circle in which it had moved. The ancients used slings as a means of warfare, and the precision acquired in their use may be inferred from the fate of Goliah, as well as from what is said in the 20th chapter of Judges, "Among all this people there were seven hundred chosen men, left-handed, every one could sling stones at a hair breadth, and not miss." The fragments of a millstone, or grindstone, that bursts from its rapid circular motion, fly off in a tangent from the circle they described before the breaking. The expression "off at a tangent" is often used facetiously in common discourse to signify an abrupt departure.

LECTURE XIV.

THEORY OF WHEEL CARRIAGES.

ALTHOUGH the use of *Wheel Carriages* shows considerable advancement in science, the invention was made at a period too early for the light of history.

The savage, and even the brute beast, will drag his burden upon the ground when he is unable to carry it, and this seems to be the simplest form of transporting burdens by draught. But it is exceedingly objectionable in several respects. It would injure the body in many instances, and would produce a great amount of friction. Very soon the idea of a sled or slide would occur, as protecting the body from injury and diminishing friction. The hunter might drag home the fruit of the chase, as the beast of prey would drag his victim to his lair, but it would be neater and easier perhaps to place it upon a pole or other implement that would raise it from the ground; and soon he would learn to construct such a carriage permanently, especially in latitudes where snow favors its use. We accordingly find the Laplander and the rude inhabitants of high latitudes traversing their snowy regions for hundreds of miles, drawn upon sledges or light sleighs, by dogs or reindeer. But though such a conveyance in such a country is admirably adapted to comfort and convenience, it serves but an indifferent purpose where there is not snow.

Except rolling the body itself, the simplest form of circular motion to overcome friction, and that most likely first to occur to the inventor is the roller, such as we sometimes see used in removing buildings and other heavy burdens. The roller answers an excellent purpose where the burden is very heavy and rapidity of motion is not important; but it would serve the purpose badly in the expedition necessary in travelling. The roller being placed between the burden and the ground, the burden resting upon the upper side, the sliding friction is entirely removed; but then the roller being introduced under the forepart, gradually passes to the rear and must be again carried forward to renew its service. The body will pass constantly over twice the surface over which the roller passes; for it will be carried with the roller, as far as the lower portion rolls upon the ground; and will roll as much farther upon the upper surface. Try the experiment.

There is evidently a difference in the friction of a locked carriage wheel upon the ground, and the friction of one that rolls or turns round. The former drags or slides, the latter merely touches the ground without any dragging motion. The latter is the kind of friction upon a roller that is allowed to pass in the way we have mentioned, from the front to the hinder part of the burden; but rollers are sometimes made to revolve in a notch that holds them to their place, and then there is a sliding friction in the notch and an impinging one upon the ground.

Let the roller A F be embraced by the collar B C, which is notched to receive it, on which any burden, as a building, rests, and moves in direction D E; then the roller A being kept in place by the collar, impinges on the ground D E, but it slides in the collar, and according as the force is applied to the burden and the roller thus caused to turn, or is applied to the roller and the burden carried onward by the revolution, will the principal friction be at the back of the roller as at A, or the forepart as at F.

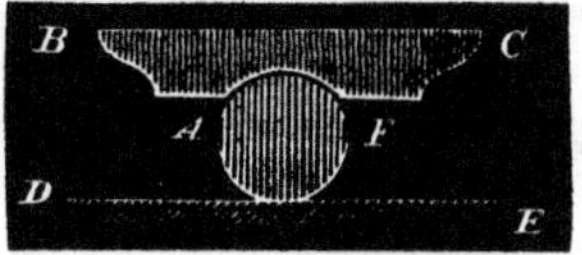

In this case the friction is very considerable, but it is transferred from the rough and uneven surface of the earth to the collar, which may be oiled or otherwise lubricated so as to reduce the friction.

But the inconvenience of operation is still great and the friction is yet very considerable, and the idea of improving the carriage by an axle passing through the centre of the wheel may be next in the order of time. The wheel may be a board cut into a circle, or it may be of frame work, as our present wheels are, having hub, spokes, fellies, and tire, but it is not likely that the wheel or its axle would for some time assume the form in which our best wheels are now made. The axles would be without droop or gather, and the wheels without dish, and of equal size. Of these properties it will be time enough to speak hereafter; at present let us examine the advantage gained by means of the wheel in overcoming the sliding friction as well as in surmounting obstacles.

Let C B represent the semi-diameter of the wheel, and C A the semi-diameter of the nave or axle. It is obvious that if the wheel does not turn round, it must slide on the ground D E, but if it turn it will only impinge upon the ground, and the sliding is transferred to the lower part of the nave at A.

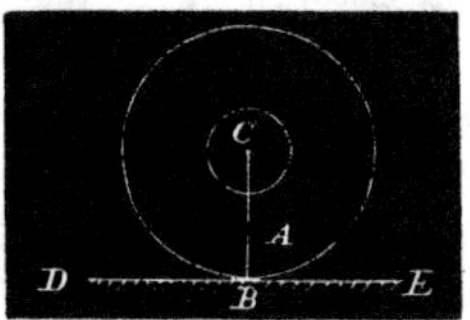

Now if C A, the semi-diameter of the nave, be 2 inches, and C B the semi-diameter of the wheel be 2 feet, then C B becomes a lever of the 2d order for the purpose of overcoming friction at A, and as C B is 12 times as long as C A, the power necessary to overcome friction will be only $\frac{1}{12}$ of what it would have been had there been no lever. To this some add the advantage of the rubbing surface moving with only $\frac{1}{12}$ the velocity of the progressive motion of the wheel, but experiment has not shown that friction is materially affected by change of velocity; though in proportion as the velocity is less, so will the wear of the axle and box be. The rubbing surfaces can be polished too, and substances, such as black-lead, tallow, &c., introduced yet farther to reduce the friction.

The impinging friction upon the ground cannot be very great, since with a wheel perfectly round, and a track perfectly smooth and hard, the wheel would touch at an infinitely small point, and as the centre of gravity would not be raised by the wheel's advancing, an infinitely small force would put the wheel in motion either forward or backward, the size of the wheel making no difference. But in practice these perfections do not exist, and the centre of gravity does not move forward in a right line, but rises and falls in its progress, and the wheel by sinking slightly is constantly encountering an ascent; and though that ascent may yield before the wheel, it requires force to press it down.

If therefore we consider the impinging friction as produced by constantly recurring obstacles, which though small must be overcome by the mechanical agency of the wheel, we may estimate its power for this purpose, just as we would its power to overcome a prominent obstacle. For though the height of the wheel could make no difference while an infinitely small power only was necessary to give motion, (*i. e.* while the wheel was perfectly round, and the plane perfectly smooth and hard) yet the moment that obstacles, however minute, present themselves, we derive advantage from the leverage of the wheel; either in pressing the obstacles down or raising the weight over them.

When the wheel does not turn, no advantage results from its use, and indeed by causing an elevation in the line of traction it would be easier to draw the burden forward by attaching the trace directly to the wheel at the point of contact with the ground, than to draw it in the ordinary way with the wheel locked.

In order to understand how a wheel enables us to surmount obstacles, let us consider the annexed figure, in which C represents an obstacle 6 inches high, which is to be surmounted by drawing the wheel in the direction A B. The semi-diameter A F or A C of the wheel being 24 inches. If the plane G H be entirely level, hard, and free from obstacles, and the wheel perfectly round, it is obvious, as has been already stated, that it will touch the plane at F at an infinitely small point and, as the centre of gravity of the wheel will be neither elevated nor depressed by the revolution of the wheel, it will be moved by an infinitely small force, and it is evident that the best direction for the operation of the force, or line of traction as it is called, will be in a line parallel with the base, which will be at right angles with the perpendicular A F.

Suppose on the other hand, as in Fig. 4, an obstacle C D presents itself, as high as the centre of the wheel, it is equally evident that no power exerted directly forward can enable the wheel to surmount it, since it draws with dead force against the obstacle. Here then are

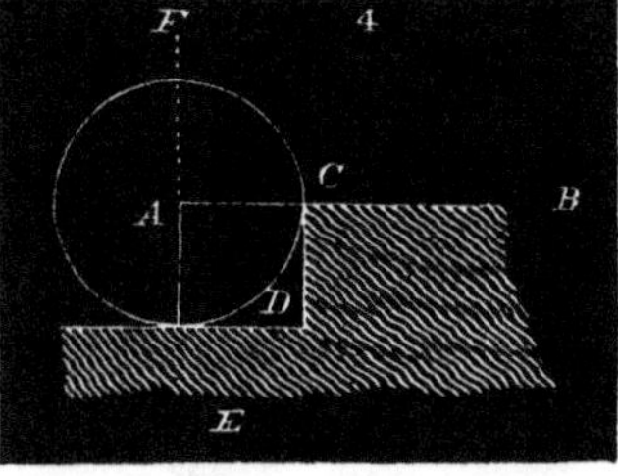

the two extremes when the power operates in a line parallel with the base. In one an infinitely small power effects the purpose, while in the other an infinitely great power will not effect it. It is certain that the wheel or the obstacle might be crushed, but that is not "surmounting," it is breaking down. Between these extremes every grade of obstacle is liable to be presented, and the question to be settled is, the law by which the power necessary to overcome them is regulated.

One position is obvious, viz; that the most efficient direction for the operation of power is at a right angle with a line drawn from the centre of the wheel to the summit of the obstacle, as A I in Figure 3, or A F in Figure 4. If the angle be less, a portion of the power is wasted against the obstacle; if greater the wheel is lifted from it; but the power should be constantly exerted at right angles and will consequently change direction at every moment: the point of traction, where the power is exerted, sinking gradually to a horizontal line, which it reaches just as the wheel reaches the height of the obstacle, when it is again carried forward upon a plane with an infinitely small force. This change of direction is necessary, since the height of the obstacle above the base of the wheel diminishes constantly as the wheel rises, and becomes nothing as the wheel reaches its higher level. In Fig. 3, the horizontal line A B would rise until when C A is vertical, and the wheel is upon the level of the obstacle, it will coincide with the upper line A B, and at that moment the line of traction A I will fall into it, A having moved through the circular arc A *a* to *a*, and I having sunk to B.

In Fig. 4, the first effort would be to lift the wheel directly upward, and its whole weight would be the measure of power necessary, but A F would gradually rise and fall over until it would coincide with A B, or rather with a line parallel with it, at the distance C D above it; as *a* B was above A B in Fig. 3.

This is on the supposition that the wheel having surmounted the obstacle continues on that level. But suppose it has again to sink to the level of the plane, it might be eased down by the mechanical force of the same power, reversed however in direction,. so as to draw it against the obstacle and thus let the wheel down gently. When the obstacle is so abrupt that the wheel at once strikes the upper part of it, the curve A *a*, Fig. 3 will be an arc of a circle, of which A C is the radius; but if the obstacle is so much inclined that the wheel rolls up its side, the curve will partake of the properties of a cycloid.

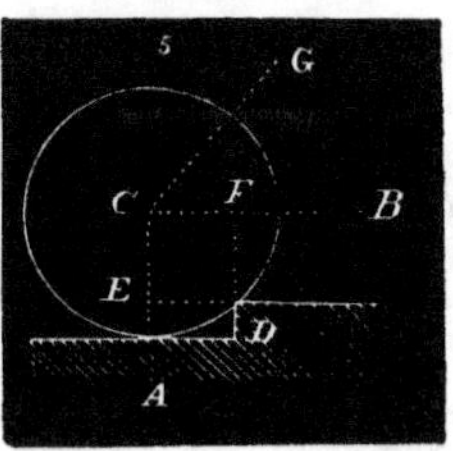

In Fig. 5, let the radius of the wheel be 2 feet, and the height of the obstacle D be 6 inches; then if the power be exerted in the direction C D, and be raised as C rises in surmounting the obstacle, so as to remain at right angles with the downward pressure of the force of gravity, the following proportion will be found to exist.

As the Power,
Is to the Weight,
So is the line D E,
To the line D F.

And as the line D E is constantly shortening, and D F lengthening, the necessary power will be constantly diminishing.

If the power at B sinks towards the base of the wheel there is a manifest loss, but if it be made to rise towards G there will be a gain until the angle, D C G is a right angle, which is the point of maximum power. The power exerted in the direction C G necessary to raise over the obstacle at D, a wheel whose radius is C E, would if exerted in the direction C B require a wheel of a radius C A. In other words a wheel of 18 inches radius with the power exerted at right angles with C D, would be equivalent to a wheel of 24 inches with the power exerted horizontally. It is evident from this that the larger the wheel, the greater the mechanical advantage in overcoming obstacles; but then other difficulties that will be hereafter noticed, would soon fix a limit to the size of the wheels.

So far as overcoming friction at the axle is concerned, the absolute size of the wheel is unimportant, *it is the relative size of the wheel and axle.* Reducing the size of the axle has the same effect as enlarging the wheel, but in a degree proportionate to the ratio of the radius of the axle, to the radius of the wheel. Suppose that an iron axle of 1½ inches in diameter be substituted for a wooden axle of 3 inches, the wheel being 5 feet in diameter. With the wooden axle the power will be as 20 to 1, but with the iron axle it will be as 40 to 1, without reference to the greater accuracy and reduced friction of the iron axle. To make the wheel an inch or two greater or less is a small matter, but to make that difference in the size of the axle is very important. Could the axle be reduced to a mathematical line, there would be no friction, the fulcrum and the resistance would fall in the same line and the nearer we can approximate to that and preserve strength, the greater will be our capability of overcoming friction. While we might on the other hand increase the size of the axle until it would coincide with the diameter of the wheel, and the wheel and all mechanical advantage would vanish together. These are the two extremes.

We have heard an experienced coachmaker urge the lighter running of coaches with iron axles, compared with wooden axles, as an argument to show that friction was reduced by diminishing the rubbing surfaces, but a much better solution of the fact is found in the diminished size and more accurate adjustment. To place the effect of reducing the axle in a stronger light, let the line A B represent the lever, with a weight of 50 lbs. suspended at C, 6 inches from B; if the hand at A, which is 6 feet from C be moved 3 inches towards C it will make but little difference, but move the weight 3 inches towards B and it will make a difference of one half, for 3 inches are half of 6.

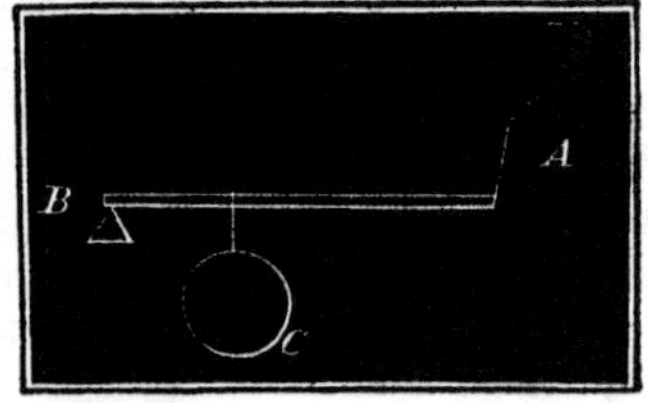

Experiments on friction may be conducted with as much certainty as on any other subject, and with more than on almost any thing else connected with wheel carriages. Treatises on Practical Mechanics abound with details of such experiments, in which various metals and other substances have been used and their friction compared, both with and without tallow or other such intervening substance. The general conclusion is that the amount of friction is but little affected by extent of surface or change in velocity; but the various details of operation, and facts brought to light, must be sought for in treatises on that subject.

Line of Traction.—The next circumstance to be considered as we now have our wheel mounted on its axis, is the *Line of Traction* or the line in which the draught of the horse or other power should be exerted, in order to produce the greatest effect. It is evident that it must not be directly upward, for that would lift the wheel from the ground, nor downward, for that would draw directly against the ground; neither of these directions could however be given by the draught of a horse. If the wagon were to be moved onward upon a level plain, without encountering obstacles upon its surface, there can be no doubt but the traces should be perfectly horizontal instead of rising from the single-tree to the horses' shoulders, as they are generally made to do. Elevation in front less than a horizontal line would create friction by drawing towards the ground; and greater elevation would diminish the pressure upon the ground by the draught bearing a part of the burden.

It is very generally agreed that in practice it is necessary to elevate the traces or line of traction, as tending to diminish friction and assist the wheels in surmounting obstacles or rising out of holes, as well as to keep the line from falling below a level when the horse sinks into mud or any depression, or depresses his chest in exerting his strength; but how great the angle of elevation should be is a question not very easily settled. Dr. LARDNER, in his treatise on *Mechanics*, says: "By mathematical reasoning it is proved that the best angle of draught is exactly that obliquity which should be given to the road in order to enable the carriage to move of itself. This obliquity is sometimes called the *angle of repose*. The more rough the road is the greater will this angle be; and therefore it follows that on bad roads the obliquity of the traces to the road should be greater than on good ones. On a smooth Macadamized way, a very slight declivity would cause a carriage to roll by its own weight; hence, in this case, the traces should be nearly parallel to the road. On railroads for like reasons, the line of draught should be parallel to the road, or nearly so."

This remark cannot be meant to apply to deep miry roads, since on such, scarcely any angle of elevation would cause a wagon to move forward, until it would approach so nearly vertical as to fall over; but on ordinary roads it might do as a general rule.

By referring back to Figure 3 it is shown that the most efficient line of traction is at right angles with the line from the centre of the wheel to the top of the obstacle, and this line is approximated by elevating the line of traction so as to form a considerable angle with the horizon; such an arrangement is therefore always best when obstacles are to be surmounted.

M. CAMUS, a French writer, attempts to prove that the line of traction should be horizontal; but this view is successfully controverted by a later writer, M. COUPLET, of the same country. He remarks that if a horizontal line were best for heavy draught, it would be necessary to elevate the line of draught, that it might not sink below a level when the horse lays out his strength.

M. DEPARCIEUX contends that horses draw principally by their weight, and not by muscular exertion. He states that they plant the hind feet as a fulcrum, and that sinking the forepart of the body so as to bring the trace horizontal, produces draught; for the trace, which before might be considered the hypotenuse of a triangle, is brought down so as to coincide with the base, and being longer than the base, its extremity must reach beyond the extremity of the base.

Thus let A C represent the trace, and C B a perpendicular let fall from the trace; if C be brought to a level it will coincide with A B, reaching beyond its termination to *b*. In that case however there could be no draught if the line A C were not inclined to the horizon. The same writer has shown by experiment that the fore feet of a horse bear less of the horse's weight when he is drawing than when he is standing unemployed. He arrives at the conclusion that the angle of elevation should be 14 or 15 degrees, and adds that experiment satisfied him that at that angle a horse drew with most effect.

Dr. Brewster, of Edinburg, says: "When I first compared Deparcieux' theory with the manner in which horses appear to exert their strength, I was inclined to suspect its accuracy: but a circumstance occurred which removed every doubt from my mind. I observed a horse making continual efforts to raise a heavy load over an eminence. After many fruitless efforts, it raised its forefeet completely from the ground, pressed down its head and chest, and instantly surmounted the obstacle."

This circumstance however does not in my mind settle the question, for though when urged by the driver, the horse might rear up as represented, and by a desperate plunge surmount the obstacle, it does not follow that it was his natural mode of drawing; and the increased effect was perhaps promoted also by increasing the angle of the line of traction; for that would increase the effect. A gentleman of close observation and much experience, to whom this was mentioned by myself, stated that horses draw very differently, but always he believed by both weight and muscle. He stated that one of his own drew so much by weight: that when in ploughing, the plough met with an obstruction every thing was wrenched; while another of his own drew so much by muscular strength, that when the plough met with an obstacle, he yielded without any injury to the plough. His opinion in reference to draught is that for the ease of the horse, the trace should form a right angle with the shoulder; the slant of which however is very different in different horses; and such a rule furnishes no general guide.

Any one who has observed a horse attached to a *dray*, must have been struck with the very heavy loads thus drawn, and yet the wheels are very small, affording therefore less mechanical advantage than cart wheels. But the line of traction is much raised, and to that no doubt a large part of the advantage is to be attributed. In carts of ordinary construction, and in two wheeled vehicles for travelling, the wheels are large and the line of traction nearly or quite horizontal. These for rough and uneven roads are better than a small wheeled vehicle could be ; the dray being principally used on the streets of cities; and there preferred for its convenience in loading and unloading.

In all two wheeled vehicles, the horse should bear a part of the burden; for if the centre of gravity of the load be placed directly over the axle, the motion of the shafts will be unsteady, and if behind the axle the carriage will tilt and choke the horse. The proportion of the burden that should be borne by the horse, is matter of judgment for the driver, and probably no general rule could be given. Much would depend upon the animal and upon the road. It is obvious that in proportion as the load is placed before the axle, the wheels will be relieved, but carrying this to excess would break the horse down: and if he could support all the burden the cart or riding vehicle would be a useless encumbrance.

If the axle could be made to pass through the centre of gravity, it is obvious that the load would rest equally upon the axle in all positions, whether ascending, descending, or horizontal ; but if the centre be above the axle, as it must of necessity be, an increased burden will be thrown upon the horse in going down hill and the reverse will occur in ascending. Every boy who has used a cart knows the tendency of the shafts to rise by the tilting of the cart in ascending a steep hill. But of that hereafter.

Size of Wheels.—Having discussed the power of the wheel and the most efficient mode of applying the line of draught, we come now to discuss the proper size of wheels for practical purposes.

It is entirely obvious that the larger the wheel in proportion to the diameter of its axis, the greater will be its mechanical efficiency in overcoming friction at the axis; and its size will also aid in surmounting obstacles by causing the load to rise and fall more gradually, as well as from obstacles thus bearing a less ratio to the semi-diameter of the wheel. They would besides, often sink less between obstacles.

But whatever might be the theoretical advantage of large wheels, they must have a limit in practice. If the wheels were made very high in proportion to the length of their connecting axle, the carriage would be very liable to be upset, by having its line of direction thrown out of its base ; and any considerable increase of height above what is now generally used would subject us to the necessity of attaching the draught far below the centre of the wheel, unless our horses were tall as giraffes. But while on one hand it would be found ill-advised

19*

to mount our wagons upon mammoth wheels, it would on the other be quite practicable to make the wheels too low, thus increasing the difficulty of overcoming friction and surmounting obstacles.

In determining the size of wheels much depends on the use for which they are designed. Timber wheels being designed to have pieces of timber suspended below the axle, are made high; while dray wheels are made low; each being thus made more convenient for receiving its load.

Wagon wheels are of various heights, but it is usual, let the height be what it may, to make the fore wheels smaller than the hind wheels; the difference varying from 4 to 12 inches. This practice seems general in all countries, and though sometimes defended on false grounds, must have better reason for its adoption than vulgar error. Mr. FERGUSON, who embraced this subject in his Lectures on Select Subjects, had a model wagon, and in order to show that the greater height of the hind wheels had no tendency to push the carriage forward as some supposed, he was accustomed in his lectures to load his little wagon, and he always found that the same weight which drew it with the small wheels foremost would do the same when the wagon was reversed; a conclusive proof that difference in height has no effect of this kind on the draught. This experiment we have always thought perfectly satisfactory, for a hard and level plane would be the proper place to test the principle; but we think other experiments of the same gentleman in reference to the shape of the wheel entirely inconclusive, for he experimented on a hard surface, while wagons, &c., must be made for roads as they are.

One circumstance to be provided for in constructing a carriage is capability of being turned, for in using wagons it is necessary to turn them in all directions. If the wheels of a wagon were made of equal height, and the body set firmly upon the axles, it could only be turned by dragging the forepart around. But in the ordinary construction the fore wheels are made low, and a bolster is placed upon the axle, turning upon the body bolt as a pivot; thus permitting the fore wheels to be turned freely, and giving to the tongue the properties of a helm. This facility is increased by the wheels being low and turning partially under the body. In order still farther to increase facility in turning, wagons are sometimes constructed with a joint in the coupling pole: and when properly made thus they will turn short, and the hind wheels will track the fore wheels.

The fore wheels being low admits of that inclination in the line of traction which has been shown to promote the draught;

while the hind wheel having no advantage from this circumstance is made larger that it may make up the deficiency in mechanical advantage. This can be done too without inconvenience, and indeed it is necessary for convenience, as the body then rests about level on the bolster before and the axle behind. In going down hill the wagon is held back more readily from the better direction of the resistance; it is however in changing direction that the greatest advantage is found.

Disposition of the load.—In reference to the loading of two wheeled vehicles, some remarks have been already made; and the same principles will be found to apply in loading four wheeled carriages or wagons. On level hard ground the wheels being properly proportioned for the purpose, and the load bearing equally on both axles the inclination of the line of traction would about compensate for the difference in the size of the wheels, and the force which would take the fore wheels over an ordinary obstacle would do the same with the hind ones. The same remark would apply where the wheels sink proportionately into a soft road, but here the fore wheels would labor under disadvantage from sinking not only deeper in proportion to their diameter but *absolutely deeper*. This evil then would be increased by placing too much of the load upon the fore axle.

It would seem fair to infer that on level ground, if the wheels are proportioned as we have supposed, the load should be disposed equally upon the axles; but if not so proportioned then let the axle having the advantage have the greater load.

A manifest inconvenience results from an unequal distribution of the load when the heavily loaded wheels have to surmount obstacles. If 40 cwt. are equally placed on four wheels, and one rises over an obstacle, only 10 cwt. is lifted; but if the weight is thrown upon two wheels and one is to be raised, 20 cwt. must be raised.

Some suppose that if the load be thrown too much on the fore axle, and especially if top heavy, that it will greatly increase the difficulty of holding back; and on the other hand that if placed on the hind axle the difficulty of ascending hills will be increased. But no good reason is discoverable for this notion, which seems near akin to the opinion that making the hind wheels larger than the fore ones helps to push the wagon forward. There is no doubt but that a very unequal distribution of the load might affect the draught anywhere, and going up hill this would be felt more than on a plane. In going down hill the rising of the tongue in holding back occasions a waste of power, for though it presses the wheels against the ground, the friction does not generally make up for the waste of power.

When a load is high, any inequality in the level of the road, throws greatly increased stress on the lower wheel; and this depends as every boy knows, on the elevation of the centre of gravity. If we gently raise one wheel, or the wheels on one side, if the vehicle have more than two wheels, we may soon discover the elevation necessary to throw all the weight on the lower wheels, and the slightest additional elevation will upset the wagon.

Effects of Springs.—Springs have a decidedly beneficial effect in relieving the abruptness of irregularities upon a road's surface; and especially when a vehicle is moving with considerable velocity. Not only is the load less violently agitated, and hence the comfort of riding in such vehicle increased, but the power necessary to draw the carriage is less, since when the wheels strike an obstacle the load is not suddenly and violently raised over it, but the springs yield and the wheels *might pass over* by the yielding of the springs, without raising the load at all: though this would not be the case generally. The effect however would be to soften the violence of such concussions, and change abrupt elevations and depressions, to gentle undulations.

Stage drivers point out another advantage from the manner in which stage bodies are suspended. When the fore wheels strike an obstacle the weight of the body is thrown upon the hind wheels, while the fore ones pass over, and generally by the time the hind wheels reach the obstacle, the roll of the body will throw the weight on the fore wheels while the hind ones pass over. This applies to bodies hung on "thorough braces" more than to elliptic springs.

The advantage of elasticity in a load is so great, that wagoners who transport lead or iron, frequently place spring poles in their wagon bodies, the extremities resting over the axles, and the load being placed upon them.

Forms of Wheels and Axles.—This has been a fruitful source of cavil, and some diversity of practice amongst workmen.

The most natural form would seem to be a perfectly flat or plane circle, formed from a board or of frame work as our wheels are generally; the opening for the axle being at right angles to the plane of the wheel, and thus causing the radius of the wheel, (the spoke,) to stand perpendicularly to the axle. If wagons always moved on level plains, this would probably be the best form: but for the vicissitudes of ruts, chucks, stumps, rocks, and all the *et ceteras* which wagons "are heir to" such a form is, by general consent of workmen and wagoners, pronounced not the best.

In the first place then the wheels are *dished*, that is the spokes are not placed perpendicularly in the hub, but inclining outward, a form given to the wheel by the shape of the end of the spoke inserted, rather than by the mortice formed to receive it. This dish is varied by workmen from half an inch to two or three inches on the face or outside of the wheel; but as the spoke is made broad at the hub and tapering towards the extremity, the part of the wheel next the wagon would show very considerable dish, though the front should be without any.

Having formed the hub and inserted the spokes, the ends are then formed by the workman to receive the fellies, which form the wooden rim surrounding the wheel. This again is surrounded by the iron tire, which being adjusted to fit closely when cold, is heated, by which it is increased in size, and and having been put upon the wheel is then cooled, and by its shrinkage the wheel is drawn closely together, and the dish somewhat increased.

The wheel being formed, is then placed upon the axle in such manner as to throw the upper part of the wheels so much farther apart that the lower spokes will stand vertical on level ground. This position of the wheel is given by drooping, if we may so speak, the end of the axle. The axle is also so dressed as to throw the fore part of the wheels nearer together than the back; this is called the *gather* of the wheels, and is necessary to keep the wheel from rubbing too hard against the linch pin; and to promote ease in turning.

Against the dishing of wheels it is urged that the stress is upon the wheel when it is not perpendicular, and of course not in its strongest position; but this objection is obviated by the position given to the wheels, by which the lower spokes are made to stand upright. In favor of this position it is urged that the upper part of the wheels being farther apart than the track gives greater room to the body; this difference would amount to a foot if the wheel had 3 inches dish and the lower spokes were brought to an upright position. It is also urged against the practice, that the wheel instead of being cylindrical and disposed of itself to roll forward in a straight line, is thus converted into a frustum of a cone, and if left to itself, when rolled forward would describe a circle round a point which would be the apex of the cone, if it were complete; and hence if kept in a right line by the axle it must drag, and thus greatly increase friction.

This tendency to curvilinear motion is greatly diminished if not removed by the position of the wheel on the axle, and when we take the gather into consideration it is overcome, for a wheel properly constructed, will run for miles without a linch-pin. But it is not proved from the wheel not running off that a tendency to curvilinear motion does not exist, it only shows that it is counteracted if it does exist; though it may be at a great expense of friction. The conical shape of the wheel would cause a tendency to run off.

In favor of dished wheels it is urged that they are stronger—that they are especially stronger when the weight is thrown upon one side by inequalities in the road; but it must be admitted that this advantage is greatly reduced by making the lower spokes upright. Some light may probably be thrown on this subject by considering how a wheel generally breaks down. On this point I have conversed with different workmen, and they say that in a very great majority of cases the wheel breaks down by the lower part running out, instead of running under the wagon. This would argue that the dish does not strengthen the wheel so much as has been supposed; or that the axle had bent upward at the point.

Broad and Narrow Wheels.—The rims of wheels vary much in width, and it has been subject of discussion whether narrow or broad rims are best. In general, broad wheels are encouraged upon paved roads, both in Great Britain and America, by being permitted to travel at a reduced rate of toll. The "tread" being broad tends to press the road into a solid state rather than to cut it into ruts, and if very broad are even thought advantageous to roads.

On soft ground they may pass without sinking deep, but if they do sink they raise a great load of mud. Some suppose the friction from narrow wheels to be less upon the ground, but this can make but little difference, since well conducted experiment proves that friction depends much more on the weight of the load than the extent of rubbing surface. If the surface is large, each portion has less weight to press it down, and less friction to encounter.

Perhaps there is no subject on which theoretical and practical men differ more among themselves than that of wheel carriages. The man of science has not always looked enough to the allowances necessary in accommodating his theories to practice; while the practical man has looked at the subject too much in mass, without always attributing effects to their proper causes. We shall be gratified if our remarks shall lead any concerned to closer and more correct thought upon the subject.

Having gone through the several particulars in reference to

the construction of wheel carriages and their mechanical effects, we shall devote a few remarks to the consideration of a question that may seem of little practical importance, and yet it is not without advantage, as tending to correct thinking, and is matter of curiosity to persons who have not thought upon it.

If we suspend a wheel upon an axle and cause it to revolve as a spinning wheel, every part of the wheel equally distant from the centre, will move through equal spaces in equal times; but when we cause the wheel to roll forward the effect is very different. In the former case while the upper part of the wheel moves forward the lower part moves backward just as far, and any point will in a revolution of the wheel describe a circle. But if we take any point in the tire of a wagon wheel and watching it carefully through successive revolutions of the wheel, we shall find it describe a series of figures resembling the following.

This figure is called a cycloid, and though it resembles in some measure the arc of a circle, it has no part of the curve of that figure. It is a figure described as generated by the revolution of a circle upon a plane. Its properties are described in some mathematical works, but we desire here to call attention to one circumstance only,—the unequal motion of the point by which the figure is described. In other words it is that if, when a wagon is moving uniformly forward, we trace the motion of any point of the tire from the time it leaves the ground until it returns to it again, we shall find that it moves through very unequal spaces in equal times. The motion increases from the time the point leaves the ground until it reaches the highest point, and decreases again until it reaches the ground at the end of its revolution. *The upper half therefore of the wheel moves much faster than the lower.* This seems paradoxical, but it is strictly true, as any one may satisfy himself in a moment by setting up a stake by the side of a wheel and moving the wheel forward a few inches. The writer of this well remembers that he thought this proposition a hoax when he first heard it, and experiment alone satisfied him to the contrary. When however we come to think of the matter, we must know that such will be the case, or how could the wheel turn? Move the lower part as fast as the upper and the wheel must drag. One moment we see a given point directly at the back part of the wheel, and at the next it is in front; how did it change places but by out-traveling the other parts? In a moving wheel no part ever moves backward, as in a standing one.

If the circumference of a wheel were marked at every 10° and then rolled uniformly forward, while the point at the first degree would trace the cycloidal curve, and if the diameter of the wheel were divided into 1000 parts, the portions of the curve described in the several equal times would be (rejecting fractions less than tenths) 7.6, 22.8, 37.8, 52.4, 66.8, 80.6, 93.8, 106.2, 117.8, 128.6, 138.4, 147.2, 154.8, 161.2, 166.4, 170.4, 172.9, 174.3, which carries us to the highest point of the figure; and the same numbers reversed in regular order will carry us forward to the ground again. From this it appears that though the motion continues to increase until the mark reaches the highest point, and then decreases to the ground again, it is not in a uniform ratio. In the last 10° the generating point passes through more than 23 times the distance that it does in the first 10°, and hence on an average moves 23 times as fast, yet the last degree as compared with the first would show a far greater relative motion; and this would be increased as the parts compared are diminished.

Galileo, first treated of this figure in 1599, but he was not able to determine its properties. Mersennus, a learned Frenchman, turned his attention to it in 1615, with little better success. Other mathematicians afterwards took up the subject and succeeded, though not without labor, for we are informed that Roberval was led by the investigation to study closely the works of the Greeks, and especially Archimedes, yet it was six years after he commenced the investigation before he determined the area of the figure. The same problem engaged the attention of other philosophers, and the cycloid, and its kindred figure the epicycloid, furnished their full share of difficulties during the celebrated "War of Problems." The Epicycloid is formed by rolling a circle on the inside or outside of the circumference of another circle. This figure is useful in determining the proper shape of cogs in machinery, and also in the construction of pendulums. The curve of this figure may be seen on the surface of milk, when placed in a bright circular cup, and the light allowed to shine on the portion of the surface of the cup above the milk. Every child has noticed this figure, which nursery legends describe as the impress of the *cow's foot,* but which philosophers style the *catacaustic curve.* If the fixed circle be twice the diameter of the circle rolled within it, the resulting line will be a straight line instead of a curve.

LECTURE XV.

POSITION AND ALLIGATION.

Among the rules of Arithmetic based on the doctrine of proportion are two, generally considered especially difficult to understand; and these we shall make the subject of the present lecture. They are *Position* and *Alligation.* It is proper however to say, that both these are founded on principles which may be perfectly understood with a reasonable degree of attention; and that *they are not difficult.* To make them intelligible, however, we shall present them, and attempt their illustration, in a form considerably different from that usually found in works on the subject.

The rule called Position in modern treatises, is frequently called in the old books "Supposition," "Rule of False," "Trial and Error," &c., and from these names, as well as from the fact that the correct result cannot always be produced by this mode, many arithmeticians seem to regard it as unworthy of investigation, or of confidence. It was formerly regarded with greater favor, but the more general diffusion of a knowledge of Algebra, has caused it to be omitted in many school arithmetics.

Hutton says, in his Mathematical and Philosophical Dictionary, that "The rule of Position passed by the Moors into Europe, through Spain and Italy, along with their Algebra, or method of Equations, which was probably derived from the former." We shall now attempt briefly to show that it is not difficult to understand, or apply.

SINGLE POSITION.

When an unknown number is to be increased by some part or multiple of itself, the operation is very simple, for we have only to take any number at pleasure, and perform upon it similar operations to those indicated in the question, then say,

As the result thus found,

Is to the result given in the question,

So is the number from which the first was produced,

To the number from which the second must have been produced.

Example. What number becomes 24 by adding to it a third part of itself?

Suppose	12	Then,
$+\frac{1}{3}$ itself	4	As 16 : 24 : : 12
	16	12
		16)288
		18 *Ans.*

Here the result is only 16, but by increasing 16 to 24, and increasing 12 in the same ratio, we have 18; the number from which 24 was produced.

These operations are founded on the general principle that "*Results are proportionate to the numbers that by similar operations produce them.*"

Example 2. What number becomes 24 by being multiplied by 3?

Suppose 12. Then $3\times12=36$

And 36 : 24 : : 12 : 8, the *Ans.*

Example 3. In a certain orchard $\frac{1}{5}$ of the trees bear cherries; $\frac{1}{4}$ bear apples; $\frac{1}{2}$ bear peaches; $\frac{1}{12}$ bear plums; and the rest, 16 in number, bear pears. How many trees are in the orchard? *Ans.* 120.

Example 4. A person having a sum of money, spent $\frac{1}{3}$ and $\frac{1}{4}$ of it, and found he had $60 left. How much had he at first? *Ans.* $144.

Example 5. A lady being asked her age replied, "If $\frac{3}{6}$ of my age be multiplied by 7, and $\frac{2}{3}$ of my age be added to the product, the sum will be 292." What was her age?

Ans. 60 years.

In all the foregoing we find the results constantly proportionate to the numbers from which, by similar operations, they were produced. We might, however, by investigation, find other proportions, *e. g.* the *error* in the result is always proportionate to the *error* in the *supposition.*

What number becomes 24 by adding to it the third part of itself?

Suppose as before 12 (which is 6 too small)
$+\frac{1}{3}$ itself 4

16, this should be 24, hence the error is 8

Suppose 15 (which is 3 too small)
$+\frac{1}{3}$ itself 5

20, this should be 24, hence the error is 4.

Here we find that

The 1st error in supposition		(6)
Is to 2d error	"	(3)
As the first error in result		(8)
Is to the 2d "	"	(4)

This conclusion is perfectly reasonable, and a little examination must show that the law is invariable. If adding 3 to our first supposition reduced the error one half, adding 6, would have caused it to disappear entirely.

Again we find that—

As the whole difference between the errors in the result (4).

Is to either error in result, (say 8).

So is the whole difference in supposition (3).

To (6), the error in supposition that produced the 2nd term, or to the correction to be applied to the supposition that produced the 2nd term, *i. e.* 12; hence 12+6=18, the answer. Or in other words: The difference between the errors will be proportionate to the difference between the suppositions, as either error in the result, is to the error in the supposition which produced it. A little close thinking will make this plain.

But singular as it may appear, there are cases in which the errors in result continue proportionate to the errors in supposition, while the results themselves cease to be proportionate to the suppositions. Whenever the added or subtracted quantities bear the same ratio to the numbers from which they are produced, they will bear the same ratio to each other that those numbers do, and if the numbers and their additions bear the same ratio, their sums or differences will do the same. *e. g.* Take the numbers 12 and 18 and their thirds 4 and 6 will have to each other the same ratio; then will their sums 16 and 24 or their differences 8 and 12 have the same ratio, and this is the basis of working by a single supposition.

But to 12 and 18, add any arbitrary number, as 2, 3, 4, &c., and the results 14 and 20, 15 and 21, 16 and 22, &c.,

have no longer a similar ratio; and the same is true if we subtract.

What number becomes 20 by the addition of 4?

Suppose 12, then 12+4=16. (Error 4).

Then 16: 20 : : 12: 15, and 15+4=19.

Now if the same proportion had existed here that did in the former case, the last result would have been 20 instead of 19. Let us make another supposition and compare errors.

Suppose 14 is the number.

Then 14+4=18.

Here as the difference between errors, (2), is to the difference in suppositions (2), so is either error in the result (4 or 2), to the error in supposition that produced it. This must be true, whatever the numbers be, and is the principle which gives origin to the mode of operation in Double Position.

DOUBLE POSITION.

The proportion on which Single Position is based, exists as already remarked, only where the addition or subtraction is of some quantity whose ratio to the quantity sought is known; for then the results are proportionate to the numbers which produced them. But where the quantity sought is increased or diminished by an arbitrary quantity, we may by using two suppositions, avail ourselves of the above proportions; and *all questions that can be solved by Single Position,* can also be by *Double* Position. The following rule is deduced from the last of the preceding proportions.

Rule. Assume successively two numbers, performing on them the operations indicated by the question, and note the errors of the results. Then find the difference of the errors and say: As the difference thus found, is to the difference of the assumed numbers; so is either error, to the correction to be applied to the number which produced such error. The result will be the true number.

Example 1. What number being multiplied by 8, and having 40 added to the product, and the result divided by 6, will make 200?

Suppose 130. Then 130×8+40÷6=180, which taken from 200 leaves 20, *error* too *little.*

Suppose 136. Then 136×8+40÷6=188, which taken from 200 leaves 12, *error* too *little.* 20—12=8, difference of errors. 136—130=6, difference of supposed numbers. Then as 8, (difference of errors), : 6 (difference of suppositions) : : 20 (first error : 15 (the correction to be added to first supposition). Hence 130+15=145, the true number.

When the errors are alike *i. e.* both too great, or both too little, the difference will be found by subtracting one error from the other, but if one be too great and the other too little, then they must be added. This will be obvious if we consider that both aim at the same point, the true number; but one falls short, say 6, and the other passes beyond, say 4, their distance apart is certainly 6+4=10. It might be compared to finding the difference of latitude of two places situated on opposite sides of the equator. But if the one be 6 degrees and the other 4 from the line, and both on the same side, they will be but two asunder. We might remark, that whenever the errors are *equal* and *unlike, half* the *sum* of the suppositions is the true number.

Another rule is sometimes given as follows, Find the errors as before, and then multiply the first supposition into the second error, and the second supposition into the first error. Then if the errors were alike (*i. e.* both too great, or both too little) take their difference for a divisor, and the difference of the product for a dividend; the quotient resulting will be the number sought. But if the errors be unlike, take their *sum* for a divisor, and the *sum* of the products for a dividend, the quotient resulting will be the number sought.

In the preceding example, the first supposition was 130, the second 136. The first error was 20, the second 12, both too little.

Then, 136×20=2720
130×12=1560

8) 1160

145 *Ans.*

An example might be given and solved in which the errors would be unlike, but it is perhaps unnecessary, as the mode of operation is obvious.

The reason of the rule for multiplying the errors and suppositions crosswise, can scarcely be given without the aid of Algebra, as the rule is derived from an Algebraic formula.

Let a and b represent the two supposed numbers, r and s the corresponding errors, and x the true number sought. Then we have

As $x—a : x—b : : r : s$

20*

Mult. ex. and means $sx-sa=rx-rb$

By transposition $sx-rx=sa-rb$

Dividing by $s-r$ gives $x=\frac{sa-rb}{s-r}$, which translated into words, is precisely the rule when the errors are alike; as will be seen by substituting numbers for characters, or using words.

If both suppositions are too great, the expressions representing them will be $a-x$ and $b-x$, and the result will be the same as above. But if one be too great, and the other too small, it will resolve itself into $x=\frac{sa+rb}{r+s}$; which is the rule when the errors are unlike, *i. e.* "Divide the sum of the products by the sum of the errors."

In the above we might assign value to the several factors, and tell what the several products represent; but it would only darken the subject. The formula is fairly deduced, and that is sufficient.

We will now give a few

Questions for Exercise.

1. A son asked his father's age, the father replied, "Your age is 12 years, to which if $\frac{5}{8}$ of both our ages be added, the sum will be equal to mine." What was the father's age? *Ans.* 52 years.

2. What number is that which being increased by its half, its third, and 18 more, will be doubled? *Ans.* 108.

3. A son asking his father how old he was, was answered, "Your age is now $\frac{1}{3}$ of mine, but 5 years ago your age was only $\frac{1}{4}$ of mine." What were their ages? *Ans.* 45 and 15.

4. A and B have the same income. A saves a fifth part of his, but B by spending $50 per annum more than A, at the end of 4 years finds himself $100 in debt. What does each receive and spend per annum?

Ans. They receive $125, A spends $100, B $150.

5. A has $20; B has as many as A and half as many as C; and C has as many as A and B both. How many had they severally? *Ans.* A $20, B $60, C $80.

6. A gentleman gave his three daughters $10,000, of which the second was to have $1000 more than the first, and the

third as much as both the others. How much had each one? *Ans.* 1st $2000, 2d $3000, 3d $5000.

From the foregoing it appears that we have proportions by which we can determine numbers that have been increased or diminished by other numbers whose ratio to the first is known; or that have been increased or diminished by any arbitrary number, and thus far, the operations are as certain and based on principles as fixed, as any pertaining to mathematical science. But if the number is to be increased or diminished by some root or power of itself, or of any unknown number, then the rule fails, for the proportion no longer exists; no two numbers having the same ratio to either their root or powers respectively. Even in this case, however, we may by repeated operations, approximate the truth to any required degree of accuracy.

What number becomes 54 by adding to it the square of a third of itself?

Suppose	12	Suppose	15
Square of $\frac{1}{3}$ of 12=	16	Square of $\frac{1}{3}$ of 15=	25
	28		40
Error too little	26	Error too little	14

Then $26\times15=390$
$14\times12=168$

$12 \quad 12)\overline{222}$
$18\frac{1}{2}$

Here instead of 18, the result comes out $18\frac{1}{2}$. But why was it not accurate? Because equal quantities, (or rather quantities having the same ratio,) were not added to the suppositions, and hence the accuracy of proportion was destroyed. Still it approximates, and by taking this result and some number near the true one, the result will still more closely approximate on a second operation; and thus, step by step, we may approach the truth: like converging lines, however, which never meet, we can only approach. When the proportion is perfect, it is unimportant how wide the supposed numbers are of the truth; but where we can only approximate, the results are nearer the truth in *some degree* as the suppositions are.

Position will only approximate in questions involving powers or roots of the unknown quantity, and when the product or quotient of two or more unknown quantities is involved. The following, however, which contains several unknown quantities, can be solved by Double Position, notwithstanding many authors assert that no problem involving more than one unknown quantity can be solved in this way.

A said to B and C, give me half your money and I will have \$100; B said to A and C, give me one third of your money, and I shall have \$100; C said to A and B, give me one fourth of your money, and I shall have \$100. How much had each? *Ans.* A had $\$29\frac{7}{17}$; B $\$64\frac{12}{17}$; C $\$76\frac{8}{17}$.

Suppose A had \$24; then \$24 added to $\frac{1}{2}$ of what B and C had =\$100; and if these equal quantities be doubled, twice what A had (\$48) added to B's and C's =\$200; hence A's, B's and C's =\$176.

In like manner from the second statement, we find that three times B's added to A's and C's =\$300, and from these equalities, taking away \$176, which we before found to be the sum of A's, B's and C's, leaves twice B's =\$124; and hence B's =\$62; and \$176—62—24=\$90, =what was left for C.

Then substituting those numbers in the original statement, we have

$$\text{A's}=24+\frac{62+90}{2}=100.$$

$$\text{B's}=62+\frac{24+90}{3}=100.$$

$$\text{C's}=90+\frac{24+62}{4}=111\tfrac{1}{2}. \quad \text{Error}+11\tfrac{1}{2}.$$

Again, suppose A had \$36; then reasoning as above, what they all had =200—36=164, B's and C's share; and what they all had =300— twice what B had. Therefore B had \$68 and C \$60.

Substituting as before

$$36+(68+60\div 2)=100.$$
$$68+(36+60\div 3)=100.$$
$$60+(36+68\div 4)=86. \quad \text{Error}-14.$$

Then $24\times 14=336$
$36\times 11\frac{1}{2}=414$

$25\frac{1}{2})\ 750$

$\$29\frac{7}{17}$= A's money.

From this the other shares are readily found.

Other modes of solution, both by Position and Analysis, might be given.

The foregoing elucidation of this subject, it is hoped, will make it evident that Position is a strictly scientific rule, and worthy the attention of every one who wishes to be an accomplished arithmetician.

ALLIGATION.

The rule of calculating, by which we determine the value and proportions of compounds, is generally called in our books *Alligation;* probably from the Latin verb *alligare, to tie* or *connect together*, the numbers in some of the calculations being tied or linked together in the ordinary mode of solution. It is an ancient rule, and is supposed by Hutton to have been a part of the classification used by the Arabians. It involves no new principles, but like Barter, Fellowship, &c., is merely an application of the general principles of proportions; though from the arbitrary form on which the rules are generally given and the mysterious process of *linking*, it is seldom well understood.

Such calculations as aim to find the value of mixtures, from having the prices and quantities of materials, are usually classed under the head of Alligation Medial, the object being to find a *medium* or average price of the whole. There seems to be, however, no very good reason why such an operation should be called *alligation*, since there is no tying about it; but it has probably received that name because, like Alligation Alternate, it relates to mixtures. The two operations are the reverse of each other.

The subject of Alligation Medial is so perfectly simple that a single example may be sufficient to show its nature.

A merchant mixes 20 lbs. of Sugar, worth 8 cents per pound, with 20 lbs. at 9 cents; what is a pound of the mixture worth?

20 lbs. at 8 Cents=\$1.60
20 lbs. at 9 Cents= 1.80

40 pounds are worth \$3.40; and 1 pound is worth 8½ Cents. The reason of this is so entirely obvious, that we shall pass on at once to the reverse operation of ascertaining the proportionate quantities from knowing the value of the simples. Before, however, taking up the subject in the mode in which it is usually treated, we will briefly investigate, in a different manner, the law of mixtures.

What proportion of 10 cent Coffee must be mixed with 15 cent Coffee, that the mixture may be worth 12 cents per lb?

By Proportion. Assume some quantity, say 10 lbs. of the lower priced Coffee, which will be worth 20 cents less than the same amount of the proposed mixture, this deficiency must be neutralized by putting in the 15 cent article, each pound of which is 3 cents too valuable.

c. c. lb. lbs.

Hence: 3: 20 : : 1: 6⅔, the quantity of the highest priced necessary to be added.

By Analysis. As each pound of the higher priced will make the compound 3 cents too valuable, and each pound of the lower priced will reduce the value 2 cents, a pound and a half at 2 cents will reduce the value of the mixture 3 cents, and will neutralize one pound of the higher priced. Therefore, 1 lb. of the 15 cent and 1½ lb. of the 10 cent will make the mixture required.

Required to determine what 9 cent compounds can be made of Sugar at 5 cents, 7 cents, 10 cents and 12 cents per lb.

We may here form several single compounds.

1. Of the 5 and 10.
2. Of the 5 and 12.
3. Of the 7 and 10.
4. Of the 7 and 12.

I. Let us take, say 10 lbs. of the 5 cent, which will fall 40 cents short of the required value, and as each lb. of the 10 cent will exceed the average 1 cent, it will require 40 lbs. to neutralize the 10 lbs. of the cheaper article.

II. Into another vessel we will throw 10 lbs. of the 5 cent and neutralize it with the 12; and as each lb. of the 12 is 3 too rich, 13⅓ lbs. will neutralize the 10 lbs. of the 5 cent article.

III. We will then take 10 lbs. of the 7 cent and it will fall 20 cents below the proper amount, and this will require 20 lbs. of the 10 cent, as each lb. of that kind affects the value one cent.

IV. Again, taking 10 lbs. of the 7 cent, we will neutralize it with 6⅔ lbs. of the 12 cent, for reasons already stated.

We then have four mixtures, each of the right value, and we may mix or combine them as we please; and the mixtures thus formed will be of the right value.

For convenience the following mode of calculation is generally adopted.

Rule.—"Place the mean rate at the back of a brace, and the prices of the several simples in front, then connect each rate that is less than the mean rate with one or more that is greater, and set the difference between each rate and the mean rate opposite the number with which it is linked. If more than one difference stand opposite to any number, add the differences together; then will such single difference or sums of differences express the necessary quantities of the several simples." If any article is of the mean rate, it need not be linked, for whether much or little be put in, the value of the mixture will not be affected.

What proportions of 6 and 9 cent sugar must be mixed to form a compound worth 8 cents?

$$8\left\{\begin{array}{l}6\\9\end{array}\right\}\begin{array}{l}=1 \text{ lb.}\\=2 \text{ lbs.}\end{array}$$

Ans. 1 lb. of 6 to 2 of 9.

For by 1 lb. of the 6 the mixture is reduced as much as it is raised by 2 of 9. It is at once obvious that this operation gives only the ratios of the several ingredients; the absolute quantities vary according to the amount of the mixture. By setting the difference opposite to the number with which each is linked, the quantity is reversed as compared with the difference.

When there are but two simples, one greater and one less than the mean rate, they admit of but one combination, but when there are several simples at different prices, they may be variously combined, and hence several answers to such a question may be found, and all will be correct; for various mixtures of the same value may be made. If a merchant have sugar at 4, 6, 8, and 10 cents per lb. and wish to form a mixture of the whole worth 7 cents, it is entirely obvious that he may vary the quantity of either ingredient that is less or greater than the mean rate by using less or more of the other pair of ingredients. He may form a mixture of the 4 and 8 cent, that shall be worth 7 cents, and another of the 4 and 10 and still others of the 6 and 8 and 6 and 10; and these mixtures being all of the same value, he may form of them an infinity of combinations, by using various proportions of the several mixtures.

What proportions of 5, 6 and 10 cent sugar will make a mixture worth 8 cents?

$$8\left\{\begin{array}{c}5\\6\\10\end{array}\right\}\begin{array}{l}=2\\=2\\=3+2=5 \text{ lbs.}\end{array}$$

Here we have one kind above and two below the mean rate, and we must hence form two primary mixtures, by using 2 of the 5 and 3 of the 10; and 2 of the 6 to 2 of the 10. We then mix all together and it makes 9 pounds. We may consider in the practical operation 2 lbs. each of the 5 and 6 cent are thrown into a vessel, and then 3 pounds of the 10 to neutralize the 2 pounds of 5 cent; and 2 more to neutralize the 2 lbs. of 6 cent. One thing must be distinctly understood and remembered, that *every pair of numbers linked together, forms of itself a compound of the right proportions;* and if they do so separately, they will do so when taken together.

When there is one number greater or less than the mean rate, and two of a contrary character, but one result can be obtained, unless we vary the quantities of the primary mixtures. But as the numbers increase, the modes in which they may be alligated increase very rapidly. If there be two greater and two less than the mean rate, 7 different compounds may be produced by alligating the numbers differently. Take the case supposed a moment since in which a merchant wished to form a 7 cent compound of 4, 6, 8 and 10 cent sugar. These we may link or combine as follows—

First.

Mean	Rate	Quantity
7	4	=3 lbs.
	6	=1
	8	=1
	10	=3

Second.

Mean	Rate	Quantity
7	4	=1
	6	=3
	8	=3
	10	=1

Third.

Mean	Rate	Quantity
7	4	=1+3=4
	6	= 1
	8	=3+1=4
	10	= 3

Fourth.

Mean	Rate	Quantity
7	4	= 3
	6	=1+3=4
	8	= 1
	10	=3+1=4

Fifth.

Mean	Rate	Quantity
7	4	= 1
	6	=1+3=4
	8	=3+1=4
	10	= 1

Sixth.

Mean	Rate	Quantity
7	4	=1+3=4
	6	= 3
	8	= 3
	10	= 4

Seventh.

Mean	Rate	Quantity
7	4	=1+3=4
	6	=1+3=4
	8	=3+1=4
	10	=3+1=4

Instead of linking we may designate in any other way the kinds of which we would form our primitive mixtures, either by similar marks, or otherwise.

If a given quantity of a composition is to be made, we find the ratio of the ingredients as already shown, and then say: As the sum of the numbers expressing the ratios, Is to the number expressing the ratio of each ingredient, So is the required quantity, to the quantity of such ingredient. Or divide the required quantity by the sum of the ratios found by alligating as before, and multiply the resulting quotient by the ratios found by linking.

Required to make 80 lbs. of a 7 cent mixture out of 4, 6, 8 and 10 cent sugar. How much of each must we use?

We find by linking as before, that 3, 1, 1 and 3 lbs., respectively will answer the purpose. The sum of these is 8 lbs. instead of 80; we may therefore say,

As 8 lbs. : 80 lbs. : : 3 lbs. : 30 lbs.—the quantity of 4 cent sugar required. And this operation repeated for each kind will give the several quantities desired. Or 80÷8=10, showing that the whole quantity, and consequently each ingredient, must be multiplied by 10 to produce the required quantity.

Sometimes one of the ingredients is limited, and the rest must be adapted to it.

Required to mix 20 lbs. of 4 cent sugar, with other sugar at 6, 8 and 10 cents, so that the mixture may be worth 7 cents.

Alligate as before, then say

As 3 lbs. (the quantity of 4 cent found by linking,)

Is to 20 lbs. (what it should be;)

So is each of the other proportionate quantities,

To what each proportionate quantity should be.

If both quantity and price of some ingredients be given, and they are to be used with other ingredients of which only the price is given, find the value of the portion of the mixture made of the materials of which both price and quantity are given, and then link such mixture with the articles of which the price only is given.

Mix 36 gallons of wine at 24 cents per gallon, 8 at 52 cents, and 4 at 88 cents, with wine at $1.25, 86 cents, and 90 cents per gallon; and have the mixture worth $1 per gallon.

36	gallons at	24	cents=	$8.64
8	"	52	"	4.16
4	"	88	"	3.52
48				48)16.32

Average price per gallon 34 cents.

Then 100	34	=	25 gallons
	125	=66+14+10=	90
	86	=	25
	90	=	25

Then as we were requred to take 36+8+4=48 gallons to form the 34 cent mixture, and the linking gives but 25 gallons, we must increase the several quantities 25, 90, &c., in the ration of 25 to 48, thus—

As 25 : 48 : : 90 : 172$\frac{4}{5}$ gal. the quantity of wine at $1.25 per gallon. The quantity of the others will of course be 48, as the ratios of the quantities are all alike.

Sometimes a compound is required of a value inferior to any of the ingredients named, which strictly speaking would be impossible. But we may according to the admonition of the deacon in the story, "sand the sugar," "gravel the coffee," and "water the rum," and as these additions cost but little, we may set them down at 0 in the alligation, and proceed in every respect as directed in other cases.

In how many ways may the articles be combined when there are 4 greater and 5 less than the mean rate? Say the average price is 15 cents per lb., and the several simples are worth 8, 10, 12, 13, 14, 17, 19, 20 and 22 cents per lb.

Let us first combine of the 8 and 17
Then of the 8 " 19
" " 8 " 20
" " 8 " 22

Then of the 10 and 17
" " 10 " 19
" " 10 " 20
" " 10 " 22

Again of 12 and 17, 19, 20 and 22 respectively
And of 13 and 17, 19, 20 and 22 "
And of 14 and 17, 19, 20 and 22 "

Proceeding in the solution of the preceding problem, we may determine what proportions of each ingredient will constitute these twenty mixtures; and we may at the same time satisfy ourselves that there may always be as many *primitive* mixtures as there are units in the product of the number of simples less than the mean rate by the number that is greater. For with each one that is less we may make a mixture with each one that is greater respectively. It is proper however to remark that these compounds are formed of but *two* ingredients, and not of a portion of *all* the ingredients. The *mixing* of the *mixtures* is an operation that numbers can in no way calculate. But the answers usually obtained by *linking* suppose a portion of each simple to enter into the mixture.

The foregoing are all the principles that occur to us as involved in this subject; and we have endeavored to give them all the explanation necessary. It is not of much practical importance, though the principle may be sometimes applied very conveniently in the solution of problems.

HIERO, king of Syracuse, gave orders for a crown to be made entirely of pure gold, but suspecting the workman had debased it, by mixing with it silver or copper, he recommended the discovery of the fraud to the famous ARCHIMEDES, and desired to know the exact quantity of alloy in the crown. It is said that he was sorely puzzled on receiving the king's order, and as was his practice, retired to his bath to study upon it. On immersing his body a portion of water ran over the side of the bath; and the idea was at once clear to his mind. In the ecstacy of the moment, he sprang from the water, and ran naked through the city, exclaiming, "Eureka! Eureka!!" I have found it! The exclamation is often used in ridicule of a trifling discovery.

ARCHIMEDES, in order to detect the imposition, procured two other masses, one of pure gold, and the other of silver, or copper, and each of the same weight with the former; and by putting each separately into a vessel *full* of water, the quantity of water, expelled by them, determined their respective bulks; from which, and their given weights, it is easier to determine the quantities of gold and alloy in the crown by this case of Alligation, than by an Algebraic process.

Suppose the weight of each mass to have been 5 lbs., the weight of the water expelled by the alloy, 23 oz.; by the gold, 13 oz.; and by the crown 16 oz.; that is, that their respective bulks were as 23, 13 and 16; then, what were the quantities of gold and alloy respectively in the crown?

Here, the rates of the simples are 23 and 13, and of the compound 16, whence,

$$16\left\{\begin{array}{l}13=7 \text{ of gold}\\23=3 \text{ of alloy}\end{array}\right\}$$

And the sum of these is 7+3=10, which should have been but 5, hence,

$$\text{As } 10 : 5 :: \left\{\begin{array}{l}7 : 3\tfrac{1}{2} \text{ lbs. of gold}\\3 : 1\tfrac{1}{2} \text{ lbs. of alloy}\end{array}\right\} \text{ the } \textit{Answer}.$$

If the specific gravity of the compound was proportionate to that of the ingredients; and if there were but two metals in it, the above would be strictly correct.

Questions like the following admit of easy solution by Alligation, but it is not always easy to avoid fractions, even though integral numbers may answer the condition of the question.

Buy 100 head of Cows, Hogs and Sheep, and give \$10 apiece for Cows; \$1 for Hogs, and $16\frac{2}{3}$ cents for Sheep.

How many must there be of each to make 100 animals for $100.

Mean rate $1 { Cows, $10 ; Hogs, 1 ; Sheep, $\frac{1}{6}$ } Multiplying by 6 to clear the numbers of fractions before taking the differences, we have

Mean rate $6	Cows, $60	= 5 Cows	=	$50
	Hogs, 6	=41 Hogs	=	41
	Sheep, 1	=54 Sheep	=	9
		100 Animals worth		$100.

Having found the number of Cows and Sheep that will make an average of the proper price, the number of Hogs is determined by taking these numbers from 100. That this must produce the correct result, we may perceive by considering that the Cows and Sheep being linked together, must produce a combination that shall average $1 each, and hence will be worth just as many dollars as there are animals; and as the Hogs are worth *a dollar* each, as many hogs as will make the number 100 will make their value $100. The hogs being at just the mean rate cannot be linked, but as many can be "thrown in" as suits convenience. If 5 and 54 were not prime to each other, we might have as many additional results as there would be common measures of the numbers.

The question may be varied by supposing the Cows worth $10, Hogs $3, and Sheep 50 Cents, and $100 to purchase 100 animals.

Mean rate $2	Cows, $20	= 1
	Hogs, 6	= 1
	Sheep, 1	=18+4=22
		24

By this we know that 18 Sheep and 1 Cow will make a combination of 19, worth a dollar each on an average; and that 4 Sheep and 1 Hog will make 5 animals worth $5; and together they will make 24 animals worth $24. Then 100÷24=4$\frac{1}{6}$, the ratio in which they must severally be increased to make 100 animals.

1 Cow×4$\frac{1}{6}$=4$\frac{1}{6}$	Cows worth	41\frac{2}{3}$	at $10
1 Hog×4$\frac{1}{6}$=4$\frac{1}{6}$	Hogs worth	12$\frac{1}{2}$	at 3
22 Sheep×4$\frac{1}{6}$=91$\frac{2}{3}$	Sheep worth	45$\frac{5}{6}$	at 50 cents.
Ans. 100	Animals worth	$100	

Fractions of animals in such a question are absurd, but this is the only form in which these numbers can be linked, and if we take the numbers naturally resulting, fractions are inevitable. But we may take what multiple we choose of the *pairs*, and thus make up a hundred animals, since each combination is ever perfect in itself. Suppose then, instead of taking 1 Cow to 18 Sheep, and 1 Hog to 4 Sheep, we take 5 times the first combination, and 1 time the latter.

1st Com.	1 Cow×5=	5	Cows worth $50
	18 Sheep×5=	90	Sheep worth 45
2d Com.		1	Hog worth 3
		4	Sheep worth 2
		100	Animals worth $100

The value of an individual cannot be affected by multiplying both quantities of a combination, and we may very soon go through the whole range of results not exceeding 100 animals, and thus determine whether an answer in whole numbers is practicable; but it would be rather a matter of experiment than calculation based on any fixed rule. To obtain an integral answer to the above is a problem that has exercised the ingenuity of many, but I have never seen any mode by Algebra or otherwise, that seemed so simple as the foregoing.

LECTURE XVI.

PROBABILITIES, &c.

Our attention for the present will be occupied with the subject of Probabilities; only the main features of which we can hope to notice even hastily, in the compass of a single lecture.

When we look around us at results happening daily, of the causes of which we are ignorant, we are led to regard them as isolated incidents, subject to no rule or law; but could we see and understand the secret workings and connections exist ing between cause and effect, we might frequently discover that all works by rule. As it is, we may readily mark the boundaries, within which events must happen in very many instances; and do much to estimate their probability. We speak of *Chance* as something without plan or design, but taking in a large range, our calculations will approximate closely to the truth. When we throw a copper into the air, the chances of "heads or tails," as the boys say, are equal, and though one or the other may occur most frequently for a few throws, in a large number, say a thousand, the results will be about equally divided. In this case the sides of the coin must be equal in weight, else it will be like the grumbler's bread and butter:

"I never had a piece of bread,
Particularly good and wide,
But fell upon the sanded floor,
And always on the buttered side."

Had he put on less butter, perhaps the sides would have been more equal in weight, and the probability of the buttered side being uppermost would have been increased. Disturbing causes, unknown to us, may often shape the result; but in the absence of these, we may pretty accurately estimate our chances.

We see accidents from fire and flood, happening at times and points least expected; but the insurer has learned by observation to estimate probabilities, and by taking a wide range of country and a period of years, he does a comparatively safe business. Death takes the young and the old, but the life insurer has conned the bills of mortality, and studied the ages of those who have died, until he can estimate at once the probability of duration of life, and determine what he can afford to pay for an annuity contingent on life, or engage for a present sum, or an annual sum paid for life, to pay the heirs at the death of the insured. In one instance his estimate may fall short, and in another exceed, but the average will be about right.

So too the man who deals in lotteries and games of chance, knows the data and calculates carefully the probabilities, and though "luck" may sometimes be against him, his estimates of probabilities are based on mathematical principles, and he is secure in being ultimately the gaining party.

How these chances are calculated, depends on the data in each case, and it is not within the range of our present plan to attempt more than giving a general idea of the subject; and this with any one of ordinary prudence, will be sufficient to prevent all intermeddling with lotteries and every other species of gambling. The probabilities are always against the casual operator, even if all be conducted fairly; what then must they be when fraud and dishonesty are superadded? It is downright swindling!

In lottery schemes generally, fifteen per cent. is reserved as profit, but this is a small part of what may be secured; yet even this amounts to a great deal. If a man were to draw a prize nominally of $100,000, fifteen thousand would be deducted at once, and he would be entitled to only $85,000. It is true that in his good fortune he would not probably regard the abatement, but that does not change the principle.

In order to make the general principles of the subject intelligible, we will now take up briefly, *Variations* and *Combinations*, which form the basis of lottery schemes; and give also some estimates of Chances, that may impart an idea of that subject. He that would investigate these things thoroughly however, must look to full treatises written expressly on Probabilities.

First then, in regard to

VARIATIONS.

It is obvious that if we have a number of single things arranged in any order, we may change the arrangement into a variety of forms, and in doing so, we may take all together, or we may take only part at once. For instance, we may arrange the six vowels, a, e, i, o, u, y, in a great number of ways, as a e i o u y, a i e o u y, e a i o u y, &c., &c., or we may form them into groups, as ae, io, uy, ai, eu, oy; &c.; or, we may take three, four, five, or, as above, all at a time; and it is reasonable to suppose that the number of possible changes may, in all cases, be calculated.

When all are taken together, the operation is called *Permutation;* but if a part only be taken, it is called either a *Variation* or a *Combination;* a e, i o, u y, are distinct combinations, and are also considered one of the variations of two of which those six letters are susceptible; e a, o i, y u, are three other variations, but they are the same combinations; for a change of order will constitute a new variation, but not a new combination; hence the number of variations will always exceed the number of combinations.

The doctrine of variations and combinations forms the basis of many forms of Lotteries, and of other calculations used in practical life. We shall commence with the simplest form of variations in which all the articles are taken at once and which is called

PERMUTATION.

To determine the number of permutations, commence with unity and multiply by the successive terms of the natural series 1, 2, 3, &c., until the highest multiplier shall express the number of individual things. The last product will indicate the number of possible changes.

Example 1. How many changes can be made in the arrangement of 5 grains of corn, all of different colors, laid in a row? *Solution.* $1\times2\times3\times4\times5=120$, *Ans.*

This may seem improbable, the number being so great, but if there were but a single grain more, the possible changes would be 720; and another would extend the limit to 5040; and so onward in a constantly increasing ratio. The reason, however, will be obvious on a little scrutiny. If there were but one thing, as *a*, it would admit of but one position; but if two, as *a b*, it would admit of two positions, *ab*, *ba*. If three things, as *a b c*, then they will admit of $1\times2\times3=6$ changes, for the last two will admit of two variations, as *a b c*, *a c b*, and each of the three may successively be placed first, and two changes made to each of the others, so that $3\times2=6$, the number of possible changes. In the same way we may show that if there be four individual things, each one will be first in each of the six changes which the other three will undergo, and consequently, there will be 24 changes in all. In this way we might show that when there are 5 individual things, there will be 5 times as many changes as when there were but 4; and when 6, there will be 6 times as many changes as when there are only 5; and so on *ad infinitum*, according to the same law.

Example 2. In how many ways may a family of 10 persons seat themselves differently at dinner? *Ans.* 3628800.

When we consider that this would require a period of $9935\frac{55}{487}$ years, the mind is lost in astonishment. The story of the man who bought a horse at a farthing for the first nail in his shoe, a penny for the second, &c., is thrown into the shade; and we incline to doubt whether there is not some mistake; and yet on just such chances as one to all these do gamblers constantly risk their money!

Example 3. I have written the letters contained in the word N I M R O D on 6 cards; being one letter on each, and having thrown them confusedly into a hat, I am offered $10 to draw the cards successively, so as to spell the name correctly. What is my chance of success worth? *Ans.* $1\frac{7}{18}$ cents.

Case 2.

When several of the individual things are alike.

Rule. Find the number of permutations of individual things as above, as though all were different. Then find the number of permutations that could be made of the individuals of each separate kind that is repeated, then multiply together the several partial permutations, and divide the permutations of the whole by this product of the several partial permutations; the quotient will be the answer sought.

We may make the reason of this rule apparent by considering that three things, as *a b c* will admit of $\frac{1\times2\times3}{1}=6$ changes; but if two of them be alike, as *a a c*, then it will admit of but three changes, aac, aca, caa, $=\frac{1\times2\times3}{1\times2}$; and thus we might extend the series to any number.

Example 1. In the preceding question the name *Nimrod* is composed of six letters, all different, and we find that the probability of all the letters coming out in their natural order is as only 1 to 720; the word *persevere* is composed of 9 letters, and ought by the same rule to admit of 362880 changes, but, as some of the letters are alike, the number of changes is greatly reduced. How many can be made? $1\times2\times3\times4\times5\times6\times7\times8\times9=362880$, the number of changes if all were different. $1\times2\times3\times4=24$ permutations of e's; $1\times2=2$ permutations of r's, and $24\times2=48$ their product; $362880\div48=7560$ the number of possible changes.

It is clear if we had these letters placed on cards, the probability of drawing out an *r* would in the first instance be doubled, because there would be two cards with that inscription; and the probability of drawing out an *e* would be quadrupled; indeed the chances would be increased still more, for not only are the sought for cards increased, but the number of an opposite description is diminished.

Example 2. How many different numbers can be formed of the following figures, 1223334444? *Ans.* 12600.

Case 3.

When part only of the things are taken out at once.

Rule. Take a series of numbers, beginning at the number of things given, and decreasing by 1, as many times as the number of quantities to be taken at a time; the product of all the terms will be the answer required.

Example 1. How many different numbers of four figures each can be formed of the figures 1 to 8, inclusive, no two figures in the same number being alike?

$8\times7\times6\times5=1680$ *Ans.*

Example 2. How many different words of 8 letters each can be made of the 26 letters of the alphabet, allowing every different arrangement to make a distinct word, without regard to vowels or consonants? *Ans.* 62990928000.

We may illustrate this rule by giving the various arrangements of two letters each in the word H A R D.

Here we have $4\times3=12$ *Ans.* And by actual experiment we have HA, HR, HD; AH, AR, AD; RH, RA, RD; DH, DA, DR; each letter successively combining with each of the other three; or as we might say, each of the 4 leading in 3 changes, and hence making 4 times $3=12$ changes. If there were 5 letters, as H A R D Y, it is obvious that each of the five would combine with each of the other four, making $5\times4=20$ changes, and so on with any number of individuals. Hence the reason of the rule is manifest.

Case 4.

Variations with Repetitions. In this case every different arrangement of individual things, including repetitions, is called a Variation, and, like Combinations, the *class* of the variation is denoted by the number of individual things taken at a time.

Rule. Raise the number denoting the individual things to a power whose exponent is the number expressing the class of the Variation.

Illustration. Let us make a variation of the second class of three things, *a b c*. Here we have aa, ab, ac, ba, bb, bc, ca, cb, cc. Each one of the letters leads in a variation with the others, including itself, and of course there must be $3\times3=9$ variations. So if we make a variation of the third class of three things, *a b c*, we shall find that *a* will lead in 9 variations, and as each of the others will do the same, there will be $3\times9=27=3^3$ variations. The law of increase is hence manifest.

Example 1. How many variations with repetitions of the 4th class can be formed out of 5 individual things. *Ans.* $5^4=625$.

Example 2. How many numbers of 9 places of figures each can we form out of the 9 digits, provided we are allowed to make a repetition of figures? *Ans.* $9^9=387420489$.

COMBINATIONS.

If we have a number of things, it is obvious that we may parcel them out into groups or combinations, and our present object is to determine the number of such combinations that may be made of any given number of things taken in given parcels. We shall first consider,

Combinations without Repetitions. In this case the repetition of an individual thing, as *aa*, is not considered a combination. It should also be premised that the number of things taken at a time indicates the class of a combination. If two things be taken, the combination will be of the second clsss; if three, of the third class, &c.

Second Class. When two things are combined.

Rule. Find the sum of the natural numbers, 1, 2, 3, &c., to as many terms *less one*, as there are things to be combined.

Reason. Two things admit of one combination, add another and it will unite with each of the others, making two additional combinations; so four will make three more; five, four more, &c.

Third Class.—Rule. Find the sum of the series, 1, 3, 6, 10, &c., to as many terms, less *two*, as there are things to be combined.

Reason. Three things will form one combination; four will form three additional ones; five will make six more, and so on. We subtract *two*, because the first three individuals only make one combination.

General Rule. From the number of individual things subtract one less than the class of the combination, and find the sum of as many terms of the series belonging to the class as there are units in the remainder, such sum will express the number of combinations.

We might easily show the correctness of the rule in any particular case, but the foregoing illustration is deemed sufficient. The following are a few of the series referred to above.

2d Class,	1,	2,	3,	4,	5,	6,	7,	8,	9,	&c.
3d "	1,	3,	6,	10,	15,	21,	28,	36,	45,	&c.
4th "	1,	4,	10,	20,	35,	56,	84,	120,	165,	&c.
5th "	1,	5,	15,	35,	70,	126,	210,	330,	495,	&c.
6th "	1,	6,	21,	56,	126,	252,	462,	792,	1287,	&c.

Here the law of formation is obvious, each term being formed of the sum of a corresponding number of terms of the preceding series. Thus the fourth term of the series belonging to the third class is the sum of the first four terms of the series belonging to the second class; and the same law holds good *ad infinitum*.

To save calculation we may, instead of finding the sum of the series, take the corresponding term of the series belonging to the next higher class.

Example 1. How many different numbers, of 4 figures each, may be expressed by the figures from 1 to 8, inclusive; no two of the numbers having all their figures alike?

Here 8—(4—1)=5, the number of terms of the 4th class whose sum, 70, or the 5th term of the 5th class series is the answer sought.

Example 2. How many combinations without repetitions of the 5th class can be formed of 12 different things?

12—(5—1)=8. Here 792, or the 8th term of the 6th class series is the number.

Where large numbers are concerned, the following rule is more convenient in application than the foregoing, but it rests on the same general principle.

"From the number of individual things, subtract the number denoting the class of the combination, *less one;* multiplying this remainder by the successive increasing terms of the series of natural numbers until we reach the term denoting the number of individual things; then divide this product by the number of permutations of a number of individual things denoted by the class of the combination."

Example 3. In order to form a lottery scheme I have put into the wheel as many cards as I can put 4 letters of the word Charleston on, without having the same letters on any two cards. I offer $100 to him who draws the cards having on them the first 4 letters of the word in any order whatever; what is a chance of drawing the prize worth?

Ans. Only 47$\frac{13}{21}$ cents.

Combinations with Repetitions. In this case the repetition of an individual is considered a new combination. Thus *ab* admit of but one combination if we do not repeat, but if we do we can form three combinations, viz, *aa*, *ab*, *bb*. The following rule is deduced in a manner similar to the one in the preceding case.

Rule. The number of combinations will be denoted by the sum of as many terms of the series belonging to the class as there are individual things.

How many combinations of the 5th class can be formed of 9 individual things?

According to the rule it will be the sum of 9 terms of the

series belonging to the fifth class; or what is the same the 9th term of the sixth class series=1287.

Example 2. How many different numbers of four places of figures each can be formed out of the nine digits?

Ans. 495.

The following rule is more readily applied to large numbers than the preceding, but it is not so easily illustrated.

To the number of individual things add the number denoting the class of the combination, *less one*, multiply the sum successively by the decreasing terms of the series of the natural numbers until we reach the term denoting the number of individual things; then divide this product by the number of permutations of a number of individual things denoted by the class of the combination.

Example 3. How many different combinations of six things at a time can be formed out of 11 individual things?

Ans. 8008.

In regard to the number of combinations that can be made out of any given number of single things, it increases with the number taken at a time until you reach half the whole number of things, after which the number of combinations will decrease, since the multipliers of the divisor will be larger afterwards than the corresponding multipliers of the dividend. And if we pass on increasing the number in a combination until the whole are taken at a time the divisor and dividend will be the same and there will be but one combination.

The number of variations will continue to increase as the number in the group increases, for there is no divisor to counteract by its greater increase, the increased product of what becomes the dividend in calculating combinations. The number of variations will be greatest when the whole number of things is taken at once, and the number of combinations will then be least, for whether the number be great or small it can form but one combination.

The preceding are the most important portions of the doctrines of Variation and Combination; and though there are portions of the subject which we have entirely omitted and others which have been but cursorily noticed, we shall now proceed to consider another field of inquiry in which the doctrines of Permutation, Combination, &c., find their most important applications. We allude to the subject of Probabilities.

PROBABILITIES.

In considering any future event, we are generally unable to determine whether or not it will happen; yet we can often ascertain the cases that are possible, and of these how many favor the production of the event in question. In our uncertainty we say that there is a *chance* it will happen; and thus our idea of chance arises from our wanting data, which might enable us to decide whether or not the event will take place. If for instance a bag contain one white and two black balls, it is impossible to decide whether or not a black ball will be drawn at one trial; but we know that there are three cases possible, of which two favor the appearance of a black ball, and one the contrary, and of this we have no reason to think one more probable than another.

Simpson defines the probability of an event to be the ratio of the chances by which the event in question may happen to all the chances by which it may happen or fail. If a bag contain no white and ten black balls, the probability of drawing a white ball is evidently 0; if on the other hand, the bag contain 10 white and no black balls, the probability of drawing a white one is unity or is a certainty; and whatever be the number of black balls the probability of drawing a white one must be some fraction between 0 and 1, which are its limits. When the fraction which expresses the probability of an event is little different from unity, we say the event is very probable or nearly certain; when it is but little greater than $\frac{1}{2}$ we say it is probable; when $\frac{1}{2}$ doubtful; when rather less than $\frac{1}{2}$ improbable; when much less than $\frac{1}{2}$ very improbable; and when 0 impossible.

The first author who is known to have written upon the subject is Galileo, who died in 1642. After him Pascal, Fermat, and other continental mathematicians bestowed some attention upon the subject. From the earliest periods a prejudice has existed against it on account of its ready applicability to games of cards and dice. An anonymous writer who in 1692 published the first English essay "Of the Laws of Chance," deemed it necessary to protest in his preface that the design of his book was "not to teach the art of playing dice, but to deal with them as with other epidemic distempers, and perhaps persuade a raw squire to keep his money in his pocket."

In later years Demoivre, James and Daniel Bernouilli, Leibnitz and otners lent their aid to the calculus of chances, and the extensive application of its principles to the calculations connected with Life Insurances, Annuities, and indeed to every kindred risk, has given the subject an importance that demands for it a closer and more extensive investigation than our present plan will permit.

If $m+n$ be the whole number of cases and m is the number of cases favorable to an event the probability of such event happening may be expressed as $\frac{m}{m+n}$, and the chances against it $\frac{n}{m+n}$; the sum of the fractions expressing the chances for and against an event being always unity.

Suppose a coin be thrown up, having two faces, what is the probability that the obverse (heads) side will fall upward, and what the reverse?

Here there are only two possible cases, and one favors each of the contingencies the probability of each will be $\frac{1}{1+1}=\frac{1}{2}$; there being no reason why one side should fall uppermost rather than the other.

What would be the probability of either side presenting upwards twice in two throws?

Here we have 4 possible cases, viz:

Obverse and reverse
Obverse both times
Reverse and obverse
Reverse both times.

Of the 4 possibilities there is only one which favors the turning up of the obverse twice in succession, and the same is true of the reverse, hence the probability of either is only $\frac{1}{4}$.

In like manner we might show that the probability of the obverse presenting upwards three times in succession will be $\frac{1}{8}$, or $\frac{1}{2}\times\frac{1}{2}\times\frac{1}{2}$; the general principle being to multiply successively together the independent probabilities of an event for the fraction expressing the chance of all the events happening.

It is required to determine the probability of an event happening *once* and *no more* in two trials.

It may happen the first time and fail the second, or fail at the first and happen at the second, and if m denote the probability of success, and n that of failure at each trial, the chance for its happening the first time is $\frac{m}{m+n}$, and of failing the second $\frac{n}{m+n}$. Hence the chance of its happening the first and failing the second is $\frac{mn}{(m+n)^2}$ and the same for failing the first and happening the second, therefore the chance of the event happening once, and only once in two trials, is expressed by $\frac{2mn}{(m+n)^2}$.

It is sometimes required to find the probability of an event happening at least a given number of times, without limiting the number, the chances of happening and failing each time being given.

What is the chance of an event happening once in 2 trials, when the chances of its happening each time are m and the chances of failure n.

This may take place in three ways; it may happen the first and fail the second; happen the second and fail the first, or happen both times. Hence the chances is $\frac{mn}{(m+n)^2}+\frac{mn}{(m+n)^2}$ $+\frac{m^2}{(m+n)^2}$ or $\frac{m^2+2\,mn}{(m+n)^2}$.

We might extend the solution of the last two questions so as to embrace the recurrence of the event more than once, and also increase the supposed number of trials, but it would exclude more valuable matter.

The investigation of this portion of our subject might be pursued farther and show how the probability of an event may be calculated, whether dependent on contingencies similar to the foregoing or still more complex. But we prefer directing the attention of the reader for a brief space to the subject as connected with Annuities and Lotteries.

We have remarked in another lecture, when speaking of the subject of Insurance, &c., that the calculations connected with Life Insurance, Life Annuities, and the value of reversionary claims, were based on tables of mortality, as they are styled, in which the ages of persons who die are registered. As a necessary consequence, the value of such claims as those mentioned above varies with the age of the individual, and we shall now give a few problems in relation to this subject.

Example 1. A person 30 years of age has an annuity for 10 years, the present worth of which is $1000, provided he lives but the 10 years, for, if he dies, the annuity ceases. What is the annuity worth, as it is ascertained that about 75 out of every 4385 persons die annually between the ages of 30 and 40 years? *Ans.* $826 $\frac{598}{877}$.

If 75 persons die in one year, in 10 years 750 would die, and 4385—750=3635 would probably be living. Hence,

As 4385 : 3635 : : \$1000 : $\$826\frac{598}{877}$.

Example 2. A, who was 70 years of age, had an annuity which was to last 10 years, provided he lived until the end of that time. B gave him for it as a fair price \$1250; but he has forgotten what A was to receive annually. Now between the ages of 70 and 80, 80 persons die out of 832 on an average. What then was A's annuity worth, in hand, provided his life had been secured 10 years? *Ans.* $\$1129\frac{193}{208}$.

Example 3. In order to form a lottery scheme, I have put into the wheel as many cards as I can put 4 letters of the word Charleston on, without having the same letters in the same order upon any two cards. I offer \$100 to him who draws the card having on it the first four letters of the said word in their natural order (Char). What is the chance of drawing a prize worth?

There are 10 letters in the word, and the combination is of the 4th class; and, according to the mode of determining combinations with repetitions previously elucidated, we find the whole number of combinations of the 4th class which the word admits of is 210. Then he has one chance in 210 of drawing the letters Char, in *some* order. The number of permutations of 4 individual things is $1\times2\times3\times4=24$, and $210\times24=5040$ his chance of drawing them in the right order, and \$100 divided by 5040 gives *Ans.* $1\frac{62}{63}$ Cents.

Suppose that the numbers from 1 to 78, inclusive, be placed upon 78 cards, and the cards placed in a wheel by which they are thoroughly mixed; and then 13 cards be successively drawn out, by a person who has no means of choosing, and the numbers on them registered. Suppose also that tickets have been issued, containing each three of the 78 numbers, but no two having *all* the same numbers, and that he who holds the ticket having on it the first three drawn numbers in their regular order, shall be entitled to \$100,000; what would the probability of drawing such a ticket be worth? *Ans.* $21\frac{5183}{5858}$ Cents.

Note.—It is usual also, to give smaller prizes to the holders of tickets having the numbers in any order, or having any two or one of the drawn numbers. Lotteries may be arranged on a great diversity of plans, and in each the probability of drawing prizes will vary.

A speaks the truth 3 times in 4; B. 4 times in 5, and C 6 times in 7. What is the probability of an event which A and B assert, and C denies? *Ans.* $\frac{140}{143}$.

If there be 4 white balls and 6 black ones in a hat, what is the chance of drawing out 2 black balls at two successive trials? *Ans.* $\frac{2}{15}$.

For further information on this subject, consult Liebnitz, Bernouilli, and some monographs on the subject, published in the Transactions of the Royal Society of London.

LECTURE XVII.

ARITHMETICAL ALGORITHM, SYNTHESIS, ANALYSIS, FORMATION OF RULES FOR SOLVING PROBLEMS, PROOFS, CONTRACTIONS, &c.

We might find ample material for a long lecture, in tracing the changes that have marked successive ages, in the algorithm or mode of arithmetical calculation; but the result would be rather speculative than practical, and be less acceptable to some than the solution and illustration of problems, which will soon occupy our attention.

We alluded in our first lecture to the primary modes of calculation by means of sensible objects, as pebbles, counters, &c., and experience has taught us that a great degree of skill and accuracy may soon be acquired in the use of such means. They are used by many nations even to the present day; and Professor Leslie, in his Philosophy of Arithmetic, treats extensively on the subject, under the head of *Palpable Arithmetic.* The Greeks and the Romans advanced a step farther and adopted letters as signs of numbers. On this also we have dwelt perhaps sufficiently in detail, and we allude to it and the palpable modes in this connection, rather as introductory to a recent change, that has revolutionized the algorithm of our forefathers, and has not been devoid of many advantages. We allude to the Pestallozian system, and that modification of it, called in our country the Analytic system. It is long since the commencement of the historic period, that the Arabian system was adopted in Great Britain, and as the circle of ordinary school studies was then very much circumscribed, being usually limited to Reading, Writing and Arithmetic, the latter was an important branch of knowledge, and for ages was communicated principally by the oral instruction of the teacher, for whose use alone, the few books that were published, were designed. In those days the doctrines of mildness and moral influence, were little thought of, and the teacher was emphatically a school *master*. Tedious and intricate problems were given out by the teacher, and pored over for days by the pupil, who was expected to solve them with the smallest possible amount of assistance. It was almost high treason against the master's dignity to ask a second time for instruction; while the application of the birch most liberally was the panacea for treachery of memory or deficiency in natural aptness. And when instruction was given, it was as to the mode of operation, without a why or wherefore. The rules of operation being such generally as were framed from algebraic formulæ, or drawn by its aid from the less obvious principles of the science, and blindly followed, by far the greater number of learners, who appeared to regard the rationale of the rule, as something with which they should not meddle; and the doctrine was common, that none could understand the rationale of arithmetic without a knowledge of algebra.

In this way the study of Arithmetic became little more than learning by rote a set of arbitrary rules, without any knowledge whatever of the principles on which they were founded; and he was the most expert arithmetician who could apply those rules most dextrously in producing results. He no more thought of investigating the rules, than the clown does of studying the internal structure of his watch. He looked only to the way the hands pointed.

The practice of putting books on the subject into the hands of learners, came gradually into use, but still they were brief treatises that gave only dogmatical rules, few of which were explained on principle; and these were the only books of the kind known, in schools, long since our school boy days.

Early in the present century, M. Pestallozi, of Switzerland, perceiving the defects of the system pursued, opened a school in that country, and engaged in the cause of practical education. He pursued this course for years, with great success, and his labors resulted in changing to a great extent, the policy pursued both in Europe and America. He aimed to take nature's method, and having orally taught his pupils to *Think*, they were then taught to *Write*, before learning to *Read;* for said he, "The first one who read, must have written before he had any thing to read." In arithmetic he commenced with the simplest possible forms of enumeration and calculation, and thus made the mind familiar with questions that it could comprehend. Proceeding from these to others more difficult, the operator learned to think for himself, and to analyze each problem, without reference to any general rule: and from the analysis of particular problems, general principles were inferred.

His notions were extremely radical, and his system not adapted in its full extent to the wants of practical life; but he had broken the old routine, and men saw that the young mind was capable of thinking and inferring for itself; and hence sprung up the system of Prussia and other European countries.

In 1805 Joseph Neef, a coadjutor of Pestallozi, was induced to emigrate to the United States, and to establish a school on the Pestallozian system near Philadelphia. But it did not suit the genius of our people, and after a few years it was abandoned; but something of its spirit had gone abroad, and the defects of the old arbitrary system had become so glaring that teachers of enterprise, were pleased with the prospect of finding something better adapted to the wants of the world.

About the year 1820, Warren Colburn, of Massachusetts, published a treatise on Arithmetic, which he called an "*Introduction to Arithmetic on the Inductive or Intellectual System.*" He adopted many of Pestallozi's notions, and his system rose rapidly into favor; securing, as success always does, hosts of imitators. Many of these have sought to combine the best features of the old system and the new, until now there is every degree of admixture, from the purely *Inductive* system of Colburn, who gives no rules, but solves every thing on its own merits; to the old dictatorial systems that deal in rules alone, without troubling the learner about whys or wherefores. We must not confound Warren Colburn with Zerah Colburn, the calculator, hereafter spoken of.

It is said that Colburn complained much in his latter years that others had robbed him of his system, and destroyed its beauties. The plan of Pestallozi and his imitators, has been frequently called the Mental system, perhaps from the fact that they all commence with oral exercises, of a very simple character, and do not resort to the pen and pencil, until somewhat advanced in study. They distinguish the subject into *Mental* and *Written* Arithmetic; and the whole system has with many taken the name of the merely introductory exercises.

We shall not attempt to institute a detailed comparison between the Analytic system and the Synthetic; but a few remarks may be profitable. In the old system, sometimes called the Synthetic, the rules of operation having been framed from general principles, are laid down for the student's use, and he is required to solve his individual problems, in accordance with those general rules. It is evident that he may become very expert in the application of his rules, without understanding the principles on which they are based; but it does not follow that he cannot become acquainted with them, neither that he should not; and if he does thoroughly investigate their principles, he must be master of his subject, and in their application afterwards the reason will in his mind follow the rule as the shadow does the substance.

Pestallozi, and even Colburn, used no rules, but left every thing to be inferred. This may be well enough for the schoolmaster, who is engaged daily in the business of calculation, but will not do in practical life, in which men will naturally forget many things; and in their active pursuits they have no time to examine a train of causes and effects. In such case rules may be remembered and promptly applied, when a long train of reasoning is impracticable. Rules may be reviewed at any time, and the memory refreshed. We have often thought that if those who prepare books and teach, had greater familiarity with practical life, it would be better for their pupils.

In the analytic system, general rules are inferred from the examination of particular cases, instead of being drawn from general principles. This mode does not seem entirely satisfactory. If an inference be drawn from a single result, certainly it is strengthened by each corroborating result; but still when ninety-nine special results have raised a very strong probability, what conclusive evidence can the operator have that the hundredth particular case may not be at war with all its predecessors?

The mind is not fully satisfied with a strong probability—it asks something conclusive—something that precludes the possibility of an excepted case. It is true that this inference of general truths from special results and facts, is the mode pursued in studying the phenomena of nature; but this arises from the necessity of the case, since we have no means of arriving at general principles, but by considering a great variety of individual cases; and it by no means follows, that because it is the best plan in that instance, it should be pursued in all others. In Arithmetic and Geometry we have the means of arriving most conclusively at general truths; and though particular facts and cases may be adduced to *illustrate* them, they are not needed to *prove* them.

Which system then, is most profitable to the student? Which most profitable to the man in after life?

These are questions on which there may well be difference of opinion. In the first place, mental, or more properly, oral exercises may be readily carried on by either system, and, to a certain extent, they are profitable, especially with young students. Either system may be used also with the pen or pencil. It is only because they who use the analytic have adopted oral exercises as introductory, that they have become identified in the view of many with that system.

The friends of the inductive mode claim that the development is from within; while instruction on the Synthetic mode is from without. On the other hand it is contended that though the principles and the rules are from the pen of an another, the student, by proper mental labor, may make them his own, and understand them as thoroughly as though he had invented them; and with an immense saving of time. Every one must understand that for each one to invent every science for himself, instead of availing himself of the accumulated wisdom of ages, is to require that man should spend his lifetime in preparing to live. Ere we could pass the threshhold of a few every day studies, the business period of life would be upon us—and soon the "sere and yellow leaf."

Some have thought that the Analytic system is best adapted to first study, and the Synthetic to review; for as the former is connected in a continued train of premises and inferences, it is scarcely practicable to refresh the memory afterwards by consulting a single point. On the other hand, the Synthetic mode admits of distinct, and to some extent, disconnected classification; so that any point may be examined at pleasure. We have sometimes thought, indeed, that some Analytic authors delighted in presenting a tangled web completely interwoven from end to end. Such a system cannot be well adapted to the wants of youth of limited school opportunities, and they form the mass of the youth of the country. They certainly require brief and clearly arranged practical treatises, adapted to their real wants, and such as they may consult with profit after leaving school. That such a system would not be as well for all does not clearly appear; for it is undoubtedly the superficial or thorough mode of study, not the system, that makes a scholar superficial or profound.

One objection to the Analytic system is the great length of time required to complete the study; thus crowding out other things of greater value. Since English Grammar, Geography and the principles of Natural Philosophy, have become every day studies, Arithmetic should not occupy the space it did when, with the Reading and Writing, it formed the entire circle of school science: and since Algebra is so much simplified, many of the more intricate parts of Arithmetic are rendered of little account. Indeed the boy was not far wrong when he concluded that some parts of Arithmetic were very much like the Irishman's horse, "Hard to catch, and worth very little when caught."

Were we to give our own opinion as to the best form of presenting the subject of Arithmetic to such as expect to be engaged in the practical pursuits of life, we would present it in the old Synthetic form, and in a shape as nearly adapted to the wants of life as practicable; and with thorough explanation of all the principles involved; but if for the youth of greater leisure, we might give the analytic and synthetic combined; but even then we would prefer to pass on early to the simpler portions of Algebra, as affording better modes of solution and greater exercise in developing a course of mental discipline. Many analytic solutions are but algebra stripped of its dress, and in a much more difficult form. For the purposes of life, the Synthetic cannot be profitably dispensed with; and should be retained if either be omitted.

We will now discuss briefly the formation of rules. It has been said that the arithmetical analyst solves particular examples, and hence infers general principles; how this is done may be understood when we come to the analytic solution of problems. Colburn gives the following as his general mode of analyzing: "In all cases, reason from many to one, or from a part to one; and from one to many or a part. If several parts be given, always reason from them to one part; and then to many parts or the whole."

Rules for the solution of problems, as we generally find them in synthetic arithmetics, are framed from the obvious properties of numbers, or from algebraic formulæ; and if having reference to geometrical quantities, the principles of geometry will be involved.

As an instance of rules, based on obvious arithmetical principles, we may cite the reduction of compound quantities, as well as the rules of fractional quantities generally. For nothing can be more obvious than that, as twenty shillings make a pound, we should, in order to change pounds to shillings, "Multiply the number of pounds by 20 for the number of shillings."

The rule for stating questions in the Rule of Proportion, or, as familiarly called, the Rule of Three, are based on principles not so simple certainly as those referred to above, but still such as need no aid in investigation beyond common arithmetic. He who looks carefully into the subject, will see the reason of each step, without the aid of Algebra or Geometry.

It is not necessary for us to spend time in the farther investigation of Proportion, as that subject was fully treated of in our seventh lecture. Fellowship, Barter, Loss and Gain, Interest and several other branches of this subject, are only an application of this principle to different business transactions, and involve no new scientific principle whatever.

The rule for extracting the square root, was explained in our eighth lecture; and in our ninth we gave a number of rules based on the relation of quantities, all of which we explained as fully as we thought proper. They involve the relation of geometrical quantities; as do also the rules referred to in our thirteenth lecture. It would be impossible, for instance, to understand thoroughly, without some knowledge of Geometry, the following simple rule for finding the area of a circle, when the diameter is given: "Square the diameter of the given circle, and multiply by .7854 for the area."

Yet admitting that circles are to each other as the squares of their diameters, the reason is obvious enough.

The Permutation and Combination of quantities, and the doctrine of chances, depend on principles fully set forth in our sixteenth lecture; and the peculiar modification of proportion, or rather the expedients resorted to sometimes in order to obtain a proportion, are set forth in our fifteenth, under the head of Position; while Alligation is but an obvious application of the same general doctrine of proportion. There is nothing in all the foregoing but what a little attention will make plain; but suppose the student were asked to frame a rule for solving such questions as the following:

Sold a horse for $56, and gained as much per cent. as the horse cost me. Required the cost.

Here the doctrine of proportion fails, neither can you institute a proportion between the errors, if numbers be supposed; and yet the following rule will solve the question:

"Multiply the selling price by 100, and add 2500 to the product; of the sum extract the square root, and from the root subtract 50. The remainder will be the prime cost."

Horse sold for \$56
100
———
5600
+2500
———
$\sqrt{8100}=90$
—50
———
Leaves \$40 cost.

Proof—
Cost \$40
Per cent 40
———
Gain 16.00
Cost 40
———
Sold for \$56
———

Human ingenuity would perhaps fail to find a reason for the above rule, by the aid of common arithmetic merely, or to explain the steps satisfactorily to a learner. It seems to be without reason, and yet it will solve all questions involving a similar principle. Take another instance;

I sold a lot for \$96, by which my gain per cent. was equal to the original cost. Required the cost of the lot?

\$96×100=9600
2500
———
$\sqrt{12100}=110$
— 50
———
Cost \$60

The above rule is formed by solving the question algebraically, and then changing the formula into words; thus—

Let x represent cost; and a the selling price.

Then $x+\frac{x^2}{100}=a$

Mult. to clear fractions $100x+x^2=100a$

By trans. $x^2+100x=100a$

Comp. sq. $x^2+100x+2500=100a+2500$

By Evolution $x+50=\sqrt{100a+2500}$

Hence $x=\sqrt{100a+2500}-50$

Which changed to words gives the rule we have laid down. But to resort to algebra to frame a rule for each class of problems, that they may be solved by arithmetic, is very much like applying to a tailor to cut a paper pattern for an old woman to cut your breeches by.

In one mode of working Double Position, you are directed

to multiply the errors and suppositions crosswise. This is based on an algebraic formula, given under the head of Position.

Let a rule be framed for solving such questions as the following:—

A certain field is 15 rods wide, and how long we know not; but we know that if 100 square rods were added to the side, the field would be square. How many acres does the field contain?

Let x represent the side of the field; a the given width; and c the proposed addition.

Then by the question $x^2=ax+c$

By trans. $x^2-ax=c$

Comp. sq. $x^2-ax+\frac{1}{4}a^2=c+\frac{1}{4}a^2$

By evolution $x-\frac{1}{2}a=\sqrt{c+\frac{1}{4}a^2}$

By trans. $x=\sqrt{c+\frac{1}{4}a^2}+\frac{1}{2}a$, the formula sought; which may be thus expressed in words:—"To the given addition add the square of half the width, and extract the square root of the sum; then add to the root one half the width, and the sum will be one side of the square."

Solution.—Given addition 100 rods
$\frac{1}{2}$ the width squared 56.25

$\sqrt{156.25}=12.5$
$+7.5$

20 rods, one side of the square, and $20\times20=400$ rods$=2\frac{1}{2}$ acres.

It is useless to extend these rules, as their number would know no limit. We will close the subject of rules, therefore, by giving the "*Land Question*" with a numeral rule, found in many books.

1. A and B gave \$600 for 300 acres of land, they paying equally. In dividing the land according to quality, it was agreed that A pay 75 cents per acre more than B. How much land did each receive, and what did he pay per acre?

Ans. A got 122.8 acres at \$2.44.3+ B got 177.2 acres, and paid \$1.69.3+ per acre.

2. A, B and C each contributed \$200 to a common stock, and went to the west to buy land. They bought 300 acres, and in dividing it equitably it was determined that A allow 75

cents per acre more for his land than B and C gave. How many acres did each receive, and at what price?

Ans. A paid \$2.55.4+ per acre and got 78.3+ acres; the others paid \$1.80.4+ and got 110.849 acres each.

We give the foregoing questions because they are often alluded to as not being solvable by common arithmetic; and certainly they are not directly so by Position, since they involve a quadratic equation; but they may be by the subjoined process. In solving the second by this rule, we must consider the shares of B and C as one, its value being double that of A.

Rule. From the whole amount paid for the land, take the value at the given difference per acre; and reserve the remainder. Multiply the aforesaid value at the difference in price, by four times the whole amount paid by him who paid least per acre, and to the product add the square of the reserved remainder: of the sum extract the square root, and to it add the reserved remainder. Divide the sum by twice the whole number of acres, and the quotient will be the price per acre paid by him who paid the least per acre. The quantity is then easily found.

In either case there will be an interminable decimal, which cannot be expressed as a vulgar fraction, since it arises in extracting a root; and the same is true when the questions are wrought by Algebra. They may however be approximated to any extent.

To solve the second, we say A paid \$200, and the others \$400. Then 300 acres at 75 cents=\$225; and \$600—\$225 =\$375, reserved remainder. $\$400\times4=\$1600\times225+375^2=500625$; and $\sqrt{500625}=707.5+$.

Then $707.5+375\div(300\times2)=\$1.80.4$ price per acre paid by B and C. Then adding 75 cents, we have \$2.55.4 the price paid by A. Dividing this into the money paid by each respectively, will give his quantity.

The first is solved in a similar manner.

To explain the rule, we will let a represent the whole number of acres bought; b the whole number of dollars paid by both; c what B paid; d the number of dollars that A paid per acre more than B; y the cost of B's land per acre.

$\frac{c}{y}=$ number of acres that B got

$\frac{b-c}{y+d}=$ number of acres that A got

And $\frac{c}{y}+\frac{b-c}{y+d}=a$, the whole number of acres bought.

Multiplying the last equation by y to clear it of fractions, gives $c+\frac{by-cy}{y+d}=ay$

And by $y+d$, and canceling $cd+by=ay^2+ady$

By trans. $-ay^2-ady+by=-cd$

Ch. signs and dividing by a gives $y^2-\left(\frac{b-ad}{a}\right)y=\frac{cd}{a}$

Comp. Sq. and Mul. both terms of $\frac{cd}{a}$ by $4a$ to make den's. alike, we have

$$y^2-\left(\frac{b-ad}{a}\right)y+\frac{(b-ad)^2}{4a^2}=\frac{(b-ad)^2}{4a^2}+\frac{4acd}{4a^2}=\frac{(b-ad)^2+4acd}{4a^2}$$

And by Evolution, $y-\frac{b-ad}{2a}=\frac{\sqrt{(b-ad)^2+4acd}}{2a}$

By Trans. $y=\frac{\sqrt{(b-ad)^2+4acd}}{2a}+\frac{b-ad}{2a}$

Or, $y=\frac{1}{2a}\left(\sqrt{(b-ad)^2+4c\times ad}+b-ad\right)$

The above formula, changed to words, will be the language of the rule; and by varying the notation and the mode of working out, no doubt a dozen different rules might be given. The price paid by him who paid most per acre, could be just as readily found.

We are to distinguish between the above mode of forming rules, and the mere expression by letters and signs of rules already found. Take as an example, the following:

Let c represent the circumference of a circle; d, diameter; a, area: then the diameter being given to find the area, we have the following formula:

$d^2\times.7854=a.$ Or, $\frac{d^2}{1.2732}=a.$

The area given to find the diameter. $\sqrt{\frac{a}{.7854}}$

The following will now be readily understood.

$d\times3.1416=c$; Or, $\frac{d}{.318309}=c$

We shall now proceed to close our lecture by giving some remarks on Proofs, Contractions, Canceling, &c.

Many persons are fond of proving the correctness of work,

and pupils are often instructed to do so, for the double purpose of giving them exercise in calculation and saving their teacher the trouble of reviewing their work.

The operation of proving is generally effected by performing a counter operation, as if we add to any amount a given sum, and from the result subtract the sum added, we shall certainly have the number we commenced with. Thus we prove subtraction by addition, for when to the less of two numbers we add their difference we obtain the greater; just as certainly as subtracting the difference from the greater would leave the less; or taking the less from the greater would leave the difference. Multiplication and Division, like Addition and Subtraction, are reverse operations, and are used to prove each other; for if we divide the product of two numbers by one of the numbers the result will be the other; and if we multiply the quotient after division, by the divisor, the result will be the dividend.

These operations, and counter operations are well adapted to make the powers and properties of numbers familiar. Reduction Ascending and Reduction Descending are the reverse of each other, and hence each will prove the other; for if 5 miles changed to yards produce 8800; then 8800 yards changed to miles should make 5.

In the Rule of Three, every question may be reversed, and each term successively be made the required term: by which the operation is fully proved, on the principle to which we have alluded.

Involution and Evolution are the reverse of each other, and of course prove each other on the same principle. If a given number be squared or cubed and the square or cube root of the result extracted, the given number will be reproduced.

Most of the cases in Fractions are in pairs, one being the reverse of the other, and of course mutually proving each other, but it is not necessary to enter upon a detail of them.

This principle of proving by reversing is susceptible of various applications, which the ingenious student cannot fail to discover. The following is an example:

There is a certain number which being divided by 7, the quotient resulting multiplied by 3, that product divided by 5, from the quotient 20 being subtracted, and 30 added to the remainder, the half sum shall make 35. What is the number? *Ans.* 700.

$$35\times2-30+20\times5\div3\times7=700 \textit{ Ans.}$$

Or more plainly $35\times2=70$ whole sum, from 70 take 30 and add $20=60$, and $60\times5=300$, and $300\div3=100$ which $\times7=700$ *Ans.*

Another general mode of testing the correctness of work is to produce the result by different modes, as calculating by the Rule of Three and by Practice, or Analysis; or by Vulgar Fractions and Decimal Fractions; and if the result be the same, we may fairly conclude that the work is correct.

There are also special modes of proof of elementary operations, as by casting out threes or nines, or by changing the order of the operation, as in adding upwards and then downwards. In Addition some prefer reviewing the work by performing the Addition downward, rather than repeating the ordinary operation. This is better, for if a mistake be inadvertently made in any calculation, and the same routine be again followed, we are very liable to fall again into the same error. If, for instance, in running up a column of Addition you should say 84 and 8 are 93, you would be liable in going over the same again in the same way to slide insensibly into a similar error; but by beginning at a different point this is avoided.

This fact is one of the strongest objections to the plan of cutting off the upper line and adding it to the sum of the rest, and hence some cut off the lower line, by which the spell is broken. The most thoughtless cannot fail to see that adding a line *to* the sum of the rest, is the same as adding it in *with* the rest.

The mode of proof by casting out the nines and threes was fully explained in a former chapter.

A very excellent mode of avoiding error in adding long columns is to set down the result of each column on some waste spot, observing to place the numbers successively a place further to the left each time, as in putting down the product figures in multiplication; and afterwards add up the amount. In this way if the operator lose his count, he is not compelled to go back to units, but only to the foot of the column on which he is operating. It is also true that the brisk accountant who thinks on what he is doing is less liable to err, than the dilatory one who allows his mind to wander. Practice too will enable a person to read amounts without naming each figure, thus instead of saying 8 and 6 are 14, and 7 are 21 and 5 are 26, it is better to let the eye glide up the column, reading only 8, 14, 21, 26, &c., and still further, it is quite practicable to accustom one's self to group the figures in adding, and thus proceed very rapidly. Thus in adding the units' column, instead of adding a figure at a time we see at a glance that 4 and 2 are 6, and that 5 and 3 are 8, then 6 and 8 are 14; we may then if expert add constantly the sum of two or three figures at a time, and with practice this will be found highly advantageous in long columns of figures; or two or three

87
23
45
62
24
—

columns may be added at a time, as the practised eye will see that 24 and 62 are 86, almost as readily as that 4 and 2 are 6.

Teachers will find the following mode of matching lines for beginners very convenient, as they can inspect them at a glance.

```
Add   7654384
      8786286
      3408698
      2345615
      1213713
     --------
     23408696
```

In placing the above the lines are matched in pairs, the digits constantly making 9. In the above, the first and fourth, second and fifth are matched; and the middle is the *key line*, the result being just like it, except the units' place, which is as many less than the units in the key line as there are pairs of lines; and a similar number will occupy the extreme left. Though sometimes used as a puzzle, it is chiefly useful in teaching learners; and as the location of the key line may be changed in each successive example, if necessary, the artifice could not be detected. The number of lines is necessarily odd.

In Multiplication the cross, already explained, is the usual mode of proof, as dividing the product would be to pre-suppose a knowledge of Division before the pupil has reached it, unless he attempt to learn both at once.

Where the multiplier is not too large, and can be divided into two or more factors, there is a saving in adopting that mode, and it may be used as a proof of the common mode. If I seek to multiply, say 7864 by 24, it requires in the usual way two product lines and finding their sum by addition, making a third operation; but if I multiply by 6 and that product by 4, or by 8 and 3, or 12 and 2, the business is dispatched in two lines. But being able to multiply in a single line is still better.

The same remarks apply in Division, and hence there is often economy of figures in dividing by factors of your divisor, and if a remainder occur only in your first division it is the true one; but if in the second only then multiply such remainder by the first divisor, or all if more than one, and if there was a remainder on the first division also, it must be added in and the sum will be the true remainder.

Divide 73640 by 24—

6)73640

4)12273 and 2 over.

3068 and 1 over. $1 \times 6 + 2 = 8$ the true remainder.

Or— 4)73640

6)18410

3068 and 2 over, and $2 \times 4 = 8$ as before.

Another mode of proving Division, is to divide the dividend by the quotient, and the result, will be the divisor; the same remainder occurring in one case as in the other, but not of the same fractional value; for as the dividend exceeds some multiple of the divisor and quotient, just the amount of the remainder, that remainder will be the same without regard to which of the factors occupies the divisor's place, and as the divisors would differ, the value of the fraction formed by the remainder and divisor must differ. To make the dividend an exact multiple subtract the remainder from it.

Another mode of proving Long Division is to add the remainder and the several products together as they stand in the work, and the result will equal the dividend. Thus—

23)74653(3245
69*

56
46*

105
92*

133
115*

18*

74653

* These lines are to be added up perpendicularly.

23
3245

115
92
46
69
18 remainder.

74653

By comparing the numbers added in the sum wrought out, with the numbers in the common multiplication on the right hand, it will be seen that they are just the same only in reversed order; an arrangement resulting from the multiplication commencing with the highest figure, viz, 3 thousand, and proceeding towards the units' place; instead of commencing with the units and proceeding towards thousands; an arrangement in nowise affecting the result.

That it does not is easily seen by disposing the numbers and working them out, thus

Multiply	7854	7854
By	32	32
	23562	15708
	15708	23562
	251328	251328

This is shown, not because it is a better form, but to present the operation in a different light, and illustrate more fully the properties of numbers.

Multiply	4756
By	182
	9512
	38048
	4756
	865592

In this case the second product line may be conveniently found by multiplying the first by 4; but if there is an error in the first it will run into the second. A shorter mode would be to multiply the first product line by 9, which would give the product by 18 in one line; and save work. Try it.

As a large number of the problems solved in this book are solved by different modes, they furnish a great variety of modes of proof.

In calculations of measurement it is easy to vary the mode and thus effect a proof; but here as in all other cases the operator should cultivate a habit of correct calculation, for it is not desirable to travel far out of the way in common business, and it is doubtful whether it is not better to rely on care and correct habits to avoid error, than on any mode of detecting blunders after they are made.

Perhaps too this would be as proper a time as any other to urge the importance of another good habit; I mean *that of making plain figures*. Some persons accustom themselves to making mere scrawls, and important blunders are often the result. If letters be badly made you may judge from such as are known; but if one figure be illegible, its value cannot be inferred from the others. The vexation of the man who wrote for 2 or 3 monkeys, and had 203 sent him, was of far less importance than errors and disappointments sometimes resulting from this inexcusable practice.

Contractions, like modes of proof, serve to show the powers and properties of numbers, and to amuse and exercise the faculties of the student; and in many instances they are practically useful. But contractions should not be resorted to unless they are perfectly understood in theory and in mode of operation; for as we have already remarked, like by-ways in a forest, they are convenient to him who knows the whole ground, but strangers do better to keep the high way. Persons who practise mental calculation to much extent, find abbreviations very necessary, and generally adopt such as suit their convenience best. "Bringing down" ciphers in multiplication, and "Cutting off" in division, are very common abbreviations. Some of the following may be practically convenient, and all of them may be worth the student's study, as a farther means of familiarizing the subject to his mind.

1. To multiply any number by 25, add two ciphers and divide by 4.

Multiply 3969 by 25. Product 99225.

4)396900

99225

This is in effect to multiply by 100 and take one fourth the product, which is the same as to multiply by 25, the fourth of a hundred. On the same principle add 0 and divide by 2 to multiply by 5; or add 00 and divide by 8 to multiply by $12\frac{1}{2}$; or add 000 and divide by 8 to multiply by 125; or add 000 and divide by 3 to multiply by $333\frac{1}{3}$; the principle is nearly allied to that of *Practice* and needs no labored demonstration. It may be applied in multiplying by any number that is an even part of 10, 100, 1000, &c., as $16\frac{2}{3}$, 25, $33\frac{1}{3}$, &c. On the same principle we may multiply by 75 by adding 00 and multiplying by 3 and dividing by 4; $66\frac{2}{3}$ by adding 00 and multiplying by 2 and dividing by 3; and we may apply the same principle to many other multipliers and *divisors;* for what applies to the former, applies also by reversal to the latter. So we may multiply by 75 by adding 00, dividing by 4, and subtracting the quotient. The same principle can be readily applied to other numbers.

2. To multiply by 99, add 00 and subtract the given number from the result.

Multiply 31416 by 99.

```
          3141600
            31416
          -------
Product   3110184
```

As 99 times is 1 time less than 100 times, and adding 00 is in effect multiplying by 100, one time the multiplicand is deducted, which leaves 99 times, as required.

This principle is applicable where any number of 9's is the multiplier; as many ciphers being added of course as there are nines. If the multiplier were 98 we would subtract twice the multiplicand; if 97 three times, and so on.

3. To multiply by any number between 10 and 20 as 16 or 18, multiply by the units' figure and set the product under the multiplicand, but put it one place to the right; then add the lines together. The reason is evident on looking at the calculation.

Multiply 3854 by 16.

```
3854
 23124
-----
61664,    Ans.
```

On the same principle if any number of ciphers intervene, as 106, 1006, 10006, &c., set the product so many places farther to the right.

Multiply 3854 by 1006.

```
3854
   23124
-------
3877124,    Ans.
```

Cross Multiplication is a mode of Multiplying by large multipliers in a single line; and by practice the operation may be performed with great expedition. It is necessary to begin with small numbers, say of two places, and carry the calf diligently, if you would carry the ox successfully.

Here we multiply 5×2 and set down and carry as usual, then to what you carry, add 5×3 and 4×2, which gives 24; set down 4, and carry 2 to 4×3, which gives 14. It is obvious that this is just the usual mode, with the intermediate work done in the head.

```
  32
  45
----
1440
```

Here the first and second places are found as before, for the third add, 6×1, 4×3, 5×2, with the 2 you had to carry, making 30; set down 0 and carry 3; then drop the units place and multiply the hundreds and tens crosswise as you did the tens and units; and you find the thousands figure; then dropping both units and tens, multiply the 4×1 adding the 1 you carried, and you have 5, which completes the product. The same principle may be extended to any number of places; but let each step be made perfectly familiar before advancing to another. Begin with two places, then take three, then four, but always practising some time on each number; for any hesitation as you progress, will confuse you.

```
  123
  456
-----
56088
-----
```

4. To multiply by 21, 31, &c., to 91 in a single line, multiply by the tens' figure and set the product one place to the left underneath the multiplicand, then add.

Multiply 3854 by 21.

```
 3854
7708
-----
80934,    Ans.
-----
```

If ciphers intervene, as 201, 3001, &c., multiply as before, but set the product as many additional places to the left as there are ciphers.

Multiply 3854 by 6001.

```
   3854
23124
-------
2316254,    Ans.
-------
```

5. The following is a convenient mode of multiplying by any two figures, and is not difficult to apply.

```
         Multiply 3754
         By         27
                ------
Product         101358
                ------
```

I here multiply 27 by 4, setting down the first product figure, and carrying the others; I then multiply by 5, and set down and carry in the same way: so proceeding to the highest place of the multiplicand.

6. In the Rule of Three, when the first term and either the

second or third are alike, (*i. e.* when a dividing and multiplying term are alike,) both may be rejected; and if not alike they may be divided by any number that will measure both, and the resulting quotients may be used. This process is called canceling.

If 12 yards of cloth cost $108, what will 35 yards cost?

	y'ds		y'ds		$
As	~~12~~	:	35	: :	~~108~~
	1		9		9

$315, *Ans.*

If 12 yards of cloth cost $108, what will 48 yards cost?

	y'ds		y'ds		$
As	~~12~~	:	~~48~~	: :	108
	1		4		4

$432, *Ans.*

If 120 yards of cloth cost $490, what will 180 yards cost?

	y'ds.		y'ds.		$
As	~~120~~	:	~~180~~	: :	490
	2		3		3

2)1470

$735, *Ans.*

In this sum one term is not the multiple of another, but 60 will measure the first and second, and they will thus be proportionately reduced; for if 120 yards cost $180, 2 yards will cost $3, *i. e.* 120 bears the same ratio to 180 that 2 does to 3. But it is plain that it would not do to divide the 2d and 3d, by which both the multiplying terms would be reduced, and of course the dividend which their product is, while the divisor would remain unaffected. If one is reduced, the other must be reduced in the same ratio.

7. In reducing compound fractions, similar terms in the numerators and denominators may be canceled.

Reduce $\frac{3}{4}$ of $\frac{6}{7}$ of $\frac{2}{3}$ of $\frac{5}{6}$ of $\frac{4}{5}$ to an equivalent simple fraction. *Result* $\frac{2}{7}$,

$$\frac{\not{3}}{\not{4}}\times\frac{\not{6}}{7}\times\frac{2}{\not{3}}\times\frac{\not{5}}{\not{6}}\times\frac{\not{4}}{\not{5}}=\frac{2}{7}$$ *Ans.*

We may proceed by canceling or striking out similar terms, and we will find only 2 remaining of the numerators, and 7 of the denominators, these united form a fraction equivalent to the whole expression; for had we multiplied all the numbers together that would have increased both terms proportionately, and the overgrown numbers would have shrunk to $\frac{2}{7}$ at last.

The operation of canceling may sometimes, in the hands of an expert operator, be employed to advantage, but like other labor saving expedients and by-ways, it requires intimate familiarity with the subject; and is not adapted to the common purposes of life. Some have attempted to build up a system and call it a universal mode of solving all Arithmetical questions by one rule, on the Prussian Canceling System. Though occasionally well enough, as a labor saving expedient, it is as a system, with numbers got up to show it off, a perfect humbug; and well calculated to bring the whole matter into disrepute. Any one familiar with cross multiplication, canceling, &c., and possessing an oily tongue, with a good amount of impudence, may astonish (if he does not benefit) an audience.

8. In order to obviate tedious multiplication of decimals, some adopt this mode: "Invert the figures of the multiplier, and place them so that the tenths may be under that order of decimals of the multiplicand to which it is proposed to limit the product; then multiply each figure of the multiplier into the figures of the multiplicand, beginning at the figure immediately above it, and taking in the carriage from the right hand. Place the first figure of each partial product in the same column, and the amount of the whole will give the total product restricted to the number of certain places sought."

Multiply 18.7568925 by 13.256825 limiting the product to four certain decimal places.

```
18.7568925                          18.7568925
528652.31                            13.256825
------------                       -----------
18756892 .                           937844625
 5627067 . .                        375137850
  375137 . . .                    1500551400
   93784 . . . .                 1025413550
   11254 . . . . .               937844625
    1500 . . . . . .            375137850
      37 . . . . . . .         562706775
       9 . . . . . . . .      187568925
------------------          ------------------
248.65680 . . . . . . . .   248.6568414163125
```

Multiplied in full the product would be 248.6568414163125. The points are placed merely to show the figures not counted. By carrying one where the rejected figures exceed 5 the result would be more nearly accurate; but as it is rather curious than useful, it is not necessary to dwell upon it. Compare the work with the solution in full and the reason will be plain. A method very similar is used to shorten division of decimals, but an illustration of it is not worth the space it would occupy.

9. Logarithms furnish a short mode of Multiplying, Dividing, Extracting Roots, Raising Powers, &c.; but their explanation here would be out of place; unless tables of them could be furnished.

10. To multiply a decimal or a mixed number by 10, 100, 1000, &c., remove the decimal point one, two or three places to the right; and to divide by such numbers remove the point as many places to the left; and if there be not a sufficient number of places on the left, ciphers must be prefixed.

11. To save trouble some persons add ciphers to the remainders and divide in extracting roots, as they would in common division, but this is not accurate. See prop. 50.

12. To multiply or divide by any composite number; multiply, or divide, as the case may be, by the factors of the numbers successively.

This was sufficiently illustrated under the head of Proofs.

The French mode of operating in Long Division has some advantage over ours. They place the divisor on the right of the dividend, as we do the quotient, and place the quotient underneath the divisor, by which the figures to be multiplied together are brought near each other. Thus—

```
Divid.  Divis.
 3936)96
 384  ‾‾
      41 Quotient.
 ————
   96
   96
   ——
```

For sake of brevity they frequently omit the product figures, setting down only the remainders, which they find as they pass along. Thus—

```
3936)96
  96 ‾‾
     41, Ans.
  ——
```

This, however, applies to our mode as well as theirs.

The following mode of multiplying by large multipliers in a single line, is both curious and useful. It is the same that is used by PETER M. DESHONG, a calculator of some notoriety in the United States.

Multiply 7865 by 432 in a single line.

On a slip of paper, separate from that on which the multiplicand is written, place the multiplier in inverted order: thus, 234 and close to the upper edge of the paper. Then bring the multiplier so that the 2 shall be directly under the 5, or units' place of the multiplicand: multiply those figures, set down 0 and carry 1. Slide the paper to the left one place, that 2 may be under 6, and 3 under 5; and to the 1 you carried add the products of the 2 by 6, and 3 by 5, making 28—set down 8 and carry 2. Again move your paper one place to the left, and to the 2 you carried, add the several products of the multiplicand figures with the figures of the multiplier that are under them, viz. 8×2, 6×3, 5×4, and the result will be 56; set down 6 and carry 5. Slide again and you have 5 (that you carried) +14+24+24=67. Thus proceed towards the left until the multiplier passes from under the multiplicand, each time adding what you carry, to the several products of the figures that stand one over the other, the result will be 3397680. These additions will soon be performed at a glance, as the products are obvious *without the formality of naming the factors.*

To understand these directions clearly, factors must be placed upon slips of paper, and the directions strictly complied with; by which the mode of operation and the reason will be better understood in ten minutes, than in three hours without them. When familiar with the slide, the operator may proceed without it, and perform operations astonishing to the uninitiated; the largest numbers being multiplied together readily in a single line.

The mode used by Mr. DESHONG for dividing in a single line by a large divisor, is somewhat similar to the French mode, the product figures not being set down, but it is of no practical importance, and not worth the space its illustration would occupy. We shall allude to his Addition hereafter. His subtraction is worthless; and the same is true of his giving the roots of even squares and cubes.

LECTURE XVIII.

PROBLEMS PURELY ARITHMETICAL.

Having elucidated the general properties of numbers, and explained the modes of framing rules for the solution of problems, we shall now present for solution such problems as may seem best adapted to serve our purpose. In order to save space and to present the subject in a condensed form, the details of division, multiplication, &c., are omitted; the solutions being rather skeletons than otherwise; but we presume they will be found intelligible to the attentive student. We might suggest to the operator that he must not expect problems of intricacy to yield to a hasty glance, they require to be dwelt upon with the whole energies of the mind; for it is in this way that the tangled mass, which at first view seemed without system or design, is made gradually to disentangle itself, and to become perfectly simple under the intense gaze of the mind. Solutions often seem very plain when once performed, even by another; as the Spanish nobles thought it a very simple matter to make the egg stand on the end, after Columbus had shown them the mode. But the ambitious student will not be satisfied with following in the wake of another, he will make the egg stand on its end at his own bidding.

Solving problems is a very useful exercise to give practical skill, but for this purpose they must be solved understandingly, for following arbitrary rules will no more develop the reasoning faculties of the human mind, than turning a mill will give sight to a blind horse. You might as well expect to become an engineer by turning a grindstone; or a musician by turning the crank of a hand organ.

It is scarcely possible to give general rules that would be of much service to the student; though he will generally find an advantage, if his problem contains large numbers, and he cannot make them clear to his mind, in taking small numbers that can be borne in memory, until their relation becomes familiar. He may often too find advantage in taking something familiar, or of a practical character, and involving the same principle from which he may understand that which is less familiar. Or if complicated he may separate the parts; but on no account let him fail to investigate the problem thoroughly. He should begin with problems that are simple, and gradually proceed to the more difficult; and he may be greatly benefited, especially if he has no teacher, by closely investigating solutions made by others.

To furnish such an opportunity, we shall now proceed to present a variety of problems, with synthetical and analytical solutions. The student should not only examine the solutions, but perform the work.

1. If 5 yards of cloth cost $13.50, what will 75 yards cost?

By Analysis. If 5 yards cost $13.50, 1 yard will cost one fifth as much, and $\frac{1}{5}$ of $13.50 is $2.70. Then if 1 yard cost $2.70, 75 yards will cost 75 times $2.70=$202.50, the answer. The reason of the process is obvious.

By Proportion. As 5 yds : 75 yds : : $13.50 : $202.50.

It was shown when treating of Proportion, that where the rate is the same, one quantity will be proportionate to any other quantity, as the price of the one is to the price of the other. The reason for multiplying the 2d and 3d terms together, and dividing by the 1st, was also fully explained under the same head.

Or we might explain the process by showing that as $13.50 is the price of 5 yards, we obtain five times too much by multiplying by 75, and hence we divide by 5 for the true amount.

We may save labor by dividing the 1st term into the 2d or 3d before multiplying, and then multiplying the quotient and the remaining term together. In the above case, after stating the question, divide 75 by 5, and the quotient, 15, multiplied into $13.50, will give the answer. A little thought will show that whether we multiply two numbers together and divide the product by a given divisor or divide either of the terms by the given divisor, before multiplying, the result will be the same. Furthermore to constitute a proportion, the 4th term must be as many times greater or less than the 3d, as the 2d is greater or less than the 1st; so that we have but to multiply or divide the 3d by the ratio of the 1st and 2d to find the 4th.

If the division involves no fractions, the operation is rather simplified than otherwise, by dividing before multiplying; but if dividing leaves a fraction, the operation becomes more difficult. It is for this reason indispensable to become intimate with fractional quantities at the very outset. Suppose that in the above case the price of 5 yards had been 13.57\frac{1}{2}$, the operation would be as follows:

If 5 yards cost $\$13.57\frac{1}{4}$, 1 yard will cost $\frac{1}{5}$ of $\$13.57\frac{1}{4}=\$2.71\frac{9}{20}$; then if 1 yard cost $\$2.71\frac{9}{20}$, 75 yards will cost 75 times as much$=\$203.58\frac{3}{4}$. As remarked above, we may however divide by 5, and obtain 15 as a multiplier; for as 75 is 15 times as much as 5, the price of 75 must be 15 times as much as the price of 5.

2. F, G and H freight a ship with 108 tuns of wine; F's share was 48, G's 36, and H's 24 tuns. A storm arising, the seamen threw 45 tuns overboard; how much should each merchant sustain of the loss?

By Analysis. F had 48 tuns of 108, or $\frac{48}{108}=\frac{4}{9}$, and having $\frac{4}{9}$ of the stock he should sustain $\frac{4}{9}$ of the loss. One ninth of $45=5$, hence $\frac{4}{9}=4$ times $5=20$, what F lost.

The same process with 36 and 24, the shares of G and H, will show that the former lost 15, and the latter 10.

By Proportion. As 108 tuns (the whole freight) : 48 (F's share of freight) : : 45 (the whole loss) : 20, F's share of loss. And so of the others.

3. Three graziers pay among them $120 for a grass lot, into which L put 80 oxen, N 100, and C 120; how much should each person pay?

By Analysis—

L $80=\frac{80}{300}=\frac{4}{15}$	and $\frac{4}{15}$	of $120	$=\$32$,	what L should pay.		
N $100=\frac{100}{300}=\frac{1}{3}$	" $\frac{1}{3}$	"	$= 40$,	"	N	"
C $120=\frac{120}{300}=\frac{2}{5}$	" $\frac{2}{5}$	"	$= 48$,	"	C	"
300		Proof	$120 whole cost.			

By Proportion. $80+100+120=300$, the whole number grazed. Then, As 300 oxen, (the whole number grazed) : 80 oxen, (what L owned) : : $120 (the whole cost) : $32 (what L must pay.)

Were explanation necessary, we might say that by summing up the oxen we find there were 300 pastured; of which L had 80, or $\frac{80}{300}=\frac{4}{15}$ of the whole number; and as he had $\frac{4}{15}$ of the cattle, he must pay $\frac{4}{15}$ of the cost of pasturage. The same reasoning applies to the others; and it is equally obvious that the amount to be paid by each, should be proportionate to the oxen owned by each.

4. L, N and C bought a lot of pasture at $50, into which L put 80 oxen for 3 months, N 100 for 2 months, and C 120 for 1 month; what should each pay?

By Analysis. L's 80 oxen for 3 mo.=240 for 1 mo.
N's 100 " " 2 mo.=200 " 1 mo.
C's 120 " " 1 mo.=120 " 1 mo.

The whole will equal . . . . 560 " 1 mo.

Ten $\frac{240}{560}=\frac{3}{7}$ and $\frac{3}{7}$ of \$50=\$21$\frac{3}{7}$, what L pays.
$\frac{200}{560}=\frac{5}{14}$ " $\frac{5}{14}$ of 50= 17$\frac{6}{7}$, " N "
$\frac{120}{560}=\frac{3}{14}$ " $\frac{3}{14}$ of 50= 10$\frac{5}{7}$, " C "

Proof \$50

By Proportion. Find, as before, how many months' pasturage there will be; then say, As 560 : 240 : : \$50 : \$21$\frac{3}{7}$, L's share. The others are to be found in the same way.

Note.—The 2d and 3d questions belong to the class usually placed under the head of *Single Fellowship;* the 4th belongs to *Double Fellowship*, or *Fellowship with Time*, as arranged in books on Arithmetic generally.

5. If by a pole that's 10 feet long,
A shade of 6 is made,
What is the steeple's height in yards,
That's 90 feet in shade?

By Analysis. If 6 feet shade is from 10 feet pole, 1 foot shade must be from $\frac{10}{6}$ feet pole, and 90 feet shade from $\frac{10}{6}\times 90=\frac{900}{6}=150$ feet=50 yards. *Ans.*

By Proportion. As 6 ft. sh. : 90 ft. sh. : : 10 ft. pole : 150 ft. steeple=50 yards.

6. C owes D \$100 due in 6 months; \$100 due in 9 months; and \$700 due in 12 months; what would be an equitable time for the payment of the whole at once?

By Analysis.

It is obvious that the interest on \$100 for 6 months will be the same as the interest on \$1 for 600 "
And \$100 for 9 months . . . .=\$1 " 900 "
Furthermore, \$700 for 12 months=\$1 " 8400 "

Hence the whole . . . .=\$1 " 9900 "

But if we have, as above \$900, it will require only $\frac{1}{900}$ part of the time that \$1 would, and $\frac{1}{900}$ of 9900 months=11 months, the equated time.

By the usual mode.

$$\begin{array}{r}100\times 6 = 600\\ 100\times 9 = 900\\ 700\times 12 = 8400\\ \hline 900 \quad)9900(11 \text{ months, } Ans.\end{array}$$

Note. This belongs to the doctrine of Equations, or finding a single time at which a debt due at different times may be paid, without loss to either party. In other words, such a time that the debtor will gain as much by keeping some payments after they are due, as he will lose by paying others before they are due. The nature of this rule and the objections to this mode of calculation have been explained in the lecture on Interest, and the kindred rules.

7. If 16 men finish a piece of work in $28\frac{1}{3}$ days, how long will it take 12 men to do the same work?

By Analysis. If 16 men do the work in $28\frac{1}{3}$ days, 1 man will do it in 16 times $28\frac{1}{3}$ days$=453\frac{1}{3}$ days; and 12 men will do the same in $\frac{1}{12}$ the time,$=37\frac{7}{9}$ days.

By Proportion. This question is of the class whose terms are in Inverse Proportion, for it is obvious that fewer men will require more time. If the question had reference to the pay, then the amount would be directly proportionate to the number, and so it would be if it were asked what amount of work they could do; but the amount of work is here a fixed quantity, and 12 men are required to do as much as 16 had done; they must therefore have more time. In the amount of work and the pay, the answer would be $\frac{3}{4}$ of what 16 would give, for 12 is $\frac{3}{4}$ of 16; but in *time* the answer will be one-third more, for 16 is $\frac{1}{3}$ more than 12. By the old mode of stating we would say,

As 16 men : $28\frac{1}{3}$ days : : 12 men : $37\frac{7}{9}$ days.

And we would multiply the 1st and 2d terms together, and divide by the 3d, for the 4th term, or answer. By the general rule we would set $28\frac{1}{3}$ days in the 3d place, because the answer is days; then because the answer must be greater than this, we would set 16 in the 2d place and 12 in the 1st; thus,

As 12 men : 16 men : : $28\frac{1}{3}$ days : $37\frac{7}{9}$ days.

The numbers are here in Direct Proportion; the 1st is to the 2d, as the 3d to the 4th; but in the former statement they are in Inverse or Reciprocal Proportion, the 1st is to the 3d, as the 4th to the 2d.

8. If 7 horses consume $2\frac{3}{4}$ tons of hay in 6 weeks, how many tons will 12 horses consume in 8 weeks?

By Analysis. If 7 horses consume $2\frac{3}{4}$ tons in 6 weeks, 1 horse will consume $\frac{1}{7}$ of $2\frac{3}{4}$ tons$=\frac{11}{28}$ tons in 6 weeks; and in

1 week will consume $\frac{1}{6}$ of $\frac{11}{28}=\frac{11}{168}$; and if 1 horse in 1 week eat $\frac{11}{168}$ of a ton, 12 horses will eat 12 times $\frac{11}{168}=\frac{132}{168}$, and in 8 weeks will eat 8 times $\frac{132}{168}=\frac{1056}{168}=6\frac{2}{7}$ tons, *the Answer.*

By Proportion. This question offers 5 terms for the purpose of ascertaining a 6th. It is equivalent to two questions in the single rule of three, and hence such problems are classed under the head of the *Double Rule of Three.*

To solve it by two statements of the single rule, we may first find what 12 horses would eat in 6 weeks; thus,

As 7 horses : 12 horses : : $2\frac{3}{4}$ tons : $4\frac{5}{7}$ tons.

Then knowing what 12 horses would eat in 6 weeks, it is easy to find what they would eat in 8 weeks; thus,

As 6 weeks : 8 weeks : : $4\frac{5}{7}$ tons : $6\frac{2}{7}$ tons. *Ans.*

To solve this by a single statement, we might arrange the numbers variously, but the following form may be as good as any:

$$\text{As} \left\{ \begin{matrix} 7 \text{ horses} \\ 6 \text{ weeks} \end{matrix} \right\} : \left\{ \begin{matrix} 12 \text{ horses} \\ 8 \text{ weeks} \end{matrix} \right\} : : 2\tfrac{3}{4} \text{ tons.}$$

42   96

$2\frac{3}{4}$

42)264($6\frac{2}{7}$ tons. *Ans.*

This is an exemplification of the doctrine of compound proportion. Seven horses are to 12 horses, as $2\frac{3}{4}$ tons, what 7 will eat, are to what 12 will eat. And 6 weeks are to 8 weeks, as $2\frac{3}{4}$ tons, the food for 6 weeks, are to the food for 8 weeks; and being so proportioned separately, their products are proportionate. By multiplication it is resolved into—

As 42 : 96 : : $2\frac{3}{4}$: $6\frac{2}{7}$, *the Answer.*

As questions of this sort are as readily solved, in all cases, by two statements of the single rule, as by a single statement of the double rule, it is unnecessary to do more than present a single additional problem.

9. A man and his family, numbering 5 persons in all, did usually drink $7\frac{4}{5}$ gallons of cider in a week; how much will they drink in $22\frac{1}{2}$ weeks, when three persons more are added to the family?

By Analysis. If 5 persons drank $7\frac{4}{5}$ gallons in 1 week, 1 person would drink $\frac{1}{5}$ of $7\frac{4}{5}=1\frac{14}{25}$ gallons; and if 1 person in 1 week drink $1\frac{14}{25}$ gallons, 1 person in $22\frac{1}{2}$ weeks will drink $22\frac{1}{2}$ times $1\frac{14}{25}$ gallons$=\frac{45}{2}\times\frac{39}{25}=\frac{1755}{50}$, and 8 persons (3 more than 5) will drink 8 times $\frac{1755}{50}$ gallons$=\frac{14040}{50}=280\frac{4}{5}$ gallons. *Answer.*

By Proportion—

As 1 week : $22\frac{1}{2}$ weeks : : $7\frac{4}{5}$ gal. : $175\frac{1}{2}$ gal.

Then,

As 5 per. : 8 per. : : $175\frac{1}{2}$ gal. : $280\frac{4}{5}$ gal. *Ans.*

Or,

$$\text{As} \left\{ \begin{array}{l} 1 \text{ week} \\ 5 \text{ persons} \end{array} \right\} : \left\{ \begin{array}{c} 22\frac{1}{2} \text{ weeks} \\ 8 \text{ persons} \end{array} \right\} : : 7\tfrac{4}{5} \text{ gal.} : 280\tfrac{4}{5} \text{ gal.}$$

Converting $7\frac{4}{5}$ into 7.8 in the above solutions will greatly simplify the calculation, by clearing it of vulgar fractions. It will be observed that in making the double statement, the number of the same name as the answer is placed in the third place, and the others are arranged with reference to it, as in the single rule; which places the two antecedents in the first place and the two consequents in the second; and the same numbers thus ultimately become divisors and multipliers, as in the single statements. It is not necessary, perhaps, to consume space by explaining the old mode of shifting the inverse terms, as the general rule is preferred by most persons, and that avoids the necessity of changing the terms.

10. If 7 masons can build a house in 30 days, in what time can 11 masons build a similar one?

By Analysis. If 7 masons require 30 days, 1 mason will require 7 times=210 days; and 11 will require $\frac{1}{11}$ of 210 days $=19\frac{1}{11}$ days. *Ans.*

By Proportion—

As 11 masons : 7 masons : : 30 days : $19\frac{1}{11}$ days.

Some may suppose that because the general rule for stating dispenses with the old mode of multiplying the 1st and 2d terms, and dividing by the 3d, the proportion does not still exist; but such a notion would be entirely incorrect, for it exists in the very nature of things, and must continue to exist, so long as increasing the operating means lessens the time necessary to produce a given effect; or as when surface is narrow, its length must be greater to produce a definite area, than when the surface is broad.

11. If 10 lbs. of cheese are equal in value to 7 lbs. of butter, and 11 lbs. of butter to 2 bushels of corn; and 11 bushels of corn to 8 bushels of rye, and 4 bushels of rye to 1 cord of wood; how many pounds of cheese are equal in value to 10 cords of wood?

When the series runs through several *antecedents* and *consequents*, as in the above question, it is called by some *Conjoined Proportion*, by others the *Chain Rule;* but all such problems might be wrought by the single rule of three, or by analysis, should such a problem ever occur. For the sake of brevity, however, we may adopt the following rule: "Place the numbers alternately, the antecedents at the left hand, and the consequents at the right, and let the last number stand on the left hand; then multiply the left hand numbers continually together for a dividend, and the right hand for a divisor; and the quotient will be the answer."

Thus,		
	10	7
	11	2
	11	8
	4	1
	10	
	112)48400	112
	$432\frac{1}{7}$ lbs.	*Answer.*

By Proportion—

As 7 lbs. : 11 lbs. : : 10 lbs. : $15\frac{5}{7}$ lbs. ch.=2 b. of corn.
And, As 2 b. : 11 b. : : $15\frac{5}{7}$ lbs. : $86\frac{3}{7}$ lbs. ch.=8 b. of rye.
And, As 8 b. : 4 b. : : $86\frac{3}{7}$ lbs. : $43\frac{3}{14}$ lbs. ch.=1 cord w.
And, As 1 c'd : 10 c'ds : : $43\frac{3}{14}$ lbs. : $432\frac{1}{7}$ lbs. ch.=10 c. w.

Here we see that the numbers which were multiplied together in the first operation to produce dividend and divisor, are in the latter operation made dividends and divisors separately; and the reason of the first operation is made plain. The latter needs no explanation.

12. What number is that, which being multiplied by 8, and the product divided by 6, the quotient will be 200?

Problems of this kind are readily solved by reversing the proposed operation. If the unknown number must be multiplied by 8 and divided by 6 to produce 200; then, as multiplication and division are the reverse of each other, if 200 be *multiplied* by 6 and *divided* by 8, it will give the unknown or required number. Thus 200×6÷8=150. *Ans.* This process is not strictly *Analysis*, but it is such solution as this class of questions admits of, and is often very convenient, as will more clearly appear hereafter. This question and several that follow, belong to *Single Position;* but they may be all solved by other modes, without supposing any unknown number. The solutions by Position are generally omitted, as that subject has been fully discussed.

13. The third part of a number is 120; what is the whole of it.

Solution. If *one* third is 120, *three* thirds, or the whole of it, will be three times as much=360.

14. To a certain number we add one fourth of itself, and it makes 150; required the number?

Solution. The number and its fourth, equal five fourths, and if five fourths is 150, one fourth will be $\frac{1}{5}$ of 150=30; and four fourths, or the number itself, is 4 times 30=120.

15. In a certain orchard $\frac{1}{5}$ of the trees bear cherries; $\frac{1}{4}$ bear apples; $\frac{1}{3}$ bear peaches; $\frac{1}{12}$ bear plums; and the rest, 16 in number, bear pears. How many trees are in the orchard?

Solution. We may consider the whole orchard as a unit, as 1 orchard, from which $\frac{1}{5}+\frac{1}{4}+\frac{1}{3}+\frac{1}{12}=\frac{52}{60}$, being subtracted, $\frac{8}{60}=\frac{2}{15}$ remain, which by the question=16, hence *one* fifteenth is 8, and 15 fifteenths are the whole=15×8=120. Or, Having found as above that $\frac{8}{60}$=16, say,

As 8 (sixtieths) : 60 (sixtieths) : : 16 trees : 120 trees.

16. A laborer contracted to receive 75 cents for every day he wrought, and to pay 25 cents board for every day he was idle. On settlement at the end of 60 days he received $30. How many days did he work, and how many was he idle?

By Position—

Suppose he wrought 40 days, then 40×.75=	$30.00
He was idle then 20 days at .25=	5.00
Received on our supposition	$25.00
But we know he received	30.00
Error too little	$5.00
Suppose he wrought 50 days, then 50×.75=	$37.50
Then he was idle 10 days at .25=	2.50
Received on second supposition	$35.00
Actually received - - -	30.00
Error too much	$5.00

5+5=10, difference of errors; and 50—40=10, difference in suppositions.

As $10 (diff. of er.) : $10 (diff. of sup.) : : $5 (1st er.) : $5 (correction.)

Then 40+5=45, days he wrought; and 60—45=15, days idle. 45 days work=$33.75; 15 days idleness $3.75; and $33.75—$3.75=$30 Proof.

By Analysis. It is obvious that had he wrought all the time he would have received 60×.75=$45. He forfeited his wages and 25 cents, making $1, every day he was idle; and as he received but $30 instead of $45, he must have been idle

15 days; and consequently wrought 60—15=45 days, as before.

17. A young lady being asked her age, replied:

"My age if multiplied by 3,
Two-sevenths of that product tripled be;
The square root of $\frac{2}{9}$ of that is 4,
Now tell my age or never see me more."

By Reversal. $4^2 \times 9 \div 2 \div 3 \times 7 \div 2 \div 3 = 28$, *Her Age.*

Proof. $28 \times 3 \times 2 \div 7 \times 3 \times 2 \div 9 \surd = 4$.

This question cannot be solved by Position, unless you omit the extraction of the root, and use 16 as the $\frac{2}{9}$.

18. What number multiplied by half itself will make $4\frac{1}{2}$?

Solution. If multiplying by half itself produces $4\frac{1}{2}$, multiplying by itself, or in other words squaring, must produce 9. Hence 3, the square root of 9, is the number sought.

19. The distance from Zanesville to Columbus is 53 miles. Suppose that a stage coach leaves Columbus at 6 in the morning for Zanesville, running 5 miles per hour; and at 7 a light wagon leaves Zanesville for Columbus, running 3 miles per hour: where and at what time will they meet?

The coach will be out an hour, and of course have advanced 5 miles, when the wagon starts. Then,

As the sum of their travel per hour, (5+3=8,)
Is to the whole distance to be travelled by both, (48 miles,)
So is the distance per hour traveled by either, (say 5, when the coach travels,)
To the whole distance which such one will travel, before they meet.

This will give 30 miles, the distance traveled by the coach before they meet: which will require 6 hours, or from 7 A. M. to 1 P. M.; to which we may add 5 miles, traveled before the wagon starts. The wagon in the same 6 hours will travel 18 miles. They meet therefore, at 1 o'clock P. M. at the 18th milestone from Zanesville.

20. The following occured some time ago in business, and is of a character that may occur again. A owed B $500, for which B was willing to wait a year longer, provided A would pay a part and the interest at 6 per cent. in advance on the remainder: A paid $200, and it is required to determine what part is to be credited on the principal, and what part will be required to pay the year's interest on the unpaid portion of the principal?

To know the amount to be credited we must deduct from $200, the interest for one year on the balance of debt remaining. This remainder will consist of $300, with its interest for one year, and an infinite series of interest on the several additions of interest; in other words, of $300 principal and 6 per cent. on its own amount.

If we add the interest on $100, which is $6, to $94, we shall have two numbers, $94 and $100, that bear the same ratio to each other as $300, and a sum that shall have its interest added to $300. Hence, As $94 : $100 : : $300 : 319.14\frac{42}{47}$, the sum for which the note is to be given. The interest on this sum, 19.14\frac{42}{47}$, being settled for in advance, the note bears no farther interest for a year. This sum being deducted from $200, leaves 180.85\frac{5}{47}$ to be credited on the note.

Any other number may be as well assumed as $100, since it is only to procure a proportion, by taking two numbers bearing the required ratio.

We might find the amount by the summation of the infinite series of sums of interest accruing, thus:

Interest on	$300 -	=	18.00
"	$ 18 -	=	1.08
"	$ 1.08	=	.0648
"	$ - .0648	=	.003888

This series might be extended onward forever, but the above being summed up gives $19.148688; differing very little from the true result found above.

To find the sum of such a series accurately, we must consider the last term as nothing, since it is to that it approximates. The ratio as a multiplier in this case is .06, or what is the same thing, 16$\frac{2}{3}$ as a divisor, the extremes are $18 and 0; and the rule in such cases is to divide the difference of the extremes by the ratio less 1, and the quotient increased by the greater term will be the sum of the series. 18—0÷(16$\frac{2}{3}$—1) =1.14$\frac{42}{47}$, and 18 being added gives 19.14\frac{42}{47}$, as before.

Perhaps the present may be an appropriate time to make a few remarks on the doctrine of *approximation*. In the above series of interest on interest, it is obvious that at every step the accruing interest diminishes in the ratio of 100 to 6, or 16$\frac{2}{3}$ to 1, and though the amount would soon be inconceivably small, there would still be an accruing interest. In summing the series, however, we suppose that it really reaches the point to which it approximates, viz: 0, and we find the amount accordingly. We have a familiar instance of approximation in reducing vulgar fractions to decimal ones. In reducing one third to a decimal, the first figure is .3, which differs one thirtieth from the truth; the second is .33, which differs one three hundredth, thus approaching ten times nearer at every step, without ever quite reaching the precise value of one third. In determining the ratio of the diameter of a circle to its circumference, this approximation has been carried so far, that according to the estimate of Mr. Grund in his Plane Trigonometry, "in a circle whose diameter is 100000000 times greater than that of the sun, the error would not amount to the hundred millionth part of the breadth of a hair."

Kindred with this is the position that two lines may approach each other forever without coming in contact. For if we imagine a line extended through half a given intervening space; and then through half the remaining, and so on, through half of each remainder, indefinitely, there will always be a portion remaining to be divided. So any part of the curve of a circle constantly approaches to a straight line, as the circle is enlarged; but it can never become entirely straight. This may be illustrated by drawing two parallel lines near each other, and connecting them by straight lines drawn across; then placing one foot of your dividers on the lower line at the left hand extremity, draw arcs from the upper end of each cross line to the lower line. At first the curve will depart entirely from the right line; but as the radius increases they approximate, and would finally bid defiance to the finest instruments to show the difference; but in truth they still differ, *for one is curved and the other is not.*

We might illustrate it by supposing a trough of indefinite length, and a slider placed as one end. If the trough be filled with fluid, and the slider moved in the direction of the indefinite extension, the fluid will follow, but theoretically the surface of the fluid could never coincide with the bottom of the trough. It is true that in practice the fluid would become so attenuated that it would cease to flow; but in theory the plane of its under side and surface could never coincide. Or imagine two lines to issue from a point, and to pass through holes in a board, standing at right angles to their direction. As the board recedes, the lines must approach, but can never coincide. The conic sectional curve, called an asymptote, is another instance. Problems are sometimes found that can be wrought only by approximation. The standing decimals .7854, 3.1416, .5236, and many others, are but approximations.

21. Divide 100 into two such parts, that one shall be five times as great as the other.

Call the smaller number 1 part, then the larger must be 5

such parts, hence both=6 parts; and $100 \div 6 = 16\frac{2}{3}$, the smaller number, and $16\frac{2}{3} \times 5 = 83\frac{1}{3}$, the larger. Or we may solve it by Position.

22. A, B, C and D undertake to build a house. A, B, C, can build it in 20 days; B, C, D, in 24 days; C, D, A, in 30 days; and A, B, D, in 36 days. In how many days can all together build it, and in how many days will each do it separately?

Suppose each company to work 36 days:

Then will A, B, C build $\frac{36}{20} = 1\frac{4}{5}$ such houses.
And B, C, D " $\frac{36}{24} = 1\frac{1}{2}$ " "
And C, D, A " $\frac{36}{30} = 1\frac{1}{5}$ " "
And A, B, D " $\frac{36}{36} = 1$ " "

And the whole can build $5\frac{1}{2}$ " "

In order to effect this amount of work, it is obvious that each one works in 3 different companies for 36 days, and dividing $5\frac{1}{2}$ by 3 gives $1\frac{5}{6}$, what all four would do in 36 days. Then, $1\frac{5}{6}$ what A, B, C, D do—$1\frac{4}{5}$ what A, B, C do, leaves $\frac{1}{30}$ of a house in 36 days for D, or a whole house in 1080 days.

$1\frac{5}{6}$—$1\frac{1}{2}$, what B, C, D do, leaves $\frac{1}{3}$ for A, or a house in 108 days.

$1\frac{5}{6}$—$1\frac{1}{5}$, what C, D, A do, leaves $\frac{19}{30}$ for B, or a house in $56\frac{16}{19}$ days.

$1\frac{5}{6}$—1, what A, B, D do, leaves $\frac{5}{6}$ for C, or a house in $43\frac{1}{5}$ days.

Then, As $1\frac{5}{6}$ houses : 1 house : : 36 days : $19\frac{7}{11}$ days, the time in which all will build one house.

At first view there would seem to be a difficulty in this description of questions, for as A, B and D have all wrought in other combinations, before they work together, the amount each can do, appears to be fixed, and their amount of work when working together would appear inferable from what each had done before. It may be plainer to take a different question.

23. A, B, C and D undertake to make 6000 rails for E. A, B, C, can make them in 10 days; B, C, D in $7\frac{1}{2}$ days; C, D, A in 8 days; and D, A, B in $8\frac{4}{7}$ days. How many rails can each make per day?

Ans. A 150; B 200; C 250; D 350.

Now as the quantity each can make per day is fixed by the working of the first three combinations, how can we know without previous calculation what the last company can do, when they operate together? By proceeding as in problem 22, we find that A can make 150, B 200, C 250, and D 350 per day. Suppose that instead of saying that D, A, B can make the required number in $8\frac{4}{7}$ days, which they will do if they work as above stated; we had said they could make them in 6 days, which it is obvious they could not do without making more per day than when at work with their associates, according to the supposition already stated. In this case C must make a less number; but though this supposition destroys altogether the proportion in which they had wrought, the new numbers will still answer the conditions of the question; and we may so suppose the numbers as to make some of them even negative. Say that A, B, C require 10 days; B, C, D 6; C, D, A 8; and D, A, B 5; then proceeding as before, we find that C is $16\frac{2}{3}$ rails per day worse than nobody: while A makes $183\frac{1}{3}$, B $433\frac{1}{3}$, and D $583\frac{1}{3}$. It is obvious that the work done by each may be greatly varied, and still the aggregate will be the same. The subject is worth a critical examination.

24. A father dying, left his son a fortune, $\frac{1}{4}$ of which he spent in 8 months; $\frac{3}{7}$ of the remainder lasted him 12 months longer, after which he had only $410. What did his father bequeath him?

Whether these portions were spent in 8 months, 12 months, or 7 years, has nothing to do with the question. He spent $\frac{1}{4}$, and of course had $\frac{3}{4}$ left, $\frac{3}{7}$ of which $\frac{3}{4}=\frac{9}{28}$ of the whole, he then spent. This added to $\frac{1}{4}$ makes $\frac{16}{28}=\frac{4}{7}$, and the remaining $\frac{3}{7}=\$410$: hence $\frac{7}{7}$, or the whole, was worth $\$956\frac{2}{3}$.

25. Suppose 2000 soldiers had been supplied with bread sufficient to last them 12 weeks, allowing each man 14 oz. per day; but on examination they find 105 barrels, containing 200 lbs. each, wholly spoiled; what must be the allowance to each man, that the remainder may last them the contemplated time?

We might find what all would consume in the whole time, and from this deduct the whole loss, and then find what the remainder would be per day; but it would be briefer to find the per diem deduction from the whole amount lost.

Two modes, by *Analysis*, might be as follows:

1st. If 1 man ate 14 oz. in a day, he would eat 7 times $14=98$ oz. in a week; and if he ate 98 oz. in a week, he would eat $12\times98=1176$ oz. in 12 weeks; and 2000 men would eat 2000×1176 oz.$=2352000$ oz. in 12 weeks.

105 barrels of 200 lbs. each$=21000$ lbs. destroyed; and $21000\times16=336000$ oz., which deducted from the whole quantity 2352000 oz. leaves 2016000 oz. to be consumed. Then if 2000 men consume 2016000 oz. in 12 weeks, 1 man

will consume $\frac{2016000}{2000}$=1008, oz.; and if 1 man consume 1008 oz. in 12 weeks, he will consume $\frac{1008}{12}$=84 oz. in 1 week, and $\frac{84}{7}$=12 oz. in 1 day. *Ans.*

By the 2d Mode. 105 barrels of 200 lbs.=21000 lbs. destroyed, and of course be deducted from the allowance; and if 2000 men lose 21000 lbs., one man will lose $\frac{21000}{2000}=\frac{21}{2}$ lbs. $=\frac{21}{2}\times16=168$ oz., and if 1 man in 12 weeks or 84 days lose 168 oz., in 1 day he will lose $\frac{168}{84}$=2 oz. to be deducted from their former allowance of 14 oz. which will leave 12, the *Answer* as before.

By Proportion, we may state as follows:

$$\text{As}\left\{\begin{array}{l}2000 \text{ soldiers : 1 soldier} \\ \quad 12 \text{ weeks : 1 week}\end{array}\right\} : : 105\times200 \text{ lbs. : 14 oz.}$$

per week=2 oz. per day, as before.

Or by two statements,

As 2000 soldiers : 1 soldier : : 21000 lbs. whole loss : $10\frac{1}{2}$ lbs. what each soldier loses in the whole time.

Then, as 12 weeks : 1 week : : 10.5 lbs. : 14 oz. what each loses in a week; and 14÷7=2, what he loses per day, as before.

26. If $\frac{5}{8}$ yard cost \$$\frac{5}{7}$, what will $\frac{9}{15}$ of an Ell English cost?

By Analysis. If $\frac{5}{8}$ y'd cost $\frac{5}{7}$\$ $\frac{1}{8}$ will cost $\frac{1}{5}$ of $\frac{5}{7}=\frac{5}{35}$ of a \$, and $\frac{8}{8}$ being 8 times as much as $\frac{1}{8}$, will cost 8 times as much,$=\frac{5}{35}\times8=\frac{40}{35}$ the price of 1 yard; but as an English Ell is $\frac{1}{4}$ of a yard greater than a yard, so will the price be $\frac{1}{4}$ greater, and $\frac{1}{4}$ of $\frac{40}{35}=\frac{10}{35}$ which added to $\frac{40}{35}=\frac{50}{35}$ the price of an E. Ell. Then if 1 E. E. cost $\frac{50}{35}$, $\frac{1}{15}$ will cost $\frac{1}{15}$ of $\frac{50}{35}=\frac{50}{525}$, and $\frac{9}{15}$ will cost 9 times as much, or $\frac{450}{525}=\frac{90}{105}=\frac{18}{21}=\frac{6}{7}$\$ $=85\frac{5}{7}$ cents, *Ans.*

By Proportion. $\frac{5}{8}$ y'd$=\frac{20}{8}$ qr. and $\frac{9}{15}$ Ell$=\frac{45}{15}$ qr. Then stating and inverting the terms of the divisor, we have,

As $\frac{8}{20}$: $\frac{45}{15}$: : $\frac{5}{7}$: $\frac{1800}{2100}=\frac{6}{7}=85\frac{5}{7}$ cents. The *Ans.*

27. If my age were doubled, and the sum increased by 8, the cube root of that number would be 4. What is my age?

By Reversal. 4^3=64, and 64—8÷2=28. *Ans.*

The following question we have seen in Sir Isaac Newton's Universal Arithmetic, published two hundred years ago. He was probably its author.

27. If 12 oxen eat up $3\frac{1}{3}$ acres of grass in 4 weeks, and 21 oxen eat up 10 acres in 9 weeks, how many oxen will eat up 24 acres in 18 weeks, the grass being at first equal on every acre and growing equally?

We are to understand that the oxen eat equally, but while they eat, the grass grows, and hence as the second lot of oxen fed 9 weeks, more oxen will be fed to the acre than when they fed but 4 weeks. It is to determine the comparative amount of this growth, that two sets of terms are necessary, and our first effort will be to ascertain the ratio of growth weekly to the quantity originally standing on the land.

By Analysis. If 12 oxen eat $3\frac{1}{3}$ acres of grass and its *growth for 4 weeks*, in 4 weeks, 10 acres, being 3 times as much, would serve 36 oxen for the same time; but if they be allowed 9 weeks to eat it in, without further growth of grass, only $\frac{4}{9}$ of 36 oxen, which is 16, will be necessary. 16 is therefore the number of oxen that will eat 10 acres of grass, *with 4 weeks' growth*, in 9 weeks.

By the 2d condition of the question, we learn that 21 oxen were necessary to eat 10 acres of grass, *with 9 weeks' growth*, in 9 weeks. It follows then that 21—16=5 oxen were fed for 9 weeks on the additional 5 weeks' growth, or 1 ox for 9 weeks, on 1 week's growth, and 9 on 9 weeks' growth. But if 9 of the 21 oxen were fed on the growth, 12 must have fed on the grass originally standing on the ground; and as 1 week's growth fed 1 ox for 9 weeks, the weekly increase was $=\frac{1}{12}$ the original quantity on the land.

Then the original quantity on $3\frac{1}{3}$ acres, being increased by growth $\frac{1}{12}$ weekly for 4 weeks, will equal the original quantity on $4\frac{4}{9}$ acres; and this is eaten by 12 oxen in 4 weeks; hence 1 ox eats $\frac{1}{48}$ of $4\frac{4}{9}$ acres, or $\frac{5}{54}$ of an acre in 1 week, and in 18 weeks $\frac{90}{54}$ or $\frac{5}{3}$ of an acre.

The original quantity on 24 acres being increased by growth $\frac{1}{12}$ weekly for 18 weeks, will amount to the original quantity on 60 acres, then as 1 ox will eat $\frac{5}{3}$ of an acre in 18 weeks, $60\div\frac{5}{3}=36$, will be the number required.

By Proportion.

As $\left\{\begin{array}{l}12 \text{ oxen} : 21 \text{ oxen} \\ 4 \text{ weeks} : 9 \text{ weeks}\end{array}\right\} :: 3\frac{1}{3} \text{ acres} : 13\frac{1}{8} \text{ acres}.$

This quantity would serve 21 oxen 9 weeks, if the grass ceased to grow after the first 4 weeks, (the first lot being supposed to grow so long,) but the quantity actually required is by the question supposed to be 10 acres, instead of $13\frac{1}{8}$, so that the growth on 10 acres in 5 weeks, is equal to the grass on $3\frac{1}{8}$ acres with 4 weeks' growth.

Then as $\left\{\begin{array}{l}10 \text{ acres} : 24 \text{ acres} \\ 5 \text{ weeks} : 14 \text{ weeks}\end{array}\right\} :: 3\frac{1}{8} \text{ acres} : 21 \text{ acres},$

the growth on 24 acres in 14 weeks, compared as before; then

21+24=45 acres, the number of acres equivalent to 24, that do not grow after the first 4 weeks.

Lastly, As $\left\{\begin{array}{l} 3\frac{1}{3} \text{ acres} : 45 \text{ acres} \\ 18 \text{ weeks} : 4 \text{ weeks} \end{array}\right\}$: : 12 oxen : 36 oxen. *Ans.*

We might solve it also by Position, by assuming a quantity as the growth per acre, and thus finding the ratio of the weekly growth to the original quantity; but the process would be longer than either of the above, and at the same time less satisfactory. Various modes by Analysis might be adopted, but it is believed that none are simpler than the above.

28. A traveling at the rate of 11 miles per day, B at the rate of 8 miles, and C 5 miles, start to travel in the same direction around a lake 60 miles in circumference. How soon will they all come together?

Solution. The two hands of a watch afford a familiar instance of this kind of chasing, and by increasing the number of hands we might bring up the whole subject.

Suppose a clock face were furnished with two hands, one moving with twice the velocity of the other, and that they leave the 12 o'clock point together; then the faster will reach 12 again at the moment the slower reaches 6; and passing on will reach 12 a second time, at the moment the slower reaches there the first time. If the motions be as 3 to 1, the faster will be at 12 when the slower is at 4; and moving on will pass the slower at 6, and reach 12 a second time when the slower is at 8; passing on again they will reach 12 together, the first time for the slower, and the third for the faster. So suppose they move as 6 to 1; the swifter will be at 12 when the slower is successively at 2, 4, 6, 8, 10 and 12, and the points of transit may be readily determined by calculation. The faster gains 5 on the slower in running 6; and on starting on its second revolution it has 2 hours to gain, hence

	gained		to be gained		time		time
As	5	:	2	: :	6	:	$2\frac{2}{5}$.

The swifter then will overtake the slower in $2\frac{2}{5}$ hours' run; and we may thus determine every point of transit. Or we may do the same by multiplying $2\frac{2}{5}$ successively by 2, 3, 4, &c., since the space to be passed over in overtaking will increase in that ratio.

When the motion is as 2 to 3, 3 to 4, &c., there will not be a conjunction at every revolution of the slower, but at 2 of the one and 3 of the other, three of the one and 4 of the other, &c.

In the above problem the parties start together, and before A can overtake B, he must gain a whole revolution upon him, *i. e.* 60 miles, which he will do in 20 days, for he gains 3 miles per day. And in exactly the same time B overtakes C, for he gained upon him 3 miles per day, and hence they will all be together at the end of 20 days; C having performed $1\frac{2}{3}$ revolutions, B $2\frac{2}{3}$ and A $3\frac{2}{3}$. And if they continue they will meet next time at $3\frac{1}{3}$ revolutions, $7\frac{1}{3}$, $10\frac{1}{3}$, and the third heat would bring them together at the starting point, after 5, 8 and 11 revolutions respectively.

If they travel respectively 11, 8 and 6 miles, A will overtake B in 20 days as before; but B will not have overtaken C, for he gains but two miles per day upon him, and will therefore be 20 miles behind him, which will require 10 days further pursuit; and when he overtakes him A will have travelled $10\times11=110$ miles further, or one revolution and 50 miles. They cannot come together until the space traveled by the slowest is a multiple of the rate of the swiftest; and the same number must be a multiple of the circuit, or else equally exceed some multiple of it. This is obvious. If then we take the differences and find them all alike, such common difference divided into the circuit traveled will give us the number of days, or other periods of travel, in which all would come together; and the greatest common measure of such differences, which when they are equal will be such difference itself, divided into the space traveled by each, will give the number of revolutions each must perform, before all will come together; and this is independent of the length of the circuit. If the differences be not equal, but have a common measure; still the periodical travel of each divided by such common measure will give the number of revolutions made by them respectively, and if the quotients be whole numbers they will meet at the starting point; otherwise the fraction will show how far from such point.

If the differences are prime to each other, then the number of days or other periods of travel necessary will be just equal to the number of miles or other spaces in the circuit; and in that case it can make no difference at what rates they travel, or whether together or in opposite directions, or whether there be two or a "great multitude," if they travel integral miles they must come together at that time. For they who have traveled 1, 2, 3, &c., miles, will have made 1, 2, 3, &c., perfect revolutions, and they must all then be together at the starting point; but if any one travel a fractional space, as $2\frac{1}{2}$, $3\frac{1}{4}$; &c. miles, then he will be a corresponding fractional part of a revolution ahead of his fellows. If the length of a circuit be divided by the greatest common measure of the differences, the quotient will express the *least time* in which all will come together. The reason of these positions will be obvious on reflection, but may, perhaps, be made plainer by a few examples.

29. The length of the circuit remaining at 60 miles, and the parties travelling 8, 6 and 4 miles, they will meet in 30 days, for the common difference is two, and the first will overtake the second in 30 days, for he has 60 miles to gain, and he gains 2 miles per day: and in just the same time, and for the same reason, the second will overtake the third.

30. Suppose they travel 15, 9 and 6, then 6 and 3 will be the respective gains, and 3 being their greatest common measure, $60 \div 3 = 20$ will be the number of days occupied in the pursuit. And 6, 9 and 15 divided by 3 will give 2, 3 and 5 as the number of revolutions each will make, and they will meet as before, at the starting point.

31. Suppose they travel 14, 8 and 5 miles, they will meet as before, at the end of 20 days, for 3 will be the greatest common measure of the gains; but it will not be at the starting point, for 5, 8 and $14 \div 3 = 1\frac{2}{3}$, $2\frac{2}{3}$ and $4\frac{2}{3}$; their point of meeting will therefore be $\frac{2}{3}$ of the circuit beyond the starting point.

32. Suppose they travel 13, 8 and 5 miles, then the difference 3 and 5 have no common measure, and they cannot meet until the end of 60 days.

33. Suppose they travel $13\frac{1}{2}$, 8 and 5 miles, they cannot meet for 120 days, he who traveled $13\frac{1}{2}$ miles per day having made $13\frac{1}{2}$ revolutions when the others have made 5 and 8; but in as much more time he will have gained the other half, when they will respectively have made 10, 16 and 27 revolutions.

The usual mode is to find the time in which the first will overtake the second, the second the third, and so on, and the least common multiple of such periods will be the first time of general meeting.

Taking the foregoing distances, A 15, B 9, and C 6; A overtakes B in 10 days; and B overtakes C in 20 days, and the least common multiple of 10 and 20 is 20, the time required. In that time B overtakes C *once*, and A overtakes B *twice*, hence all are together. The reason will be obvious on examination; a plainer mode however would seem to be to take the greatest common measure of the gains per day, or other period, and dividing the circuit by such measure will give the periods of time occupied in pursuit; and the spaces in each period (say miles per day) being divided by such common measures will give the revolutions performed by each and fix the place of meeting.

34. A going 10 miles per day, B 12, C 16, D 24, and E 30, all commence on the first day of January, circumambulating in the same direction, an island 56 miles in circumference, on what day of the month will all meet and where?

A	loses on	B	per day	2	miles		
B	" "	C	" "	4	"		Com. Measure 2.
C	" "	D	" "	8	"		
D	" "	E	" "	6	"		

56÷2=28, the number of days necessary. Hence they meet January 28th.

A 10÷2= 5	
B 12÷2= 6	
C 16÷2= 8	Number of revolutions performed by the several travelers in 28 days.
D 24÷2=12	
E 30÷2=15	

Or E overtakes	D in	56÷6= $9\frac{1}{3}$	days.	Of these the smallest common multiple is 28.
D "	C	56÷8= 7	"	
C "	B	56÷4=14	"	
B "	A	56÷2=28	"	

B overtakes A once in every 28 days; C overtakes B twice or every 14 days; D overtakes C 4 times, or every 7 days, and E overtakes D three times, or once every $9\frac{1}{3}$ days: so that they are then all together, and must necessarily be whenever the number of days is a common multiple of the time of their several overtakings.

35. A travels 5 miles per day, and C 10 miles per day in the same direction, and B 8 miles per day in an opposite direction, around a lake 80 miles in circumference. How soon will they all come together, and where?

A and C will meet every 16 days at the starting point, and B will be there every 10 days; they will therefore meet at every common multiple of 16 and 10. Eighty is the *least* common multiple, hence at the end of 80 days is the first time that all will be together.

36. A prisoner escaped from prison at 6 in the morning, and at 4 in the afternoon the sheriff started in pursuit, gaining upon the fugitive 3 miles per hour. At midnight the sheriff met an express traveling at the same rate with himself, who reported that he had met the prisoner 24 minutes before 10 o'clock. In what time from commencing the pursuit will he overtake the fugitive? And supposing every thing to be as

above, except that the sheriff gained 5 miles per hour, in what time would he overtake the fugitive?

When the sheriff met the express the fugitive was 2 hours 24 minutes of the sheriff's travel ahead, and when the sheriff reached the point where the express met the prisoner, which would be in 10 hours 24 minutes from the time he started, *i. e.* at 24 minutes past 2 o'clock in the morning, the fugitive would have made another 2 hours 24 minutes, or would be 4 hours 48 minutes of his own travel ahead. But he was at first 10 hours ahead, so that he has lost 5 hours 12 minutes in 10 hours 24 minutes; and

As 5 h. 12 min. : 10 h. : : 10 h. 24 min. : 20 h., the whole time of pursuit; so that he overtook him at noon on the day after he commenced the pursuit. And strange as it may appear, the time of pursuit would be the same let the rate of gain be what it might; for the actual speed of both would be altered. That the gain makes no difference is obvious; for the question is solved without reference to it. From the difference of the rate the distance traveled is easily found, by Position or otherwise.

37. A, B and C, in company, put in \$5762: A's stock was in 5 months, B's 7 months, and C's 9 months; and they gained \$780, which was so divided that $\frac{1}{4}$ of A's was $\frac{1}{5}$ of B's, and $\frac{1}{5}$ of B's was $\frac{1}{3}$ of C's. But B absconded after receiving \$2087. What did each gain or lose by B's misconduct?

Solution. If $\frac{1}{4}$ of A's be $\frac{1}{5}$ of B's, and $\frac{1}{5}$ of B's be $\frac{1}{3}$ of C's, then A's, B's and C's are to each other as 4, 5 and 3. And $4+5+3=12$, and $780\div12=65$, which multiplied successively by 4, 5 and 3, will give:

A's share of the gain,	\$260
B's " "	325
C's " "	195

And these divided by their months respectively, will give their several monthly gains, viz.

A's, - - -	\$52
B's, - - -	$46\frac{3}{7}$
C's, - - -	$21\frac{2}{3}$

Which gains, as the gains were proportionate to the investment, give us the ratio of the share of each to the whole stock.

$\$52+\$46\frac{3}{7}+21\frac{2}{3}=120\frac{2}{21}$, the sum of the monthly gains.

As $120\frac{2}{21} : 52 :: 5762 : \$2494\frac{1118}{1261}$, A's original stock.
As $120\frac{2}{21} : 46\frac{3}{7} :: 5762 : 2227\frac{728}{1261}$, B's " "
As $120\frac{2}{21} : 21\frac{2}{3} :: 5762 : 1039\frac{676}{1261}$, C's " "

Then, when B ran away, he was entitled to his original capital, $\$2227\frac{728}{1261}$, and 7 months' profit, or \$325, making in all $\$2552\frac{728}{1261}$, of which he had received \$2087; consequently A and C gained $\$465\frac{728}{1261}$, by his rascality, and this should be divided between them in the ratio of their stocks.

Dividing it in the ratio of their investments, or as 4 to 3, we have $\$266\frac{390}{8827}$ A's share of B's unclaimed portion, and $\$199\frac{4706}{8827}$ C's share of the same.

38. A father leaves a number of children, and a certain sum, which they are to divide amongst them as follows:—The first is to receive \$100 and one-tenth of the remainder; and after this the second to have \$200 and one-tenth of the remainder; and so on, each succeeding child is to receive \$100 more than the one immediately preceding, and then one-tenth part of that which still remains. At last it is found that all the children have received the same. What was the fortune left, and how many children were there?

Solution. The several "tenths" must decrease by just the same amounts as the specific legacies respectively increase, which in this case is 100 at each step. And if the tenths differ 100, the whole will differ 1000, and the least remainder cannot be less than that sum. The several remainders after the specific legacies will then be found in "round" thousands. Furthermore, 100 is to be deducted before the first tenth is taken, 200 before the second, &c. The first tenth, therefore will be 200 short of a thousand, *i. e.* 800; and hence the undivided estate, after deducting the first \$100, must be $800\times10=8000$; and $8000+100=\$8100$ the whole estate. Then $8100-100=8000$, and $8000\div10=800$, hence $100+800=900$, the share of each, and $8100\div900=9$, the number of children.

39. A drover, having calves, sheep and hogs, was met by an inquisitive fellow, who inquired the number of animals in his drove. He replied that he might calculate for himself: that his whole drove cost him \$400; that $\frac{2}{5}$ of his drove were sheep, $\frac{3}{5}$ of the residue were hogs, and the remainder calves; and that if he could sell his sheep for $\$2\frac{1}{2}$ per head, his hogs for $\$3\frac{1}{2}$, and his calves for \$5 a head, he should make \$119 by the speculation. How many of each kind were there in the drove?

Solution. Assuming some number, we can, from the data, determine the ratio of each kind to the whole; or we may just say there were—

Of sheep $\frac{2}{5}$, leaving $\frac{3}{5}$ for the other animals. Of hogs $\frac{3}{5}$ of $\frac{3}{5}=\frac{9}{25}$, and $\frac{3}{5}-\frac{9}{25}=\frac{6}{25}$=calves.

The comparative numbers of the several kinds were, sheep 10, hogs 9, and calves 6.

10 sheep, at \$2.50=\$25.00
9 hogs, at \$3.50=\$31.50
6 calves, at \$5.00=\$30.00

Necessary to buy the above, \$86.50

But he calculated on \$400+\$119=519. Hence we have

As \$86.50 : \$519 : : 10 Sheep : 60 Sheep.
As \$86.50 : \$519 : : 9 Hogs : 54 Hogs.
As \$86.50 : \$519 : : 6 Calves : 36 Calves.

Animals 150, *Ans.*

40. A factory is divided into 32 shares, and owned equally by 8 persons, A, B, C, D, &c. A sells 3 of his shares to a ninth person, who thus becomes a member of the company, and B sells 2 of his shares to the company, who pay for them from the common stock. After this what proportion of the whole stock does A own?

Solution. There were originally 32 shares, and of course each share was $\frac{1}{32}$ part of the whole, but 2 of the shares being bought in by the company, the whole will be divided into 30 parts, and each share will be $\frac{1}{30}$ of the whole. A had 4 shares but sold 3, he therefore had 1 share=$\frac{1}{30}$ left.

41. G received of H 760 lbs. of rough tallow to try out at 60 cents per hundred pounds, clear, and was to take his pay in rough tallow at 8 cents per lb. G returned 615 lbs. net, and H paid him the balance due to him in rough tallow. Allowing 18 per cent. for waste, what was the balance due G?

Solution. Eighteen per cent.=$5\frac{5}{9}$, which deducted from the rough tallow used, leaves 615 lbs. *net*, therefore 615 lbs. $\div 4\frac{5}{9}$=135 lbs. added to 615=750, the rough tallow from which the 615 lbs. were made; and as 760 lbs. were put into his hands, there are 10 lbs. yet in his possession.

$\frac{615\times.60}{100}$=\$3.69, the price of rendering 615 lbs. at 60 cents per 100 lbs.

\$3.69÷8=$46\frac{1}{8}$, the lbs. of rough tallow that would pay him.
10 lbs. already in his hands,

Leaves $36\frac{1}{8}$ lbs. yet due H.

42. A wall was to be built 700 yards long in 29 days. Now after 12 men had been employed on it for 11 days, it was found that they had completed only 220 yards of the wall. It is required to determine how many men must be added to the former, that the whole number may finish the wall in just the time proposed, at the same rate of working?

Solution. 12 men in 11 days will do 132 days work, and if 220 yards require 132 days work, 700 yards will require 420 days work, or the remaining 480 yards will require 288 days work, which must be done in 29—11=18 days. Hence 288÷18=16 men to finish in the time required, and 16—12 =4, the number to be added.

43. Express 625 in a system of notation which shall have 4 for its radix instead of 10.

It has doubtless occurred to the student's mind that we might adopt any other number instead of 10, as the radix of our scale of notation, and it is obvious that were such change made, the same numbers would no longer be expressed by the same combinations of characters, even though the distinguishing feature of our system, viz:—the value of figures depending on their place, be retained. It is likewise clear that the characters representing numbers will vary in number accordingly as our radix varies. If 2 be the radix, then 0 and 1 will be the only characters; if 3, then 0, 1 and 2, and so on in the same manner. The names of the several places would also change and we should hear no more of *tens*, *hundreds*, *&c.*, for new terms of different import would have supplanted them, varying in signification according to the scale adopted. New characters would also be required if a radix greater than 10 were assumed.

These different scales are called Binary, Ternary, Quaternary, Quinary, Senary, Septenary, Octary, Nonary, Denary, Undenary, Duodenary, &c., accordingly as 2, 3, 4, 5, 6, 7, 8, 9, 10, 11, 12, &c., are used as radices.

We may readily transfer any number from the common or Denary scale to another, by dividing the given number and the successive quotients by the base of the proposed system, and taking the several remainders in their reversed order. Thus, in the question above, we are required to express 625 in the Quaternary scale, and it is done thus:

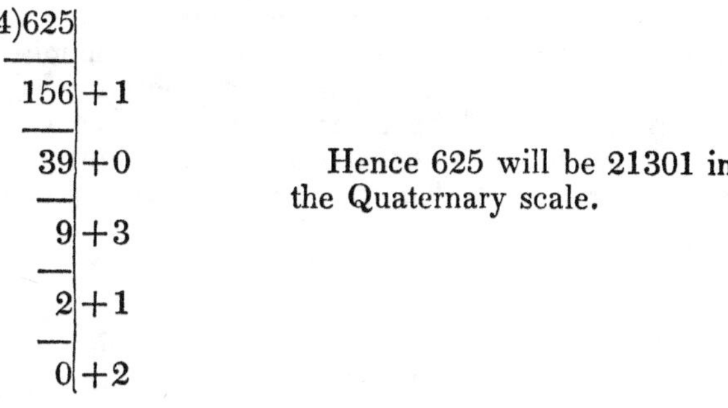

Hence 625 will be 21301 in the Quaternary scale.

We might give a number of examples, but this will serve to illustrate our meaning, and the operation is quite simple. Numbers may be changed to the Denary scale by decomposing them and then adding the several parts. Thus the result in the preceding example may be decomposed as follows:

$$(2\times4^4)+(1\times4^3)+(3\times4^2)+(0\times4)+1=625$$

To change a fractional expression, similar to our decimals, from one scale to another, it is only necessary to multiply the expression by the index of the new scale of notation, conceiving the product to be set each time a place lower, just as in the ordinary rule given for changing decimals to duodecimals, or as it is sometimes expressed "finding the value of decimals."

Many curious results arise from the peculiarities of the different scales, but we have not space to investigate them thoroughly. A few facts may however be succinctly stated.

If any prime number be assumed as a radix, every fraction when changed to a form corresponding to our decimals will circulate, as is evident from what has been shown in the lecture on Circulates, &c. The adoption of 10 as the base of the system is nearly universal, and the ten digits of the hands furnish a ready explanation of the fact, but 12 is a much better number, and it is to be regretted that it was not originally adopted, for it is divisible by 2, 3, 4 and 6, while 2 and 5 are the only factors of 10. With 12 as a radix $\frac{1}{2}$, $\frac{1}{3}$, $\frac{1}{4}$, $\frac{1}{6}$, $\frac{1}{8}$ and $\frac{1}{9}$ would terminate when changed to duodecimals, while $\frac{1}{2}$, $\frac{1}{4}$, $\frac{1}{5}$ and $\frac{1}{8}$ are the only fractions whose denominators are less than 10 that will terminate when changed to decimals. Eighteen and Sixty have many divisors but they are too large, and include too many prime numbers to be even as good radices as 10. Sixty however was formerly used to a considerable extent. The division of the circle and time, in 60ths, is probably a vestige of this scale.

The operation of extracting roots would of course be performed according to the scale in which the number is expressed, but the idea that any number which is a Surd in our present system would be otherwise in a different system is entirely groundless, as the rationality or surdity of numbers rests on other grounds than the mere scale of expression.

LECTURE XIX.

QUESTIONS OF AN AMUSING DESCRIPTION.

The questions that will occupy our attention in the present brief lecture, will not generally be of a class difficult to solve, but rather of a light and amusing character. In some of them the reader may recognize old acquaintances in the shape of puzzles, often given as tests of ingenuity; but though not all of the dignified class, they involve scientific principles as well adapted to exercise the reasoning faculties as questions of a graver cast, and they will often be studied when difficult problems would not be.

"A little nonsense now and then,
Is relished by the best of men."

1. A and B took each 30 pigs to market, A sold his at 3 for a dollar, B at 2 for a dollar, and together they received $25. A afterwards took 60 alone, which he sold *as before* at 5 for $2, and received but $24: what became of the other dollar?

This is rather a catch question, the insinuation that the first lot were sold at the rate of 5 for $2, being only true in part. They commence selling at that rate, but after making ten sales, A's pigs are exhausted, and they have received $20; B still has 10 which he sells at "2 for a dollar" and of course receives $5; whereas had he sold them at the rate of 5 for $2, he would have received but $4. Hence the difficulty is easily settled.

2. The longest side of a triangle is 100 rods; and each of the other sides 50. Required the value of the grass at $5 per acre. *Ans.* $00.

This also is a catch question, as a triangle cannot be formed unless any two of the lines are longer than the third. In this case, the base being laid down, and the two sides of 50 each, being laid down from opposite ends, they will fall upon the base and coincide with it. Imagine that you are attempting to form a triangle of three sticks, and that two of them are just as long as the third, and you will understand the matter.

3. Three men met at a caravansary or inn in Persia; and two of them brought their provision along with them, according to the custom of the country; but the third not having provided any, proposed to the others that they should eat together, and he would pay the value of his proportion. This being agreed to, A produced 5 loaves, and B 3 loaves, all of which the travelers ate together, and C paid 8 pieces of money as the value of his share, with which the others were satisfied, but quarreled about the division of it. Upon this the matter was referred to the judge who decided impartially.—What was his decision?

At first sight it would seem that the money should be divided according to the bread furnished; but we must consider that as the 3 ate 8 loaves, each one ate $2\frac{2}{3}$ loaves of the bread he furnished. This from 5 would leave $2\frac{1}{3}$ loaves furnished the stranger by A; and $3-2\frac{2}{3}=\frac{1}{3}$ furnished by B, hence $2\frac{1}{3}$ to $\frac{1}{3}$ $=7$ to 1, is the ratio in which the money is to be divided. If you imagine A and B to furnish and C to consume all, then the division will be according to amounts furnished.

4. There was a well 30 feet deep, and at the bottom a frog anxious to get out. He got up 3 feet per day, but regularly fell back 2 feet at night. Required the number of days necessary to enable him to get out?

The frog appears to have cleared one foot per day, and at the end of 27 days, he would be 27 feet up, or within 3 feet of the top, and the next day he would get out. He would therefore be 28 days getting out.

5. Two men bet which would eat the greatest number of oysters, A ate ninety-nine, B ate a hundred and won. How many did B eat more than A? *Ans.* One more.

6. Three men, Henry, Richard and Robert, with their wives, Hannah, Mary and Ann, going to a store to buy cloth, each of them purchases as many yards as he or she gives shillings per yard; each man expends 63 shillings more than his

wife; also HENRY buys 23 yards more than MARY, and RICHARD 11 yards more than HANNAH. Please to point out, from the data, the wife of each; no fractions being admitted?

{ *Ans.* ANN is HENRY'S wife, and MARY is RICHARD'S.
HANNAH is ROBERT'S.

Premising that "The product of the sum and difference of two numbers is equal to the difference of their squares," and from the question we learn that the square of the yards bought by each wife is 63 less than the square of her husband's purchase, (for the price of the whole differs that much, and the yards and price per yard in shillings being equal, the price of the whole in shillings is = the square of the yards,) we know that 63 is also the product of the sum and difference of the number of yards bought by a man and his wife. And inasmuch as the following table exhibits all the factors of 63, the sums and differences must be amongst them, the sums of course being the larger or left hand numbers.

$$63 \times 1 = 63$$
$$21 \times 3 = 63$$
$$9 \times 7 = 63$$

And the following table exhibits corresponding pairs of numbers, whose squares differ 63, calculated from the above by adding $\frac{1}{2}$ the sum to half the difference and *vice versa*, and of course the several purchases must be amongst them, the men's being on the left, and their wives' the corresponding numbers on the right.

32 and 31 for $32^2 - 31^2 = 63$
12 and 9 for $12^2 - 9^2 = 63$
8 and 1 for $8^2 - 1^2 = 63$

By examination we find that 32 and 9 differ 23, and infer that HENRY'S purchase was 32 and MARY'S 9; and as 12 and 1 differ 11, we infer that 12 was RICHARD'S purchase and 1 was HANNAH'S, and as only two remain, we infer that they were ROBERT'S and ANN'S, and that ROBERT bought 8 yards and ANN 31.

Hence HENRY (32) and ANN (31) were man and wife,
And RICHARD (12) and MARY (9) were do.
And ROBERT (8) and HANNAH (1) were do.

7. A blacksmith had a stone weight weighing 40 lbs., a mason coming into the shop, hammer in hand, struck it and broke it into 4 pieces: there, says the smith, you have ruined my weight. No, says the mason, I have made it better, for whereas you could before weigh but 40 lbs. with it, now you

can weigh every pound from 1 to 40. Required the size of the pieces?

Ans. 1, 3, 9, 27; for in any geometrical series proceeding in a triple ratio, each term is 1 more than twice the sum of all the preceding, and the above series might proceed to any extent. In using the weights, they must be put in one or both scales as may be necessary: as to weigh 2, put 1 in one scale, and 3 in the other.

8. A blackleg passing through a town in Ohio, bought a hat for $8 and gave in payment a $50 bill. The hatter called on a merchant near by, who changed the note for him, and the blackleg having received his $42 change went his way. The next day the merchant discovered the note to be a counterfeit, and called upon the hatter, who was compelled forthwith to borrow $50 of another friend to redeem it with; but on turning to search for the blackleg he had left town, so that the note was useless on the hatter's hands. The question is, what did he lose—was it $50 besides the hat, or was it $50 including the hat?

This question is generally given with names and circumstances as a real transaction, and if the company knows such persons so much the better, as it serves to withdraw attention from the question; and in almost every case the first impression is, that the hatter lost $50 besides the hat, though it is evident he was paid for the hat, and had he kept the $8 he needed only to have borrowed $42 additional to redeem the note.

9. A person remarked that when he counted over his basket of nuts two by two, three by three, four by four, five by five, or six by six, there was one remaining; but when he counted them by sevens there was no remainder. How many had he?

The least common multiple of 2, 3, 4, 5 and 6 being 60, it is evident, that if 61 were divisible by 7, it would answer the conditions of the question. This not being the case, however, let $60\times2+1$, $60\times3+1$, $60\times4+1$, &c., be tried successively, and it will be found that $301=60\times5+1$, is divisible by 7; and consequently this number answers the conditions of the question. If to this we add 420, the least common multiple of 2, 3, 4, 5, 6 and 7, the sum 721 will be another answer; and by adding perpetually 420, we may find as many answers as we please.

10. Suppose the 9 digits to be placed in a quadrangular form: I demand in what order they must stand, that any three figures in a right line may make just 15?

8	3	4
1	5	9
6	7	2

This is one variety of MASCOPULIUS' Magic Squares, of which much has been said by some writers, and which at one time received much attention, and general modes were sought for constructing them. They are infinite in their variety, but are of no practical use. Some amount to one number, and some another.

11. A gentleman making his address in a lady's family who had five daughters, she told him that their father had made a will, which imported that the first four of the girls' fortunes were, together, to make \$50000; the last four, \$66000; the last three with the first, \$60000; the first three with the last, \$56000; and the first two with the last two, \$64000, which, if he would unravel, and make it appear what each was to have, as he appeared to have a partiality for HARRIET, her third daughter, he should be welcome to her: Pray, what was Miss HARRIET's fortune?

$$\left.\begin{array}{l} A+B+C+D \quad\quad =50000 \\ \quad\; B+C+D+E=66000 \\ A \quad\; +C+D+E=60000 \\ A+B+C \quad\; +E=56000 \\ A+B \quad\; +D+E=64000 \\ \hline \quad\quad\quad\quad\quad\;\; 296000 \end{array}\right\}$$

Then, 296000÷4 the number of repetitions=74000 the sum of their fortunes.

Then,

$$A+B+C+D+E=74000$$

And

$$A+B \quad +D+E=64000$$

Ans. HARRIET's fortune=\$10000

12. A gentleman rented a farm, and contracted to give to his landlord $\frac{2}{5}$ of the produce; but prior to the time of dividing the corn, the tenant used 45 bushels. When the general division was made, it was proposed to give to the landlord 18 bushels from the heap, in lieu of his share of the 45 bushels which the tenant had used, and then to begin and divide the remainder as though none had been used: Would this method have been correct?

The landlord would lose $7\frac{1}{5}$ bushels by such an arrangement, as the rent would entitle him to $\frac{2}{5}$ of the 18. The tenant should give him 18 bushels from his own share after the division is completed, otherwise the landlord would receive but $\frac{2}{7}$ of the first 63 bushels.

13. A and B bought 200 sheep for $400, each paying $200. A pays $1.75 per head, B $2.25. How many sheep did each receive for his $200?

This question is absurd, since 200 sheep will not cost $400 at those rates. A would have $200 \div 1.75 = 114\frac{2}{7}$, and B $200 \div 2.25 = 88\frac{8}{9}$, making $203\frac{11}{63}$ sheep. A version of the "Land Question," containing an absurdity similar to the above, is sometimes met with. This question might be made fair by supposing B's sheep to be worth 50 cents per head more than A's; and then it could be wrought as question 69 is wrought. On that supposition A would receive nearly 112.35 sheep at $1.78+ per head; and B nearly 87.71 at $2.28+ per head.

A question involving a similar absurdity is sometimes given about building 100 rods of stone wall for $100. It also may be made fair.

14. How may 100 be expressed with four nines?

Ans. $99\frac{9}{9}$.

15. A, B, C and D chartered a schooner and loaded it with "*Notions;*" of the stock A owned a third, B a fourth, C a fifth, and D a sixth; and received in return 60 hogsheads of molasses, which the captain delivered according to the respective interests of the stockholders, and found he had 3 hogsheads left. How was it?

The fallacy consists in supposing that these several fractional shares will form a stock. They amount to only $\frac{57}{60}$; hence if A received $\frac{1}{3}=20$; B $\frac{1}{4}=15$; C $\frac{1}{5}=12$; and D $\frac{1}{6}=10$, there would be 3 hogsheads left. The absurdity would be more obvious if we supposed but two owners, A and B, and that the former owned one-half, and the latter one-fourth; and yet the same principle is involved.

16. Two merry companions are to have equal shares of 8 gallons of wine, which is in a vessel containing exactly 8 gallons. Now to divide it equally between them, they have only two other empty vessels, one of 5 gallons, the other of 3. The question is, how they shall divide the wine equally between them by the help of these three vessels?

Fill the 3 and pour it into the five—then fill it again and from it fill up the 5, which will leave one gallon in the 3 gallon keg—empty the 5 into the eight, and pour the one from the 3 into the 5—fill the 3 again and empty it into the 5—Then there will be 4 gallons in the 5 gallon keg and the same left in the 8.

17. A countryman having a Fox, a Goose, and a Peck of Corn, came to a river, where it so happened that he could

carry but one over at a time. Now as no two were to be left together that might destroy each other, he was at his wit's end, for says he "Though the corn can't eat the goose, nor the goose eat the fox; yet the fox can eat the goose, and the goose eat the corn. How shall he carry them over, that they shall not destroy each other?

Let him first take over the Goose, leaving the Fox and Corn; then let him take over the Fox and bring the Goose back; then take over the Corn; and lastly take over the Goose again.

18. Three jealous husbands, A, B and C, with their wives, being ready to pass by night over a river, find at the water side a boat which can carry but two at a time, and for want of a waterman they are compelled to row themselves over the river at several times. The question is how those six persons shall pass, two at a time, so that none of the three wives may be found in the company of one or two men, unless her husband be present?

This may be effected in two or three ways; the following may be as good as any: Let A and wife go over—let A return—let B's and C's wives go over—A's wife returns—B and C go over—B and wife return, A and B go over—C's wife returns, and A's and B's wives go over—then C comes back for his wife. Simple as this question may appear, it is found in the works of Alcuin, who flourished a thousand years ago. Hundreds of years before the art of printing was invented.

19. A canal boat weighing, with its cargo, 15 tons, has to pass an aqueduct of doubtful strength, the water is 4 feet deep, and the aqueduct 15 feet wide, and a waste weir adjacent to it prevents any rise in the water from the motion of the boat. How much will the pressure upon the aqueduct be increased by the boat passing over it. *Ans.* Not Any.

It is an established principle that a body floating upon a fluid displaces its weight of the fluid; and if the boat displaced a weight of water equal to its own weight, the pressure upon the aqueduct was not increased. To make the position more obvious, if the boat were placed in the aqueduct and the water permitted to freeze solid, and the boat then lifted out, the weight of the water necessary to fill the cavity in the ice, would just equal the weight of the boat and cargo. If the weight of the body exceeds the weight of its bulk of fluid it will sink; for which reason bodies will sink in spirit that will float in water; or will sink in pure water that will float in lye or brine. Every housewife knows how to try the strength of lye or brine with a new egg, which will be borne upon the surface in proportion to the strength of the fluid. The reason is that the bulk of water is not increased by having a solid *dissolved* in it; the particles of the solid being divided enter in between the particles of water, and thus make the water specifically heavier, until it becomes heavier bulk for bulk, than the egg, when the egg must swim. When it is a mechanical mixture, the bulk is increased.

20. From 1 mile, subtract 7 furlongs, 39 rods, 5 yards, 1 foot, 5 inches.

	Mile	Fur.	Rods.	Y'ds.	Ft.	In.
From	1	0	0	0	0	0
Take	0	7	39	5	1	5
		0	0	0	0	1

In this problem, instead of borrowing 1 foot, we borrow ½ a foot=6 inches, from which we take 5 inches, and 1 remains; we then carry ½ to 1, and borrowing ½ a yard=1½ feet, we have 1½ from 1½=0, and afterwards proceed as usual.

21.

	Mile	Fur.	Rods.	Y'ds.	Ft.
From	55	0	00	0	0
Take	13	7	39	5	2
	40	7	39	5	1

In this we subtract the foot as usual, and carry 1 to the yards, making 6, which cannot be taken from 5½, the yards in 1 rod; we therefore take from 11, the yards in 2 rods, and so proceed, borrowing 2 instead of 1.

Problems 20 and 21 are designed to show that we do not always borrow a unit, but more or less as circumstances may require. We prefer however, to consider the numbers added to the minuend and subtrahend, not as numbers borrowed and paid back, for that seems rather a commonplace idea, but as equal quantities added to each term, by which their inequality is not changed.

22. Jacob was by contract to serve Laban for his two daughters 14 years; when he had accomplished 10 years, 10 months, 10 weeks, 10 days, 10 hours, 10 minutes, how long had he yet to serve?

	Y'rs.	M.	W.	D.	H.	M.
From	14	0	0	0	0	0
Take	10	10	10	10	10	10
	2	11	0	3	13	50

Here it was necessary to borrow 2 weeks, and then 3 months, and then 2 years, but this was easier than to have reduced the numbers to their proper amounts, by which the subtrahend would have become 12 yrs., 0 mo., 3 w., 3 d., 10 h., 10 m.

23.

	Miles.	Fur.	Poles.	Yds.	Ft.
6)	97	7	39	4	2
	16	2	26	3	$1\frac{7}{12}$
					6
	97	7	39	$4\frac{1}{2}$	$0\frac{1}{2}$

The gist of this is that the proof line will not correspond with the multiplicand in numbers, though it does in value, for $4\frac{1}{2}$ yards and $\frac{1}{2}$ a foot added=4 yards, 2 feet. This difficulty is liable to occur whenever the ratio of value between the denominations is not a whole number.

24. What three figures, multiplied by 4, will make precisely 5? *Ans.* $1\frac{1}{4}$, or 1.25.

25. Required to subtract 45 from 45, and leave 45 as a remainder?

Solution.
$$\begin{array}{r} 9+8+7+6+5+4+3+2+1=45 \\ 1+2+3+4+5+6+7+8+9=45 \\ \hline 8+6+4+1+9+7+5+3+2=45 \end{array}$$

26. From 6 take 9; from 9 take 10;
From 40 take 50, and 6 will remain!

Solution.

SIX	IX	XL
IX	X	L
S	I	X

27. Place 10 cents in a row upon a table, thus, 1, 2, 3, 4, 5, 6, 7, 8, 9, 10; then taking up one of the series, place it upon some other: but with this condition, that you pass over just two cents. Repeat this till there are no single cents left.

Solution. Place 4 on 1, 7 on 3, 5 on 9, 2 on 6, and 8 on 10.

28. How may 13 trees be planted, so that there may be 12 rows, and 3 trees in each row?

Solution. Draw a hexagon, and plant a tree at the centre, at the middle of each side, and at each angle. A diagram will make this plain.

29. A tailor offered his customer $5 per yard for all the cloth left of his pattern; and the coat being made, the latter

asked what was left. The tailor answered, "If you had got $\frac{1}{2}$ of a yard square more, you would have had $\frac{1}{2}$ of a square yard left; and if you had got $\frac{1}{2}$ of a square yard less, you would have had $\frac{1}{2}$ of a yard square too little. How much cloth was left? *Ans.* None.

30. A ropemaker has a ball of thread, and wishes to make a rope with 31 threads, neither more nor less, and would like to have it 100 feet long; but upon forming his strands, finds he lacks just one thread: how much must he reduce the length to gain the thread?

Solution. $100 \times 30 = 3000$ feet, the length of his thread; and $3000 \div 31 = 96\frac{24}{31}$, the length it will make of 31 strands.

31. Two-thirds of 6 are 9; one half of 12 is 7—
The half of 5 is 4, and 6 are half of 11.

Solution. Two-thirds of SIX are IX=9; the upper half of XII is VII=7; the half of FIVE is IV=4; and the upper half of XI is VI=6.

32. Two men owned a mahogany board of superior beauty, 20 feet long and two feet broad at one end, but running to a point at the other. Now they desire to divide the board equally, and yet so that the share of each shall be of the same shape as the above: how can they do it?

Ans. They had better rip it flatwise.

33. A, B and C start to travel 3 miles, and have a pair of shoes to carry. The shoes are to be carried by different persons, and their several distances are to be equal. How can they arrange it?

Solution. Let A carry his shoe a mile, then give it to B, who may carry it through. Let C carry his two miles, and give it to A, who may carry it through. Each will then have carried his shoe 2 miles.

We will now add a few, and leave them to exercise the student's ingenuity.

1. Divide 45 degrees 10 miles 7 furlongs 37 poles 5 yards 1 foot 3 inches and 2 barleycorns by 4, and prove by multiplication.

2. Place the 9 digits in two different ways, so that in one case they may count 17, and in the other 31.

3. Two boys, wishing to amuse themselves by playing at snatch-apple, took a string 4 feet long, and tied it to a hook in the ceiling of a room 7 feet high, and attached an apple to the other end of the string. Now I desire to know the distance they must stand from each other, in order that the apple,

when put in motion, may touch each of their mouths; they being just $4\frac{1}{2}$ feet from the floor?

Ans. 6.2448 feet. A neat diagram might be made by the student for illustration.

4. I have a box that is 12 inches square, for which I wish to make a lid of a board which is 16 inches long, and 9 inches broad, and to have but one joint. How can I do it?

5. Three persons bought a keg of beer, containing 18 quarts: How can they divide it equally by means of a 10, an 8, and a 4 quart measure?

6. What number multiplied by 57 will produce just what 134 multiplied by 71 will do?

7. Seven out of 21 bottles being full of wine, 7 half full, and 7 empty, it is required to distribute them amongst 3 persons, so that each shall have the same quantity of wine, and the same number of bottles.

8. A gentleman owning a section of land, (which is just one mile square,) bequeathed to his wife the north-east quarter, and directed that the remainder be given in farms of precisely the same shape and size to his four sons. Required the shape and size of each?

9. Supposing there are more persons in the world than any one of them has hairs on his head, it then follows as a necessary consequence, that some two of them at least, must have exactly the same number of hairs on their heads, to a hair: required the proof.

10. Three persons bought a quantity of sugar, weighing 51 pounds, which they wish to divide equally amongst them, but having only a 4 and a 7 pound weight, it is required to find how this can be done?

11. Place 17 sticks of equal length so as to form 6 equal squares, then remove 5 sticks and leave 3 perfect squares. How is it done?

12. Let 23463 be multiplied by 12431 pyramidically, as was once the form used.

```
    23463
    12431
    -----
      3
     626
    49852
   3630389
  249723463
  ---------
  291668553
```

13. Ten times 8 is the same as 8 times 10; is $10\frac{1}{2}$ times 8 the same as $8\frac{1}{2}$ times 10. If not, why?

14. If a man 6 feet in height travel round the earth, how much farther must his head travel, than his feet?

15. A fly lighting upon a coach wheel, midway between the hub and tire, is observed to remain while the coach is running. Required the figure described by the fly, and whether it moves all the time with equal velocity?

16. If electricity travels with the velocity of light, and if the cry of *Fire* be raised at Philadelphia at 5 minutes past 12 on Sunday morning, at what hour and on what day of the week will the announcement reach St. Louis, by telegraph, allowing the difference of longitude to be 14 deg. 38 minutes.

17. Reverse the above, and allow the cry to be raised 5 minutes before 12 on Saturday night, at St. Louis, when will it reach Philadelphia.

18. A traveler went westward round the world in one week, starting on Sunday morning, and found he had one day too few in his reckoning. Which day was it that he lost? Another started and went eastward round in a week, and found he had eight days in his reckoning—of what day had he a duplicate?

19. A left Zanesville on Saturday at noon, and kept pace westward with the sun, returning to Zanesville the next day at noon. Now suppose our division of time into weeks to be used all round the earth, and our traveler to inquire the day of the week every ten minutes, where would he first be told, "It is Sunday?" And as he leaves at noon, and keeps pace with the sun, it was noon with him all the time, and he saw neither morning nor evening—and yet Saturday changed to Sunday. How could this be?

Note. The above are fair questions, involving no absurdity, and susceptible of satisfactory solutions. Though not exactly arithmetical, we trust they will be found fraught with amusement and instruction.

20. A traveler having gained the north pole on the 21st of December, on which day it was new moon, took his stand to watch the phenomena of the heavens. When would the moon rise to him—and what would be its apparent motion? When would day break, allowing it to do so when the sun is 18° below the horizon? In what direction would he first see the light? When and in what direction would the sun rise—and what would be its apparent motion, throughout its stay above the horizon? How would day disappear? No allowance being made for refraction. Let the same questions be answered in regard to a person placed at the Arctic Circle—the tropic

of Cancer and the Equator. What would be the appearance of the earth and its shadow, to an eye placed at a remote point in the line of the earth's axis extended indefinitely?

21. The following numbers are often printed on cards and used for telling ages, as high as 63, numbers thought of, &c. We insert them that the principle may be investigated. To use them, let each be handed successively to the person, and if the age or number thought of, be upon it, it is so stated; then add together the first numbers on all such cards, and the sum will be the number sought.

1	33	2	34	4	36
3	35	3	35	5	37
5	37	6	38	6	38
7	39	7	39	7	39
9	41	10	42	12	44
11	43	11	43	13	45
13	45	14	46	14	46
15	47	15	47	15	47
17	49	18	50	20	52
19	51	19	51	21	53
21	53	22	54	22	54
23	55	23	55	23	55
25	57	26	58	28	60
27	59	27	59	29	61
29	61	30	62	30	62
31	63	31	63	31	63

8	40	16	48	32	48
9	41	17	49	33	49
10	42	18	50	34	50
11	43	19	51	35	51
12	44	20	52	36	52
13	45	21	53	37	53
14	46	22	54	38	54
15	47	23	55	39	55
24	56	24	56	40	56
25	57	25	57	41	57
26	58	26	58	42	58
27	59	27	59	43	59
28	60	28	60	44	60
29	61	29	61	45	61
30	62	30	62	46	62
31	63	31	63	47	63

If the age be found on the 1st, 3d and 5th, then it will be $1+4+16=21$.

LECTURE XX.

SOLUTION OF PROBLEMS, IN WHICH ARITHMETIC IS COMBINED WITH GEOMETRY, THE PRINCIPLES OF NATURAL PHILOSOPHY, &c.

1. ASCENDING bodies are retarded in the same ratio in which descending bodies are accelerated; therefore, if a ball discharged from a gun, return to the earth in 12 seconds, how high did it ascend? *Ans.* 576 Feet.

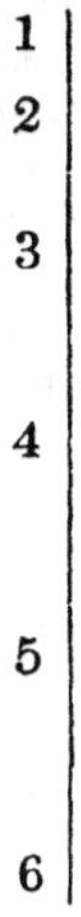

According to the usual theory, the ball was 6 seconds in its ascent, and as many in its descent; and as it was retarded in its ascent in the same ratio in which it was accelerated in its descent, it must have had the same velocity in passing any given point in its descent, as in ascending at the same point. This we might illustrate by the annexed cut, in which the points of descent at the end of each second, are marked by the figures 1, 2, 3, &c. In ascending it would pass from 6 to 5 in the first second; from 5 to 4 in the second, &c. And as the velocity, were it not for atmospheric resistance, would be the same in the ascent and descent at the same point, its power to penetrate any substance would be the same, and it would be as dangerous to encounter a ball coming down as going up. The effect, however, of atmospheric resistance on the descending ball would be great, and would increase with the increase of velocity, as is true of all solid bodies moving in fluids. The comparative resistance would depend too on the size of the ball, since the larger the ball, the less will be the resistance in proportion to the weight; the resistance being proportionate to the surface acted upon, and hence increasing as the *square* of the diameter, while the weight would increase as the *cube*. In a vacuum a cannon ball, and a rifle ball would descend with equal rapidity, but in a fluid, the larger ball would descend most rapidly, and the difference would increase with the increased density of the fluid. In water it would be greater than in air; and in quicksilver still greater than in either. Particles of water are made to float in the atmosphere, partly by their minute division, on the principle we have suggested. The question may be solved by multiplying the time in seconds by 4, and squaring the product for the answer in feet.

2. Suppose an opening to be made from any point on the earth's surface, directly through the earth's centre to the opposite side, and a cannon ball to be dropped into the abyss. Required to know where the ball would come to a state of rest?

The ball would, in passing from the surface of the earth to the centre, acquire a great degree of velocity, being constantly attracted more powerfully by the greater quantity of matter before it, than by the less through which it had passed; and the momentum thus acquired would be sufficient to carry it through to the opposite surface; by which time it would be exhausted and the ball would fall back. Passing the centre to the opposite surface just as it had done before, it would again renew its course as at its first setting out, thus continuing to oscillate from one side of the earth to the other forever. After passing the centre in each vibration, its motion would be retarded by the greater quantity of attraction tending to draw it back, than would exist to carry it forward, the forward attraction diminishing to nothing at the surface; when the opposite attraction would exert its greatest force. But if the influence of atmospheric resistance be considered, the ball at each vibration would fall a little short of the point it had before reached, and would at last come to a state of rest at the centre, as a common pendulum without sufficient maintaining power, is found to do at a perpendicular. Another question might arise, which we shall merely offer for consideration.

The earth being in motion, every thing connected with it partakes of that motion, and has a centrifugal force tending to throw it forward at a tangent to the circle in which it moves; this force is greatest at the surface, because the circle of its revolution is there greatest, and this ball partaking of that force, when detached from its connexion with the earth and dropped into the imaginary abyss, must carry this influence with it. Could it then touch the centre of the globe, and could it fall in a straight line? It certainly could not, if the theory be correct, that bodies fall in the line of an ellipse, having the earth's centre in one of the foci.

3. A B C is a triangle, the side A C being 50 rods, C B 20 rods, and the perpendicular C D 15 rods. Required the diameter of its circumscribing circle?

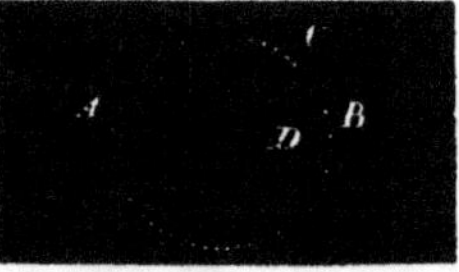

As 15 : 50 : : 20 : 66⅔ the diameter required.

This solution is based on the following position: "In every triangle the rectangle of any two sides is equal to the rectangle of the perpendicular, let fall from the angle included by such sides, and the diameter of the circumscribing circle."

4. The area of a grass plat is 36 square poles, and the sides are as 4 to 1. What is their length?

As 4 : 1 : : 36 : 9, sq're of sho'r side. } Hence { 3 sho'r side.
As 1 : 4 : : 36 : 144, " of lo'r side. } Hence { 12 lo'r side.

5. The sum of the sides of a grass plat is 15, their difference is 9; required their sides?

15÷2=7½ and 7½+4½ (=½ the difference) =12, the longer side; and 7½—4½=3, the shorter side.

6. The difference in the sides of a grass plat is 9, and of their squares 135; required the sides?

Ans. 12 and 3.

7. The surface of a ball measures 3.1416 square feet; and the ball contains .5236 of a solid foot; required the solidity of another ball having four times the surface?

The surface being 4 times the area of a great circle of the ball, the area of the circle must be 3.1416÷4=.7854; and reversing the mode of finding the area from knowing the diameter, we find the diameter to be 1: thus, .7854÷.7854=1, and $\sqrt{1}=1$; then as the supposed ball is to have 4 times the surface, it must have $\sqrt{4}=2$ times the diameter. Hence the diameter will be 2 feet.

Proof. 2×3.1416=6.2832, circumference of larger ball, and this ×2=12.5664 surface, which is 4 times the surface of the first.

8. Suppose a ball, 9 feet in diameter, to be dressed down to a cube; what would be its size?

It is evident that the diagonal of the cube will be the diameter of the ball, and that this diagonal is the hypotenuse of a right angled triangle, the base of which is the diagonal of the base of the cube, and the perpendicular the height of the cube. And furthermore, the diagonal of the base is the hypotenuse of a right angled triangle, of which the two equal sides of the base are the legs. The square of the latter hypotenuse is the sum of the squares of the length of two sides of the cube; and the square of the hypotenuse which forms the diagonal of the cube is the square of the hypotenuse just described, added to the square of the height of the cube; it is therefore made up of the squares of three equal sides of the cube. This will be obvious on examining a cube.

If therefore, we square the diameter, 9, and divide by 3, we have 27, the square of the length of one of the equal sides of the cube; the square root of which is 5.09+, the length required.

9. How large a globe may be turned out of a cube 9 feet square?

It is obvious that the globe would be formed by turning the corners off the cube, and that the diameter would be equal to the length of a side of the cube: *i. e.* 9 feet.

10. How large a square beam may be formed of a log 24 inches in diameter? *Ans.* 16.9+inch. sq.

This depends on a principle similar to the foregoing; the diameter of the log being the hypotenuse of a triangle, the legs of which are sides of the beam. Hence we have but to square 24, and extract the square root of one-half the sum, for the measure of a side.

11. How many acres are contained in a square field, the diagonal of which is 20 perches longer than either of its sides? *Ans.* 14 acres, 2 roods, 11.34 poles.

We may assume a square of any size, and find the excess of the diagonal; then institute the proportion, As the excess thus found, Is to the given excess; So is the side of the assumed diameter to the square of the true diameter.

12. A had a circular meadow, and agreed that B should have the privilege of grazing his three horses at $2 per acre, as follows: He was to drive 3 stakes in the ground, at such distances asunder that when the horses were attached severally to them by ropes, they might graze over the greatest possible quantity of ground in the circle, without encroaching on each other's premises. After the grazing was over it was found that just one acre of grass remained untouched at the centre of the meadow. It is required to give the distance asunder of the three stations—to find the length of the ropes used to confine the horses—the quantity of land in the meadow—and the amount paid by B for what his horses grazed over.

Assume some radius, say 40 rods, for the smaller circles,

and determining the area of one of them, $\frac{1}{6}$ of such area will be the space included in either sector whose centre is at an angle of the triangle A B C, and its area embraced within it, for since the triangle is obviously equilateral, either angle will measure 60°, or one-sixth the circumference of the circle. Tripling the area of such sector will give the area of the triangle, except the curvilineal space at the centre: then, as the whole area of the triangle is easily found, we have but to take the area of the three sectors from the area of the triangle, and we have the area of the curvilineal space, which in the meadow measures just one acre. Then as the area of such space in the assumed figure is to one acre, so is the square of the assumed side to the square of the true side.

Thus, $40\times2=80$ diameter, and $80^2\times.7854=5026.56$, the area of either small circle; and $5026.56\div6=837.76$, the area of the sector falling within the triangle.

Then, as A B is twice $40=80$, $80^2-40^2=4800$, and $\sqrt{4800}=69.282$, the perpendicular of the equilateral triangle; from which the area is found, $69.282\times40=2771.28$. From this deduct $837.76\times3=2513.28$, the area of the 3 sectors embraced in the triangle, and we have 258 rods, the area of the central space.

Then, As 258 rods : 160 rods, or 1 acre : : 40^2 : 992.248, the square root of which is 31.5, *very nearly*, being the proper radius of the smaller circles, and consequently "the length of the ropes used to confine the horses," twice which, or 63, being "the distance asunder of the stations," and the length of either side of the triangle A B C.

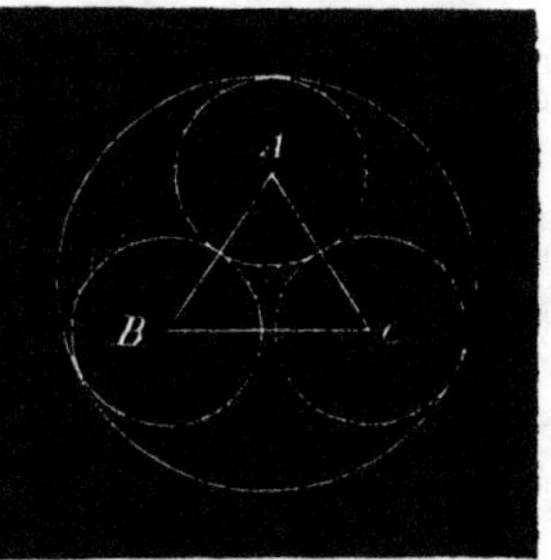

The area of the 3 circles of 31.5 rods radius, grazed over, as found by the ordinary rule, will be 58 acres, 1 rood, 31.7 poles, amounting, at \$2 per acre, to \$116.89$\frac{5}{8}$.

The perpendicular of the central triangle is the square root of $63^2-31.5^2=54.56$, and as the distance from either angle to the centre of a circle circumscribing an equilateral triangle, is just two-thirds of the perpendicular let fall from such angle to the opposite side, the distance from A to the centre is two-thirds of 54.56 rods, which is 36.37 rods, to which the radius 31.5 being added, we have 67.87, the radius of the great circle; from which its area, 90 acres, 1 rood, 31.26 poles, is readily found.

13. In turning a one horse chaise in a ring, it was observed that the outer wheel made two turns, while the inner wheel made but one; the wheels were both 4 feet high, and supposing them fixed at 5 feet asunder on the axle, what was the circumference of the track described by the outer wheel?

It is obvious that having the height of the wheel is of no importance, the *gist* of the question being to find two concentric circles, that being 5 feet asunder, the outer shall be twice the inner. As the circumferences of circles are to each other simply as their diameters, the diameter of the smaller circle must be half the greater, which by the question is 10 feet greater than the less; hence the less is 10 feet in diameter; and the greater 20 in diameter, or 62.832 in circumference.

14. What is the weight of a hollow spherical iron shell, 5 inches in diameter, the thickness of the metal being one inch; and a cubic inch of iron weighing $\frac{139}{500}$ of a pound?

$5^3 \times .5236 = 65.45$, solidity of shell, including cavity.
$3^3 \times .5236 = 14.1372$, " of cavity.

51.3128, " of metal.
$51.3128 \times \frac{139}{500} = 14.2649584$ pounds, weight of metal.

15. Suppose the earth to contain 4,000,000,000,000,000,000,000 cubic feet, and each foot to weigh 100 lbs., and that the earth was suspended on a lever, its centre being 6,000 miles from the prop or fulcrum; how far must a man be placed on the opposite side of the fulcrum, that with a force of 200 lbs., he may hold the earth in equilibrium?

The supposed weight of the earth being multiplied by its distance from the fulcrum and divided by the power the man can exert, will give 12,000,000,000,000,000,000,000, as the distance in miles of the man from the fulcrum, necessary to produce an equilibrium. From this we may see what a perfect "abstraction" was the boast of ARCHIMEDES, "Give me a fulcrum for my lever, and I will move the world," for had he adopted the above data, and had he left his fulcrum for his station at the hour Adam was placed in the garden of Eden, and traveled day and night, with the velocity of a ray of light, 12 millions of miles per minute, he would still be thousands of thousands of years from his journey's end. Of such numbers the human mind can form no conception.

16. A, B and C bought a grindstone 3 feet in diameter for $5; C paid $1.25; B $1.75; and A $2.00. Required the thickness each must grind down, C owning the central part. No allowance to be made for the eye?

36=diameter of stone; Then, as \$5 : \$1.25 : : 36^2 : 324, the $\surd$ of which is 18, the diameter of C's share.

Then C \$1.25+B \$1.75=\$3. As \$5 : \$3 : : 36^2 : : $777\frac{3}{7}$, the $\surd$ being 27.88+the diameter of C's and B's together; from which taking C's, 18, leaves 9.88, and the half of this is 4.94, the thickness B must grind down.

Then 36—27.88÷2=4.06 inches, A's share.

17. In a pair of scales, it is found that a pig of lead weighs 90 lbs. in one scale and 40 in the other, required the true weight, and the cause of the difference?

The cause of the discrepancy is that the arms of the beam, (*i. e.* the distance from the pivot or fulcrum to the points where the scales are suspended,) are not of equal length; by which any weight at the longer will evidently counterpoise a greater one at the shorter. Neither the weight indicated when the lead is at the longer or shorter end can be the true one, the former being as many times less than the true weight as the latter is greater; *hence the true weight is a mean between the weights indicated.* And it is a geometrical mean, being produced by multiplying the less extreme or dividing the greater; and not by adding or subtracting equal differences. Hence the true weight is $\surd$ (90×40)=60 lbs; which is $1\frac{1}{2}$ times 40, as 90 is $1\frac{1}{2}$ times 60. The arms must then be so divided that 60 lbs., the true weight of the lead will, when placed at the short end, balance 90 lbs. at the longer; and 60 : 90 : : 1 : $1\frac{1}{2}$ or 2 to 3, hence the arms of the beam must be as 2 to 3, which will make the virtual velocities of the lead and weight equal; 60×3=90×2; or 40×3=60×2.

This case must be distinguished from balancing 40 lbs. true weight, and 90 lbs. true weight, at the same time; for then the arms must be in the same ratio; that the virtual velocities may be the same.

Beams are sometimes defective in this way, and the defect is concealed by making the scales of unequal weight; but the defect is at once seen by weighing the same thing in both scales; or by putting at the same time two weights known to be equal, in opposite scales. The fraudulent might use such scales to buy and sell with: placing the body to be weighed in one scale or the other according to their interest.

If in buying, half the number of pounds to be weighed, be weighed in one scale, and the other half afterwards in the other, the purchaser will get more than his proper weight. Say that the beam shall be so divided that one point of suspension shall be 11 inches and the other 12 inches from the centre of motion. Putting the pound weight in the scale at the long end $1\frac{1}{11}$ lbs.

will be counterpoised at the shorter; and putting the pound weight into the shorter $\frac{11}{12}$ lb. will be balanced at the longer; add these together and the result will be $2\frac{1}{132}$ lbs. This may at first sight seem unaccountable, but it needs only a little attention to make it plain.

The beam should be so constructed that its centre of gravity may be immediately *under* the axis or centre of motion; for if the centre of gravity were itself the centre of motion, the beam would rest in any position, and would not tend to that horizontal position indispensable in a balance; while if the centre of gravity were above the centre of motion, the beam would constantly tend to upset, and indeed to turn under the axis. The centre of gravity when properly placed being below the centre of motion, the beam constantly tends to assume a horizontal position.

A line being drawn from the centre of gravity to the centre of motion and another from one point of suspension to the other, the latter line should be cut at right angles by the former and also into two precisely equal parts. If a line connecting the points of suspension do not conform to these conditions, the points must be altered until it does. If the centre of gravity be too far below the line of suspension, the instrument will not be sufficiently delicate, and hence will not weigh with nice accuracy, while on the other hand if not far enough the instrument will be unsteady.

18. Two men carry a hog weighing 200 pounds, upon a pole, the ends resting upon their shoulders; how much will each sustain if the pole be 6 feet long, and the hog hangs 6 inches from the middle of the pole?

In this case one man will be $2\frac{1}{2}$ feet, and the other $3\frac{1}{2}$ from the burden. Then as 6, the whole length : $2\frac{1}{2}$, one of the ends : : 200 lbs. whole weight : $83\frac{1}{3}$ lbs. which he that is farthest from the weight must carry. And as 6 ft. : $3\frac{1}{2}$: : 200 : $116\frac{2}{3}$ lbs. that the other must carry.

If the persons are of equal height, it will make no difference whether the weight be suspended loosely from the pole, or be firmly attached to it. But if they be not so, *and the centre of gravity* be not in the line that supports the burden, it will make a material difference whether the centre is firmly placed above or below the pole, or whether the burden *hangs* loose, so as constantly to retain a vertical position.

Two men carrying a weight upon a pole are an instance of the second order of levers, and the stress upon each is inversely as his distance from the weight. If A and B with a lever of 5 feet, carry a burden of 300 lbs. suspended 3 feet from A and 2 feet from B, A will sustain 120 lbs., and B 180 lbs.; and in making the calculation we may consider B as a fulcrum to A, and A as a fulcrum to B. But this is supposing the pole to be carried in a horizontal position, which is the simplest form. If one man be taller than the other, so as to raise one end of the pole, the proportion of their burdens will be changed.

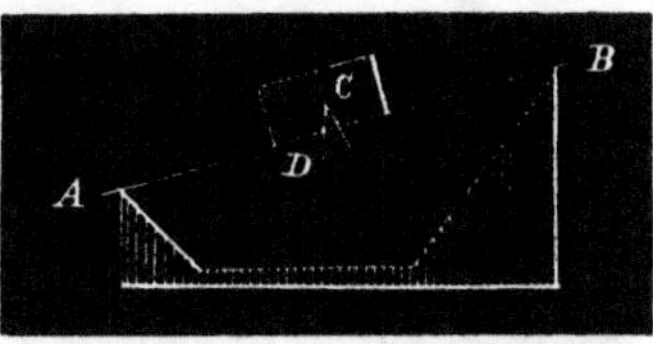

Suppose A B to represent a piece of scantling the centre of gravity of which is in the line A B, then the fulcrums at the ends will sustain equal pressure, however they may differ in elevation. But if a block be laid on the scantling, by which the centre of gravity shall be raised to C, then the perpendicular to the horizon, let fall from C, will strike the line at D, and A will sustain more than half the weight, in proportion as D B, exceeds A D. If on the other hand an addition be made to the lower side so as to throw the centre of gravity from the plane of the line to the point C, so that a perpendicular to the horizon shall cut the line at D, then B will sustain greater pressure, in proportion, as A D exceeds D B. If in these two cases A and B were persons carrying burdens, the effect on each is easily seen. Where a weight hangs loosely as supposed in the preceding question the case is different.

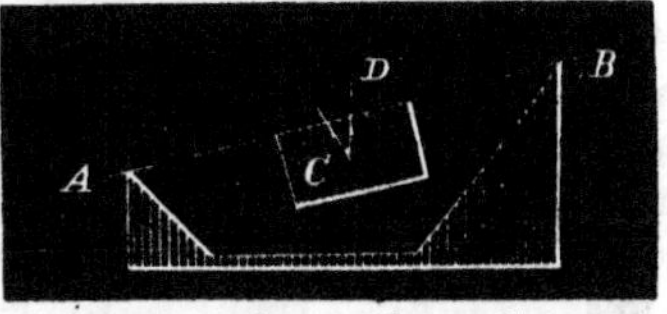

The manner in which horses draw at a double tree is another instance of the application of this principle. If three horses of equal strength are to draw abreast at a plough, the triple tree at which they draw will be attached to the plough with the clevis at $\frac{1}{3}$ its length from one end, and the longer end will be given to the single horse, while the two horses will draw at the double tree at the opposite end. If these are to draw equally, the tree will be equally divided. If it be desired that the single horse shall draw less than $\frac{1}{3}$ give him "more end," but if more, give him less. The same principle applies in dividing the double tree. I recollect when a boy, hearing a man instruct a blacksmith who was ironing a double tree for him, to put the clip a little nearer the off horse, for he was not so strong as the near horse, and he wished him to have the advantage of being nearest the load!! Dobbin perhaps owed his master gratitude for his good will, but the less he had of his scientific favors, the better for him.

In this case, each horse would be a fulcrum to the other, while the weight or draught would be between; just as would be the case of two persons carrying a burden upon a pole.

19. If a man weighing 160 lbs. rest on the end of a lever 10 feet long, what weight will he balance on the other end, supposing the prop to be one foot from the weight?

Ans. 1440 lbs.

According to the question, the man was 9 feet from the fulcrum, and the weight 1 foot; hence the man being nine times as far from the fulcrum, will balance 9 times his weight, and 9×160 lbs.=1440 lbs.

I regret that our limits will not admit, as was originally designed, a full illustration of the principles of mechanics, we must content ourselves therefore with a few principles, and refer the student to treatises written expressly upon that science. A thorough investigation of the subject cannot be too strongly recommended.

If I wish to raise 4 weights of 100 lbs. each, to a height of 3 feet, I may take them one at a time and raise them to the point I desire, and in doing so, I carry the weight 100 pounds, through a space of 12 feet. But suppose I resort to machinery instead of applying my strength to the weights themselves. I may unite the weights into one and by using a lever in which the gain is as 4 to 1, I may raise the whole to the required height by using the same strength I did before to lift one; but when I am done, it will be found that I have exercised this power through the full space of 12 feet. If I resort to the Pully or the Wheel and Axle, or any other of the six mechanical powers, I shall find that though I have raised all at once, I have traveled in the exercise of my strength over 4 feet for every foot the weight was raised. The position is correct, that to gain *Power*, you must lose *Time*; to gain *Time* you must sacrifice *Power*. Take in your hand a stick of any length, as a two feet measure for instance, and laying it across some sharp edge as a fulcrum at 12 inches. and attempt to raise a weight with it. Here you will find the arms of your lever equal—your power and the weight pass through the same space—and consequently nothing is gained: but shift your lever to the 6 inch mark, and with one pound you will raise 3, but you will find that for every inch you raise the weight, your hand will pass through 3 inches. If you balance two balls of unequal weight connected with a wire, across a sharp edge and put them in motion, and then multiply each ball by the space passed through, in any given time, the products will be equal. Say for instance

A and B are two balls, B weighing 5 lbs., and A 15 lbs., connected by a wire which sustains their centre of gravity at C, which from the proportion of the weights is necessarily 3 times as far from B as from A.

Now if the rod and balls be caused to move upon or around C as a centre, the product of A multiplied by the distance passed through in any given time, will be equal to the product of B multiplied by its distance in the same time, for it will be found that B being $\frac{1}{3}$ the weight of A, and consequently resting 3 times as far from the centre of motion, will move through 3 inches while A moves through one inch. This is called the doctrine of *virtual velocities*. If the machinery be so disposed that 50 lbs. of strength will raise the 400 lbs., the principle still holds true, for now the power, that when exercising a force of 100 lbs. moved through 12 feet, to accomplish its task, must move through 24 feet. Let us see, $50\times24=1200$ and $400\times3=1200$: the power multiplied by its distance from the centre is equal to the weight multiplied by its distance—their virtual velocities are therefore equal.

20. Suppose a railroad to have an ascent of 5 feet per mile, what power exerted parallel to the road will hold to its place a train of cars weighing 10 Ohio tons?

	mile		ft.			tons		lbs.	
As	1	:	5	:	:	10	:	$18\frac{31}{33}$	*Ans.*

If the plane were perfectly horizontal, it is certain that a body placed upon it, though without any friction or adhesiveness, would rest wherever placed; but elevate one end of the plane and the body would slide or roll off. To prevent this, force must be exerted in the direction of the elevated end; and while thus situated, two forces would be exerted to sustain the body, viz.: the plane underneath, and the force that prevents it from sliding. As the plane would be more and more elevated, less weight constantly would rest on the plane and more would fall upon the sustaining power, until reaching a vertical position, the plane would then be entirely relieved, and the sustaining power would support the whole burden.

If the sustaining power be exercised in a direction parallel to the plane, the advantage gained will be as the length of the plane exceeds the height of the plane, but if the force be exerted parallel to the base, the advantage gained will be as the length of the base exceeds the height of the plane. Let A B

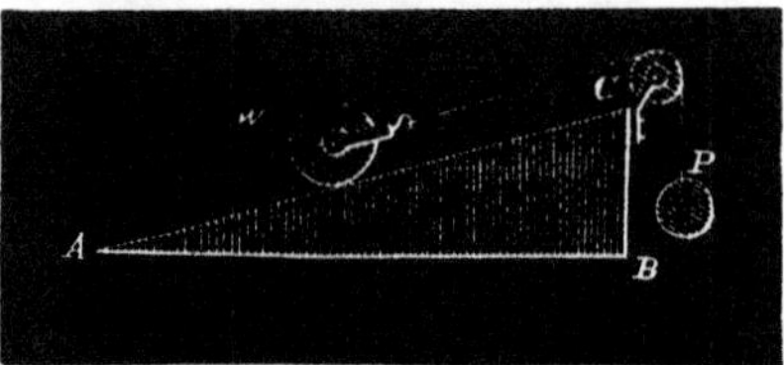

=8 feet, B C 6 feet, and A C will then be 10 feet. If then the ball W weigh 100 lbs., and it be kept in its position by the cord represented as passing the wheel at C, by which the power is exerted parallel with the plane, the length of the plane will be to its height as 10 to 6, the ball W will be to the power P in the same ratio. Hence as 10=(length of plane) : 6 (its height) : : 100 lbs. : 60 lbs. the power necessary to prevent the ball from rolling or sliding down. But if instead of exerting this power in a direction parallel to the plane, it were exerted in a direction parallel to the base, then it is evident that it would draw partially against the plane and would be less effective—it would then only gain in proportion as the length of the base A B of the plane is to its height B C.

It is clear in the above case that the whole weight 100 lbs. is supported, and that if 60 lbs. be sustained by the cord, 40 lbs. will be supported by the plane: furthermore put them in motion and the doctrine of virtual velocities applies strictly, for 60×10=600 and 100×6=600. Let a weight be suspended by a cord, it may then be moved to and fro by a very small force, and as it moves in a curve, the centre of gravity will be raised at every oscillation. This is on the principle of the inclined plane, and as there is no friction, a very great weight may thus be moved by a very small power.

As to the principle of the *Pulley*, it may be necessary to examine it more closely, since many writers on the subject refer it to the principle of the Lever, and give a false rule for finding the power of the Pulley.

Suppose it be desired to raise a stone weighing 200 pounds, from the ground to a scaffold 20 feet high, and that we can exert a power of only one man, and he can lift but 100 pounds. We might, as we have seen, use a lever of the first or second order, and by properly adjusting the fulcrum be able to raise the stone; but then any length of lever that one man could manage would raise the stone but a few inches before the operator must renew his "purchase;" and to be compelled thus to block up the stone and the fulcrum some 8 or 10 times in the required height, and to erect scaffolding at the same time for the operator, would be excessively tedious and expensive. To obviate this we might attach a cord to the stone and use a windlass or wheel and axle, by which the lever would become perpetual, and thus the weight could be raised; and if the weight were too great for the power without making the wind-

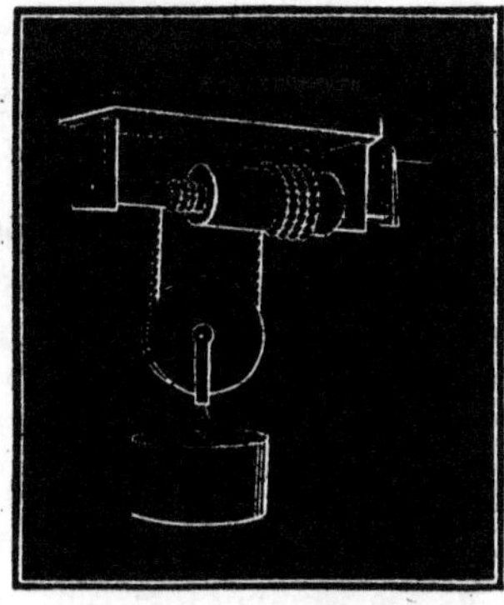

lass too small for strength, or the winch too long for the convenience of the operator, the following form could be adopted, by which any required mechanical advantage could be gained without weakening the axis or increasing the length of the handle. As the portions of the axis approach in size the power increases, for at each revolution of the axis the power will rise or fall through a space equal to the difference in their circumferences.

But there is another mode that might be adopted. We might attach a rope and the person taking his stand on the scaffold could draw up the stone *if he could lift* 200 lbs.; but as his strength is limited to 100 pounds he cannot succeed. If there were two persons each capable of lifting 100 lbs. they could attach two ropes and thus raise the stone, but here again a difficulty occurs, there is but one person If however, we can substitute one rope and so adjust it that by shortening one end, we shorten the other also at the same time, one man may manage the business, provided he can fasten one end to something capable of sustaining 100 pounds. Suppose then a ring be placed in the stone, and having attached one end of the rope to the scaffold above, we pass the rope through the ring and carry the other end up by which the man can draw; the rope will slide through the ring and each part or half of it will bear half the burden. It is then evident that the man by his strength can sustain his half, and by drawing it up, the stone will be raised to the height required. But it is not necessary that the man shall stand upon the scaffold, he may attach a ring above the scaffold and passing the rope through it, he may take his stand on the ground and by drawing the rope downward the weight will be raised; the weight will still rest on but two ropes, but the ring above will enable the operator to change the direction of his power, by which he may work to better advantage. It is true that the stiffness of any rope that could be used, and its roughness, would cause a waste of power, but that is a matter for future consideration and does not affect at all the principles which we desire to explain. If we could divide the weight amongst a hundred cords, each cord would sustain but two pounds, or a hundredth of the whole weight; and though they were all connected so that the movement of one would affect the whole, 2 lbs. suspended to the last rope would keep all in equilibrium. This is the principle of the pulley, and will be found to run through all its various combinations; though they are more numerous than those of any other simple power. But in this too we find the principle of virtual velocities, for the man raising the 200 lbs. weight with a power of 100 lbs. must shorten the end of the rope at which he works two feet for every foot the weight is raised; for each portion of the rope that bears the burden must be shortened one foot. And if the weight were suspended as suggested, on a hundred cords, every one of them must be shortened a foot before the weight can be raised a foot, and consequently the power must pass through a hundred feet to raise the weight one foot or 2000 feet to raise the weight 20 feet. Let us see whether their virtual velocities will be equal: $2000 \times 2 = 4000$ and $20 \times 200 = 4000$, just as we found in the Lever and in the Wheel and Axle.

In order to cause the rope to pass through the rings with as little friction as possible, it is customary to pass it over small wheels; but the wheel is used only to overcome the defects arising from the stiffness and roughness of the rope, and is not necessary at all to the principle of the machine. Whether the rope passes over wheels or fixed axles, the principle is the same. Some, however, contend that the principle consists in the wheel itself, and hence the wheel is often termed a pulley.

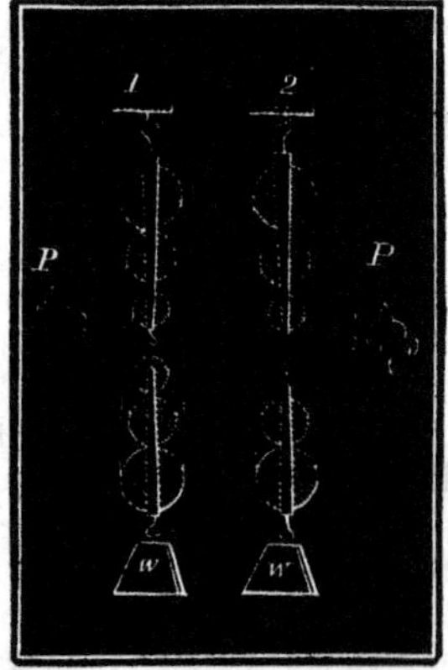

There are many forms of rigging pulleys, but the most common is represented in figure 1, though some blocks have more and some less than three sheaves. By studying any single wheel it will be evident that its cross or horizontal diameter is a lever of the second order, the fulcrum being the point of contact with the fixed part of the rope, the power is the movable part of the rope, and the weight rests on the pivot at the centre. Such a wheel would afford a mechanical advantage as 2 to 1, since the power is twice as far from the fulcrum as the weight is, and the power of a number of them would be equal to twice their number; the corresponding fixed pulleys in the upper block having no mechanical power, but only changing the direction of the power applied. As there are three wheels in the lower block of Fig. 1, and each gains as 2 to 1, the whole will gain as 6 to 1. On this view of the subject, as the wheel owes its efficiency to the lever principle, may conclude that the pulley does also; and give the general rule to find the power of a system, " Multiply the power by twice the number of wheels in the moveable block, for the weight it will balance."

But in Fig. 2, there are only two moveable wheels, and it should by this rule gain only as 4 to 1, but it will be found on trial to gain as 5 to 1; though there are only two wheels in the lower block, and consequently the same number of levers, and if that were the principle the power should be just as 4 to 1. It may be said that there are still three in the upper sheave, but that need not be so, for the "fall" may be carried upward by a power above, and the upper wheel thus be dispensed with; but whether there be two or three in the fixed block is of no importance, since they only change the direction of the power, without affording any mechanical advantage.

If then the lever principle will not apply in all cases, let us sēe what will. The rule that will apply in all cases is to multiply the power by the number of cords that support the weight. In Fig. 1, this is 6, in Fig. 2, it is 5. And this rule applies equally as well if the rope pass over wheels, slide on axles, or be fastened to the weight at detached points. It is applicable in all cases, and is based on reason. To gain in an odd ratio, as 1 to 3, 1 to 5, 1 to 7, &c., the rope must be fastened to the lower block, as in Fig. 2, while to gain as 1 to 2, 1 to 4, 1 to 6, &c., let it be fastened above. In the latter case doubling the lower wheels will do, but not in the former.

Various other forms exist, but the same principle, under proper modifications, applies throughout; and the intelligent observer may readily estimate the power of any combination.

If power be not gained by these mechanical agents, what advantage then results? may well be asked. We answer, much in many ways.

The laborer may carry small burdens one at a time, but the weighty timber cannot be hewed piecemeal that it may be transported with ease; nor could the rocks that built the Egyptian pyramids have been raised to their destined places in the wall by the unassisted hands of man. The sailor rigs his pulley, and with ease manages his sails and rigging, or the unaided builder places his heavy materials in their appropriate places. In all this he does not *create* power, neither does he annihilate gravity, but he so arranges the matter as to divide the burden amongst several agents, that he may be able to manage a division of it himself. If he uses a Lever, he throws so much of the burden on the fulcrum as will enable him to manage the remainder himself; if he adopts the Pulley, he increases the number of sustaining cords until he can with the strength at his command, control one of them; and if he uses the Inclined Plane, he reduces the inclination and throws burden on the plane, until he can sustain the remainder. So with the Wheel and Axle, the Screw, and the Wedge, the artifice, if we may so express it, is the same. It enables the operator to divide the resistance to be overcome, without dividing the body itself. A correct knowledge of these principles enables the workman to combine these powers to suit his purposes; for cranes and all other mechanical contrivances are but combinations of these simple elements. They are the A B C of mechanism, and the student cannot be too familiar with them.

21. The reservoir of the Zanesville water works is 170 feet above the forcing pump by which it is supplied with water from the river, and the length of the ascending pipe through which the water is driven is 2400 feet, the ascending pipe being 10 inches in diameter in the clear. Required the pressure on each square inch of the catch valve at the pump when the machinery is not in motion, supposing a cubical foot of water to weigh $62\frac{1}{2}$ lbs. *Ans.* $73\frac{113}{144}$ lbs.

Fluids press according to their depth, without regard to the shape or size of the vessel, if it be but large enough not to be influenced by capillary attraction. In this case the absolute weight of a column of water whose cross section is a square inch, and height 170 feet, is the measure of its pressure; but this strain upon the ascending pipe will necessarily be increased in putting the water in motion, according to the violence or velocity of the stroke, though the elasticity of the air in the air chest, will aid greatly to break the force of the stroke in its tendency to burst the pipe. The difference between the pressure and weight of water is made most obvious by inserting a small long tube into a close vessel, when the vessel and tube being filled with water or other fluid, the pressure will be as great upon each part of the vessel as though the tube were as large in diameter as the vessel. The pressure upon the base of a vessel, diminishing in size upward, is just as great as if it increased in size, and consequently capacity. The power of the Hydraulic Press is owing to this principle; and we sometimes see large rocks upturned by the operation of the same law. We have a familiar instance, and a striking one, of the force of atmospheric pressure in boring a gimlet hole into a vessel to give air when drawing off from the spile. Fluids press in all directions. If a cubical box were filled with water, the sides and bottom would of course be of equal area, and the pressure against each side would be just half what it is upon the bottom, for it is as great sidewise at the bottom as downwards, and it would diminish to nothing at the surface. Hence the amount of pressure in such a vessel would be three times the absolute weight of fluid contained. The downward pressure would be just the weight, and the sides twice the weight. But suppose such vessel be closed over, and a small tube inserted, then a few pints more of water might increase the pressure a hundred fold. In the problem above, the distance from the power house to the reservoir is of no importance, unless the machinery is to be put in motion, when the friction will be increased by the distance, as well as by any unevenness upon the inner surface of the pipe, and it is equally indifferent whether the pipe is 10 inches or 10 feet in diameter; so far as the pressure on a square inch is concerned.

22. A farmer having a bank of wheat to measure, finds himself without a vessel to measure it in. He accordingly made him a box 15 inches square, but could not tell precisely the depth necessary that it might contain just a bushel. Please assist him.

The capacity of a standard bushel is 2150.4 cubic inches, and this divided by 15×15=225 will give 9.55 inches as the depth. In this way the depth may be found to any size or shaped bottom. By extracting the cube root of 2150.4, the depth will be found of a box completely cubical, or equal on all sides. On the same principle garners, cribs, &c., may be made to contain any quantity that may be desired. The side of a cubical box that will hold a bushel is a trifle over 12.9 inches in length.

23. It is often suggested from the pulpit and elsewhere, that enough persons have lived and died in the world to cover its whole surface with bodies; and even two or three strata deep. Is this probable?

Say the earth has existed 6000 years, the population always having been 800,000,000, and the average life of man 30 years; this being the utmost that could be claimed. Allow then the State of Virginia to contain 70,000 square miles, and each grave to occupy a space of 6 feet by 2; the territory of the State would contain 162,624,000,000; while the mighty army of the dead would number only 160,000,000,000; leaving 2,624,000,000 graves yet unoccupied. How wide of truth then is the position often set forth so positively!

24. From two different sized orifices of a reservoir the water runs with unequal velocities. We know that the orifices are

in size as 5 to 13, and the velocities of the fluids are as 8 to 7; we know further, that in a certain time there issued from the one 561 cubic feet more than there did from the other. How much water then, did each orifice discharge in this space of time?

Solution. The first is $\frac{5}{13}$ the size of the second; but the velocity of the fluid is as 8 to 7 or $\frac{8}{7}$, hence taking size and velocity both into consideration, the discharge of the first will be $\frac{5}{13}$ of $\frac{8}{7}=\frac{40}{91}$ of the second. The difference between the two then will be $1-\frac{40}{91}=\frac{51}{91}$, which, by the question, is equal to 561 cubic feet. Hence

As 51 : 91 : : 561 : 1001, the amount discharged by the second, and 1001—561=440, what the first discharged.

25. A person possesses a wagon, with a mechanical contrivance by which the difference of the number of revolutions of the wheels on a journey may be determined. It is known that each of the fore wheels is $5\frac{1}{4}$, and each of the hind wheels $7\frac{1}{8}$ feet in circumference. Now when in a journey the fore wheel has made 2000 revolutions more than the hind one; how great was the distance traveled.

Solution. Hind wheel $7\frac{1}{8}$; fore wheel $5\frac{1}{4}$; difference $1\frac{7}{8}$ feet, which the hind wheel gains on the fore wheel at each revolution. $5\frac{1}{4}\div1\frac{7}{8}=2\frac{4}{5}$, the revolutions of the hind wheel while it gains a revolution upon the fore wheels; $2\frac{4}{5}\times2000=5600$ revolutions; and $5600\times7\frac{1}{8}=39900$ feet.

26. Three poles, each 50 feet long, were erected on a plain so that the upper ends met, and the lower ends were 60 feet apart. What length of rope was required to reach from their point of meeting to the ground?

Solution. The feet of the poles being connected by lines, an equi lateral triangle will be formed, and directly over the centre of such triangle will be the point of suspension of the rope, so that knowing the length of the pole having its foot at A, and its apex directly over D to be 50 feet, we have but to determine A D. We then have the base and hypotenuse to determine the perpendicular, which will be the length of rope sought.

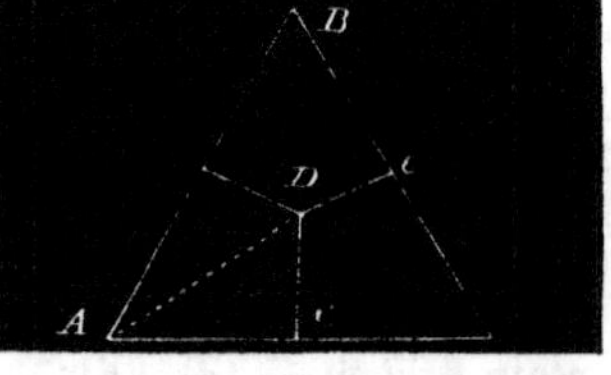

It is a general principle, that "If from any point within an equi lateral triangle, perpendiculars be let fall upon the several sides, the sum of such perpendiculars will be equal to the perpendicular let fall from either angle upon the opposite side." The perpendicular let fall to the middle of the opposite side will be $\sqrt{60^2-30^2}=51.96+$ feet, and as the perpendiculars are equal $51.96\div3=17.32$ the length of any of them as $\overline{D\ E}$, and A E=51.96—17.32=34.64=A D. Then $\sqrt{50^2-34.64^2}$ =36.05 feet. *Ans.*

27.
A landed man two daughters had,
And both were very fair:
To each he gave a piece of land,
One round—the other square.
At forty dollars the acre just,
Each piece its value had,
The dollars that encompassed each
For each exactly paid.
If 'cross a dollar be an inch
And just a half inch more,
Which did the better portion have
That had the round or square?

Ans. The Square by $239308.18.56.

Solution. Each dollar paid for the fortieth of an acre, the question then resolves itself into this: How large must a $\left\{\begin{matrix}\text{square}\\ \text{circle}\end{matrix}\right\}$ be, that each inch and a half of the periphery may just enclose $\frac{1}{40}$ of an acre?

1st. *The Square.* Suppose each side 40 rods, then the area will be 10 acres=400 fortieths of an acre. 40 rods×4 =160 rods=21120 widths of a dollar to enclose. 21120÷400=52.8 ratio. 52.8×21120=1115136 fortieths enclosed. $(52.8)^2\times400=1115136$ dollars to enclose the square, each dollar enclosing $\frac{1}{40}$ of an acre.

2d. *The Circle.* Suppose diameter of circle to be 40 rods. Then the circumference will be 125.664 rods=16587.648 times the diameter of a dollar; and the area will be 314.16 fortieths of an acre; 16587.648÷314.16=52.8, ratio.

Then $314.16\times52.8^2=875827.8144$; and 16587.648×52.-8=875827.8144, the dollars that will enclose each $\frac{1}{40}$ of an acre in a circle.

Her share who had the Square	$1115136.
" " " " " Circle	875827.8144
Difference in fortunes,	$239308.1856

28. How long will it require to travel at 5 miles per hour across an area of 256000 acres, so laid out that the longest distance across shall be the shortest possible to contain the given area?

29

Solution. The circle will be the figure whose "longest distance across shall be the shortest possible," and 256000 acres =40960000 rods, which reversing the common rule for finding the area of a circle when the diameter is given, we divide by .7854 and extract the square root of the quotient, which gives 7221.618 rods=22.5675562+ miles, and this divided by 5 gives 4 hours 30.8 minutes, the time required.

29. There is a triangular meadow 100 rods in length up on each side, and it is desired to fasten a horse by a rope to be attached to a post at one corner; required the length of rope necessary to enable the horse to graze over just half the area of the meadow.

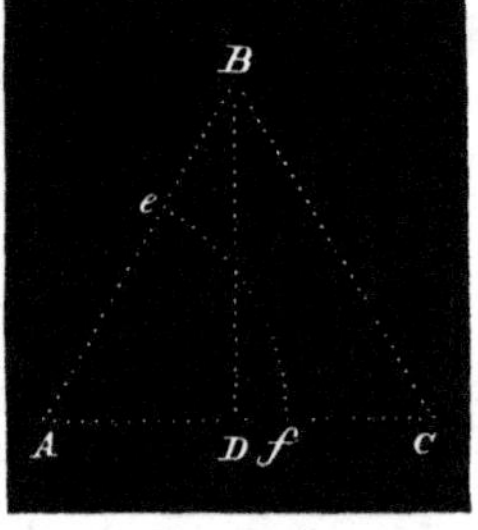

Let A B C represent the meadow, and the horse to be fastened at A. Then will the area of the meadow be found thus— $\sqrt{100^2-50^2}$=80.6+ the perpendicular D B 86.6÷2=43.3, and 43.3×100- =4330 rods, the area of A B C. As the angle at A is 60° or $\frac{1}{6}$ of a circle, we must find the radius of a circle containing $\frac{4330}{2}\times6=12990$ square rods, which we may do by reversing the ordinary rule for finding the area of a circle thus—

$\sqrt{\frac{12990}{.7854}}\div2=64.3025$ rods, the radius or length of rope required, and consequently=Ae or Af.

30. A gentleman had a circular fish pond, 100 yards in diameter, and from a rock in the centre I started my dog in pursuit of a duck that was swimming round the outskirt of the pond. The dog and duck swam with equal speed, the dog keeping constantly between my eye and the duck, or in other words swam constantly directly towards the duck. Required the distance the dog must swim before he reaches the duck; and the figure he will describe. Also required the figure described if the dog swims only half as fast as the duck; and if he swims twice as fast; and in either case the time necessary to reach the duck. Space requires that we leave the solution.

LECTURE XXI.

UNSOLVED PROBLEMS.

THAT the ingenious student may have something to exercise his ingenuity, and that he may learn to make the egg stand on end without the aid of some modern Columbus, we shall now present him with a choice collection of unsolved problems, all of which have been carefully examined, and the answers tested. They need no algebra; but admit of neat arithmetical solutions.

1. A cannon ball six inches in diameter is to be melted and cast into balls 2 inches in diameter; allowing no wastage of metal, required the number it will make? *Ans.* 27.

2. If 27 men can do a piece of work in 14 days, working 10 hours in a day, how many hours a day must 24 boys work, in order to complete the same in 45 days; the work of a boy being half that of a man? *Ans.* 7 hours.

3. Two boys, A and B, on opposite sides of a pond, which is 536 yards in circumference, set off simultaneously to go round it in the same direction. A walks 22 yards in 15 seconds, and B 68 yards in 45 seconds: how often will B circumambulate the pond before he overtakes A?

Ans. 17 times.

4. Required to find three equidifferent numbers, such that the least shall be to the greatest as 5 to 9, and the sum of the three, 63. *Ans.* 15, 21, 27.

5. What time between 4 and 5 o'clock, are the hour and minute hands of a watch together?

Ans. At $21\frac{9}{11}$ min. past 4.

6. What is the least number, which being divided by 6, 7, 8, 9, 10, and 12, shall always leave a remainder of 5?

Ans. 2525.

7. At what time between 12 and 1 o'clock, do the hour and minute hands of a common clock or watch, point in directions directly opposite? *Ans.* $32\frac{8}{11}$ min. past 12.

8. A steamboat having on board 150 barrels of sugar, pays for toll on the Muskingum river, the value of 2 barrels wanting $6 ; and another containing 240 barrels, pays at the same rate, the value of 2 barrels and $18 besides; what is the value of the sugar per barrel? *Ans.* $23.

9. A son having asked his father's age, the father replied: "Your age is 12 years; to which if five eighths of both our ages be added, the sum will be equal to mine." What was the father's age? *Ans.* 52 years.

10. A, B and C formed a joint stock of $1064; A's stock continues in trade 5 months, B's 8 months, and C's 12 months; A's share of the gain is $114, B's $133.20, and C's $165. What was the stock of each?

Ans. { A's $456. B's $333. C's $275.

11. A and B set out for the same place in the same direction. A travels uniformly 18 miles per day, and after 9 days turns and goes back as far as B has traveled in those 9 days; he then turns again, and pursuing his journey, overtakes B $22\frac{1}{2}$ days from the time they first set out. It is required to find the rate at which B uniformly traveled?

Ans. 10 miles per day.

12. What number is that which being increased by its half, its third, and 18 more, will be doubled. *Ans.* 108.

13. What number multiplied by 15 will produce $\frac{3}{4}$?

Ans. $\frac{1}{20}$.

14. What number multiplied by 57, will produce just what 134 multiplied by 71 will do? *Ans.* $166\frac{52}{57}$.

15. Required to lay out a lot of land in form of a long square, or parallelogram, containing 3 acres, 2 roods, 29 poles, that shall take just 100 rods of wall to enclose, or fence it round, what shall the length and breadth of the lot be?

Ans. 31 by 19.

16. A, B and C can do a piece of work in 6 days; B, C and D in 7 days; A, C and D in 8 days; A, B and D in 9 days; in what time can all do the work together?

Ans. $5\frac{137}{275}$ days.

17. If a heavy sphere, whose diameter is 4 inches, be dropped into a conical glass full of water, whose diameter is 5 inches, and altitude 6 inches. How much water will run over?

Ans. 26.2721536 inches.

18. A bale of cotton weighing 1667 lbs. is given to a manufacturer to be spun. The manufacturer is to be paid 13 cents

per lb. for what yarn he makes, and is to make 14 oz. of yarn to each pound of cotton. He is to keep so much cotton out of the bale as will, at 10 cents per lb. pay him for spinning the remainder on the above terms. How many pounds must he keep? *Ans.* $887\frac{20}{171}$ lbs.

19. Nine gentlemen met at an inn, and were so pleased with each other, that they agreed to tarry so long with each other as they and their host could be seated differently at dinner. Pray how long would such a frolic have lasted?
Ans. 3628800 days.

20. I agreed with a tinker whose name was DOOLITTLE,
To make for my wife a flat bottomed kettle—
Twelve inches exactly the depth of the same,
And twenty-five gallons of beer to contain—
The number of inches across on the top,
Was twice at the bottom when new at the shop;
How many inches across must the bottom then be,
Likewise the top pray show unto me?

Ans. At top 35.8096+and at bottom 17.9048+

21. A man can dig a piece of ground in 5 days, his son can do the same in 7 days; in what time can they both do it together? *Ans.* $2\frac{11}{12}$ days.

The simplest mode is to add together $\frac{1}{5}$ and $\frac{1}{7}$, the fractional parts each will do in a day, and their sum will be $\frac{12}{35}$, the fractional part of the work both will do in a day. Then, As $\frac{12}{35}$: $\frac{35}{35}$: : $\frac{1}{1}$: $2\frac{11}{12}$, *Ans.* Hence we see the reason of the arbitrary rule sometimes given for this class of questions, "Divide the product of the numbers by their sum," and we may also see the reason of the rule given where there are three or more agents, operating unequally, "Divide the product of the three given times by the sum of the products of each two taken separately." Adding the fractional parts is the plainest mode, but not the shortest.

22. A can produce a certain effect in 3 hours, B in 4 hours, and C in 5 hours; in what time can the three together produce the same effect? *Ans.* 1 h. $16\frac{28}{47}$ minutes.

23. Of Sweet wine at $\$1\frac{3}{5}$ per gallon; Port, at $\$2\frac{2}{7}$; and Madeira at $\$3\frac{5}{9}$; how much can be purchased for $\$214\frac{2}{7}$, expending $\frac{1}{3}$ of the money on each?
Ans. Sweet, $44\frac{9}{14}$ gall., Port, $31\frac{1}{4}$ gall. and Madeira $20\frac{5}{56}$.

24. If A, B and C could pave a street in 18 days; B, C and D in 20 days; C, D and A in 24 days; and D, A and B in 27 days; in what time would it be done by all of them together, and by each of them singly? *Ans.* By all in $16\frac{56}{199}$

days; by A in $87\frac{21}{37}$ days; by B in $50\frac{5}{8}$ days; by C in $41\frac{1}{79}$ days: and by D in $170\frac{10}{19}$ days.

25. A servant draws off one gallon each day for 20 days, from a cask containing 10 gallons of rum, each time supplying the deficiency by a gallon of water; and then, to escape detection, he again draws off 20 gallons, supplying the deficiency each time by a gallon of rum. It is required to determine how much water still remains in the cask?

Ans. 1.06779577 gallon, or rather more than a gallon and half a pint.

26. A merchant every year gains 50 per cent. on his capital, of which he spends $300 per annum in house and other expenses, and at the end of 4 years he finds himself possessed of a capital 4 times as great as what he had at commencing business. Find his original capital without using the rule of Position. *Ans.* $\$2294\frac{2}{17}$.

27. It is required to find a sum of money, of which, in the space of 4 years, the *true* discount, at simple interest, is $5 more at the rate of 6 than of 4 per cent. per annum.

Ans. $89.90.

28. A man leaves to his eldest child one fourth of his property; to his second, one fourth of the remainder, and $350 besides; to his third one fourth of the remainder and $975; to his youngest one fourth the remainder and $1400; and what still remains he bequeaths to his wife, whose share is found to be one fifth of the whole. Hence it is required to find the value of the whole property. *Ans.* $20,000.

29. A man travels from his own house to Wheeling in 4 days, and home again in 5 days, traveling each day, during the whole journey, one mile less than he did the preceding. How far does he live from Wheeling? *Ans.* 90 miles.

30. The men employed in a factory work 12 hours, the women 9 hours, and the boys 8 hours, each day; for laboring the same number of hours, each man receives a half more than each woman, and each woman a third more than each boy: the entire sum paid to all the women each day is double of the sum paid to all the boys; and for every five dollars earned by all the women each day, twelve dollars are earned by all the men. Hence it is required to find the number of each class employed, the entire number being 59.

Ans. 24 men, 20 women, and 15 boys.

31. One third of a quantity of flour being sold to gain a certain rate per cent., one fourth to gain twice as much per cent., and the remainder to gain three times as much per cent.:

it is required to determine the gain per cent, on each part, the gain upon the whole being 20 per cent.

Ans. The gains per cent. are $9\frac{3}{5}$, $19\frac{1}{5}$, and $28\frac{4}{5}$.

32. The less of two bales of cloth is bought at the rate of twice as many pence per yard as it contains yards, and costs £31 0 2 more than the greater, which contains 4 yards for every 3 in the less, and is bought at the rate of as many pence per yard as it contains yards. How many yards are contained in each?

Ans. 244 yards in the greater, and 183 in the less.

33. A man owes a debt due in four equal instalments, at the end of 4, 9, 12 and 20 months respectively; and he finds that discount being allowed at 5 per cent., $750 will pay the debt. How much did he owe? *Ans.* $784,74+

34. If a merchant commence trade with a capital of $5000, and gain so much, that, after paying all expenses, his capital, each year, is increased by a tenth part of itself wanting $100, how much will he be worth at the end of 20 years?

Ans. $27910 very nearly.

35. Two men, A and B, are on a straight road, on the opposite sides of a gate, and distant from it 308 yards and 277 yards respectively, and travel each towards the original station of the other. How long must they walk till their distances from the gate will be equal, B traveling 2 yards, and A $2\frac{1}{3}$ yards, per second?

Ans. 1 minute, 33 seconds, or 2 minutes, 15 seconds.

36. Every thing being supposed to be as in the preceding question, at what time will each be at the same distance from the original station of the other, as the other is from his?

Ans. In $4\frac{1}{2}$ minutes after starting.

37. If an acting partner in a mercantile concern contribute $1000 to the original joint stock of the company, and annually increase this sum by $150 saved from his salary; to how much will his share amount at the end of 11 years, on the supposition, that, after all expenses are paid, there is a clear gain of 10 per cent. per annum on the entire capital?

Ans. $5632.791+

38. Two partners, Peter and John bought goods to the amount of $1000; in the purchase of which, Peter paid more than John, and John paid—I know not how much. They then sold their goods for ready money, and thereby gained at the rate of 200 per cent. on the prime cost; they divided the gain between them in proportion to the purchase money that each paid in buying the goods; and Peter says to John, My

part of the gain is really a handsome sum of money; I wish I had as many such sums as your part contains dollars, I should then have $960000. I demand each man's particular stock in purchasing the goods.

Ans. PETER paid $600, and JOHN paid $400.

39. A young man was required as the condition of obtaining his devoted, that he should obtain a number of apples, half of which and half an apple more he should give to the father; half the remainder and half an apple more to the mother; half the remainder and half an apple more to the daughter; and retain half the remainder and half an apple more for himself. None of the apples are to be cut. Required the smallest number that will answer his purpose?

Ans. 15.

40. Not long ago says a person, a barrel of wheat was by $1, and the barrel of rye 75 cents cheaper than now; then the price of the wheat was double that of the rye, their present prices are as 20 to 11. What is the price of each?

Ans. Wheat $5, rye $$2\frac{3}{4}$.

41. A owns 720, B 336, and C 1736 rods of land. They agree to divide it into equal house lots, fixing on the greatest number of rods for a lot, that will allow each owner to lay out all his land. How many rods must there be in a lot?

Ans. 8.

42. The sum of A's and B's ages is 60, and if you divide A's age by B's the quotient will be 3. Required the age of each? *Ans.* A's age 45, B's 15.

43. A and B start at opposite points, to skate to the other's starting point: distance 8 miles. A, by having the advantage (hence B the disadvantage) of a uniform wind, performs his task $2\frac{1}{2}$ times the quickest, and 48 minutes the soonest. Required the time that each is skating, and the force of the wind per minute? *Ans.* A's time 32 min. B's 1 h. 20 m.
Force of wind per minute 396 ft.

44. A young man being asked his age, answered that if the age of his father, which was 44, were added to twice his own, the sum would be four times his own age. Required his age?

Ans. 22 years.

45. How many hills of corn may be planted on a square acre, allowing them to stand 4 feet apart, and 2 feet on every side from the enclosing lines? *Ans.* 2704.

46. The sum of four numbers is 360, and they are proportionate as 3, 4, 5, 6; what are they?

Ans. 60, 80, 100 and 120.

47. A water-tub that holds 147 gallons, has a pipe that brings in 14 gallons in 9 minutes, and a tap that discharges 40 gallons in 31 minutes; now supposing these both to be left open by mistake at 2 o'clock, and a servant at 5, finding the water running, shuts the tap, only, and is solicitous to know in what time the tub will be filled after the discovery of the accident. What is the reply?

Ans. 1 hour 3 min. $48\frac{114}{217}$ seconds.

48. What number is that which being multiplied by 6, the product increased by 18, and the sum divided by 9 the quotient shall be 20? *Ans.* 27.

49. A person went to 4 taverns in succession; upon entering each of which, he borrowed as much money as he carried to it; and upon leaving them he paid the landlords one dollar each; which done, he finds himself without money. What sum of money did he carry to the first tavern?

Ans. $93\frac{3}{4}$ cents.

50. A, B and C are employed to do a piece of work for $26,45 cents: A and B together are supposed to do $\frac{3}{4}$ of the work; A and C, $\frac{9}{10}$; B and C, $\frac{13}{20}$; or in these proportions, and are paid accordingly. What is each man's share of the money? *Ans.* A $11,50, B $5,75, C $9,20.

51. A man had four sons, whose ages differed from each other 4 years, and the youngest was half as old as the oldest; required the age of each. *Ans.* 12, 16, 20 and 24.

52. B's age is $1\frac{1}{2}$ the age of A, and C's $2\frac{1}{10}$ the age of both; and the sum of their ages is 93. What is the age of each? *Ans.* A 12 years, B 18, C 63.

53. On a certain day, 20 farmers, 30 merchants, 24 lawyers, and 24 tailors, spent at a dinner $64, which was divided among them in such a manner that 4 farmers paid as much as 5 merchants; 10 merchants paid as much as 16 lawyers; and 8 lawyers as much as 12 tailors; how much money did each class pay?

Ans. Farmers $20, Merchants $24, Lawyers $12, Tailors $8.

54. A dealer in the article drew from a barrel of whiskey of 32 gallons, 5 gallons and replaced it with 5 gallons of water; and thinking it would bear a little more water, he repeated the operation, thus drawing and replacing five times. How much whiskey was then in the barrel, and how much water?

Ans. { Whiskey, $13\frac{717419}{1048576}$ gallons.
Water, $18\frac{331157}{1048576}$ gallons

55. A, B and C have among them $135. A's+B's are to

B's+C's as 5 to 7, and C's—B's is to C's+B's as 1 to 7. How many had each? *Ans.* A 30, B 45, C 60.

56. B delivers to C $1200, to be invested in trade for one year, on condition that if C added $500 to it, and gave his time as manager, he should have $\frac{2}{5}$ of the gain; What was C's time valued at in capital? *Ans.* $300.

57. The sides of two square pieces of ground are as 3 to 5, and the sum of their superficial contents is 30600 square feet. What is the length of a side of each piece?

Ans. 90 and 150 feet.

58. G and H buy 48 acres of land at $12 per acre, of which H is to have a piece containing 12 acres, which G and H think to be $\frac{1}{3}$ better than 12 of the 36 that G is to have, the rest of G's being of the average value of $12. How much should each pay? *Ans.* G 164\frac{4}{7}$, H 411\frac{3}{7}$.

59. A, B, C, and D caught in their net 522 fishes of which A was to have a certain number; B was to have 12 more than A, C 7 less than B, and D as many as A and C. Required the share of each.

Ans. A 100, B 112, C 105, D 205.

60. In the annexed figure, suppose the ball B to weigh 20 lbs. and the distance thence to C, where the rod is supported, to be 2 feet; and from the point of support to the ball A, 6 feet. How much must A weigh to balance B. *Ans.* 6$\frac{2}{3}$ lbs.

61. A and B have apples. A said to B, if I had 2 apples more, I should have twice and half as many as you. B said to A, if I had 4 apples more I should have half as many as you. Required the number of apples that each had?

Ans. A 48, B 20.

62. A bought flour for cash, and sold it to B at an advance. B sold it to C at 10 per cent. advance, and C, on selling it to D, gained $71,28, equal to 12 per cent. profit; which was 4 per cent. more than A made, though he bought it at $5 per barrel. Required B's gain, how much C received, and the number of barrels in the lot?

Ans. { B gained $54, C received $665,28; and there were 100 barrels in the lot.

63. If the annexed lever, of the 1st order be 6 feet long, and the fulcrum is 6 inches from the end to which the weight of 66 lbs. is attached; how much power must be exerted to balance the weight? *Ans.* 6 lbs.

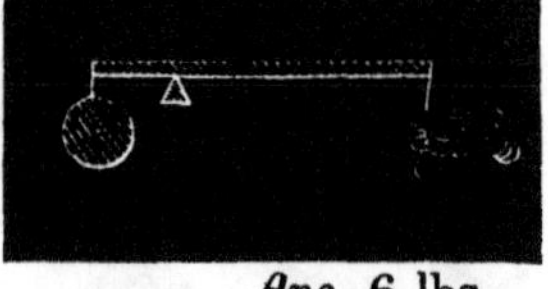

64. A, B, C, and D, found a purse of shillings and each of them took a number at a venture, afterwards by comparing them together they found that if A took 25 from B, his number would be equal to what B had left, and if B took 30 from C, his number would be three times what C had left, and if C took 40 from D, his number would be double what D had left, and if D took 50 from A, his number would be 3 times as much as A had left, and 5 shillings over. What number had each? *Ans.* A 100, B 150, C 90, D 105.

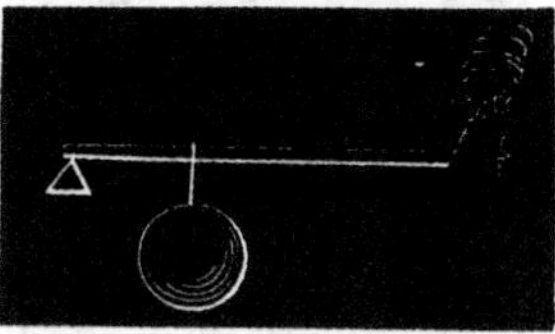

65. The lever, of the 2d order in the annexed diagram is 6 feet, and the weight 66 lbs. suspended 6 inches from the fulcrum; required the power necessary to hold the weight in equilibrum.

Ans. $5\frac{1}{2}$ lbs.

66. A merchant puts a capital of $5500 out at interest at 4 per cent. and $4\frac{1}{2}$ years afterwards, another sum of $8000 at 5 per cent. If he leaves these two capitals constantly at simple interest, in how many years will he have drawn the same interest from both?

Ans. In 10 years from the time of the first loan.

67. There are two numbers whose sum is to their difference as 8 to 1; and the difference of whose squares is 128. What are the numbers? *Ans.* 18 and 14.

68. A and B have the same income: A saves $\frac{1}{8}$ of his, but B, by spending $30 per annum more than A, at the end of 8 years finds himself $40 in debt; what is their income, and what does each spend yearly?

Ans. { Income $200 per annum; and A spends $175, and B $205 per annum.

69. A general disposing his army into a square, found he had 231 over and above; but increasing each side with one soldier, he wanted 44 to fill up the square. Of how many men did his army consist? *Ans.* 19000 men.

70. A said to B and C, give me $\frac{1}{2}$ of your money and I shall have $100; B said to A and C, give me $\frac{1}{3}$ of yours, and I shall have $100; C said to A and B, give me $\frac{1}{4}$ of yours, and I shall have $100. Determine how much money each man had, without using Position.

Ans. A had $$29\frac{7}{17}$; B $$64\frac{12}{17}$; C $$76\frac{8}{17}$.

71. If a globe 6 inches in diameter weighs 25 lbs., what is the weight of another of the like metal whose diameter is 3 inches? *Ans.* 3.125 lbs.

72. A cheese being weighed for me in one scale of a balance weighed 16 lbs., but suspecting its accuracy I transferred it to the opposite scale and it weighed but 9 lbs. Required its true weight. *Ans.* 12 lbs.

73. A, in a scuffle, seized on $\frac{2}{3}$ of a parcel of marbles; B caught $\frac{3}{8}$ of them out of his hands, and C laid hold on $\frac{3}{10}$ more; D ran off with $\frac{6}{7}$ of what A had left, and the rest E afterwards secured slily for himself. Then A and C jointly fell upon B, who, in the conflict, let fall $\frac{1}{2}$ he had, which were equally picked up by D and E; B then kicked down C's hat, of the contents of which A got $\frac{1}{4}$, B $\frac{1}{3}$, D $\frac{2}{7}$, and C and E equal shares of what was left of that stock. D then struck $\frac{3}{4}$ of what A and B last acquired, out of their hands; they with difficulty recovered $\frac{5}{8}$ of it in equal shares again, but the other three carried off $\frac{1}{8}$ apiece of the same. Upon this, they called a truce, and agreed that the $\frac{1}{3}$ left by A at first should be equally divided among them. How many marbles at least, after this distribution, had each of the competitors?

Ans. { A must have had at least 2863; B 6335; C 2438; D 10294; and E 4950.

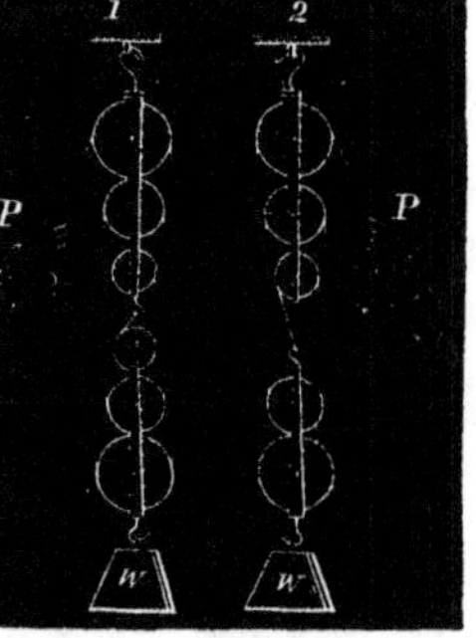

74. A block and tackle are arranged as in the annexed cut, and the power is equal to 100 pounds. How many pounds will it suspend at W, in Fig. 1, and how many in Fig. 2.

Ans. { In Fig. 1, 600 lbs. In Fig. 2, 500 lbs.

75. A had 50 dollars more than B, and the sum of the squares of the shares of both was 12500. How many dollars had each?

Ans. A 100, B 50.

76. Desiring to warm a vessel gradually, I placed it 10 feet from the fire, but it is too cool; what amount of heat will it receive if I place it within 5 feet of the fire?

Ans. 4 times as much.

77. Two ships depart from the same place, one sails due north at the rate of 6 miles per hour, the other due east, 8 miles per hour. How far will they be apart at the end of one hour? How far in 2 hours? &c.

Ans. 10 miles in 1 hour, 20 in 2 hours, &c.

78. One hundred eggs being placed on the ground in a straight line, at the distance of a yard from each other, how

far will a person travel who shall bring them one by one to a basket, which is placed one yard from the first egg?

Ans. 10100 yards, or $5\frac{65}{88}$ miles.

79. A and B have between them, a number of guineas, which are to be so divided that the sum of their squares shall be 208, and the difference of their squares 80. Supposing A's the greater number, how many has he more than B?

Ans. 4.

80. A certain gentleman at the time of marriage agreed to give his wife $\frac{2}{3}$ of his estate, if at the time of his death he left only a daughter, and if he left only a son, she should have $\frac{1}{3}$ of his property; but, as it happened, he left a son and a daughter, by which the widow lost in equity $2400 more than if there had been only a daughter. What would have been his wife's dowry if he had left only a son? *Ans.* $2100.

81. A is older than B, their ages added make 50, and the sum of their squares make 1300. Required the age of each?

Ans. A 30, B 20.

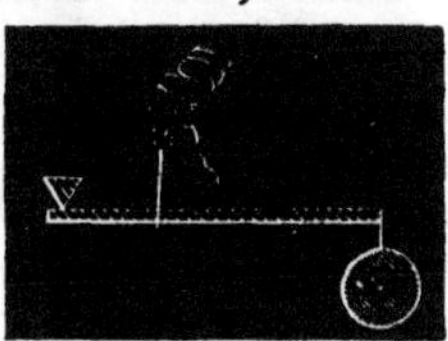

82. Suppose the annexed lever, of the 3d order, to be 6 feet long, the fulcrum at one end, the weight 66 lbs. at the other, and the power 6 inches from the fulcrum. Required the power that will hold the weight balanced.

Ans. 792 lbs.

83. From Columbus to A's tavern is a certain number of miles, thence to B's tavern is a certain other number, and just half way from Columbus to B's tavern is C's tavern. The square of the distance in miles from Columbus to A's, added to the square of the distance thence to B's is 80, and the square of the distance from Columbus to C's is 36. Required the distance from A's tavern to C's? *Ans.* 2 miles.

84. A bullet is dropped from the top of a building, and found to reach the ground in $1\frac{3}{4}$ seconds. Required the height? *Ans.* 49 feet.

85. In the annexed plane, suppose the base A B 80, the perpendicular B c 60, the plane A c 100, and the weight 66 lbs.; what power exerted parallel with the plane will hold the weight to its place, and what exerted parallel with the base A B?

Ans. Parallel with the plane 39.6 lbs.; with the base $49\frac{1}{2}$ lbs.

86. Required the pressure upon the fulcrum in each of the questions 63, 65, 82?

Ans. { In 63, 72 lbs., in 65, 61½ lbs., In 82, the upward pressure is 726.

87. A, B and C are to share $100,000 in the proportion of $\frac{1}{3}$, $\frac{1}{4}$, and $\frac{1}{5}$ respectively; but C's part being lost by his death, it is required to divide the whole sum properly, between the other two? *Ans.* A's part is \$57142$\frac{6}{7}$, and B's \$42857$\frac{1}{7}$.

88. Suppose a pole standing on a horizontal plane to be 75 feet in height. At what height from the ground must it be cut off, that the top of it may fall on a point 55 feet from the bottom of the pole; the end, where it was cut off, resting on the upright part? *Ans.* 17$\frac{1}{3}$ feet from the ground.

This question may be made much more difficult by supposing the pole to stand on the side of a hill, and that the top shall strike the ground at some distance, say 20 feet down the hill from the foot of the pole; while a line drawn parallel with the horizon from the foot of the pole, and intersecting the part broken off, shall be a given length, say 15 feet. You may then suppose it to fall up the hill, and that a horizontal line from the point of section to the side of the hill shall measure a given distance; or suppose a horizontal line drawn if practicable, from the point of contact with the hill to the stump. The object in either case being to find the point where the pole must be cut or broken off.

89. The annexed cut represents a windlass, the circumference of the larger part being 24 inches, and of the smaller 20 inches, and the circumference of the circle described by the power is 8 feet. The cord constantly winds off the smaller part, and upon the larger. See page. 331. What weight, with such a machine, will a power of 30 lbs. hold balanced? *Ans.* 720 lbs.

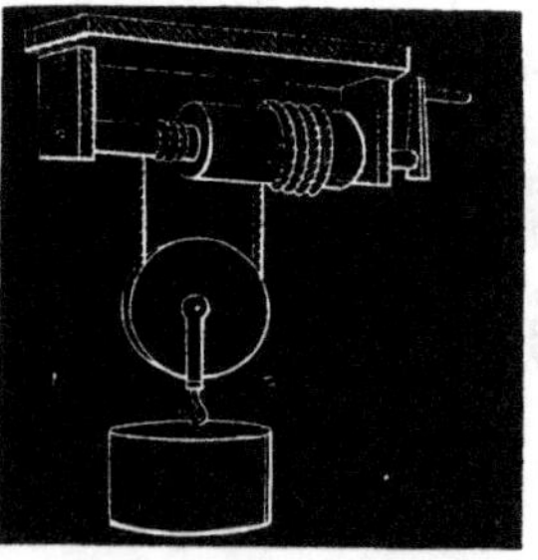

90. There is a line of paling to be constructed over a succession of hills and intervening planes, as follows: The line forms the hypotenuse of a right angled triangle, the legs of which are 60 and 80 rods; the legs being measured over level ground, and consequently the hypotenuse being a straight line; but upon it a succession of little hills put in as follows: the first 10 rods are level, then a hill puts in that covers the next 10 rods, and causes the line to rise over it in a semicircle;

then another plane and another hill of similar extent, and so on successively to the end of the line. The palings are 3 inches wide, and the space between each is just equal to the width of a paling. Required the length of the two lines of rails, to which the palings are to be nailed, allowing the distance upon the surface of the ground as measured over the hills, to be the length of one; and the number of palings necessary for the whole, supposing them to be applied continuously whether they fall on rails or posts?

Ans. { Railings a trifle over 257 rods;
And it will require 3300 palings.

91. A boy hired with a farmer for 12 weeks, on condition that he should receive $12 and a coat. At the end of 7 weeks the parties separated, when it was found that the boy was entitled to $5, and the coat. What was the value of the coat? *Ans.* $4.80.

92. What number is that which being divided into 4 or 5 equal parts, the product of all the parts in either case will be the same? *Ans.* 12 $\frac{53}{256}$.

93. In giving directions for making a chaise, the length of the shafts between the axletree and backband, being settled at 9 feet, a dispute arose whereabouts on the shafts the centre of the body should be fixed. The chaise maker advised to place it 30 inches before the axletree; others supposed 20 inches would be a sufficient incumbrance for the horse. Now, supposing two passengers to weigh 3 cwt., and the body of the chaise $\frac{3}{4}$ cwt. more, what will the beast in both these cases bear, more than his harness?

Ans. 116$\frac{2}{3}$ lbs., and 77$\frac{7}{9}$ lbs.

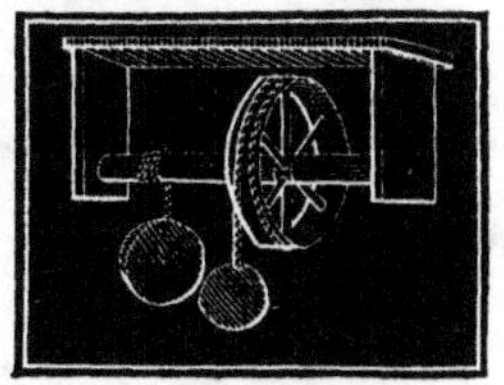

94. Suppose the annexed cut to represent a wheel and axle for raising goods. The diameter of the axis, on which the cord winds, is a foot and a half, and of the wheel upon which the power operates, 9 feet. What power at the rope that goes over the wheel will suspend 600 lbs. at the rope that goes over the axle?

Ans. 100 lbs.

95. A butcher bought a certain number of oxen, and paid $7665; and if the cost per ox, in dollars, were added to the number of oxen bought, the sum would be 386. What number did he buy and at what price?

Ans. 365 oxen, at $21 each.

96. What will be the diameter of a globe, when the square

inches upon the surface are just equal to the cubic inches in the globe? *Ans.* 6 inches.

97. Suppose there be a round column of equal size throughout, being one foot in diameter and 20 feet high, and a bean vine entwines itself around it just 12 times at equal distances, in passing from the foot to the top. How long will the vine be? *Ans.* 42.6756.

98. Five men undertake to carry a piece of scantling, 30 feet long, of equal size and density throughout, one takes post at the hind end, and the others with a long handspike, place themselves so far from the fore end, that all shall carry equally. How far were they from the man who carried alone? *Ans.* $18\frac{3}{4}$ feet.

99. Five men undertake to carry a piece of scantling, 30 feet long, of equal size and density throughout; A takes post at the hind end, B and C with a handspike at the fore end; where must D and E place themselves with their handspike that all may carry equally? *Ans.* $7\frac{1}{2}$ feet from A.

100. Determine the size of three wine casks from the following conditions. When the first empty cask is filled from the second full cask, there remains in the second only $\frac{2}{9}$ of the wine; if the second empty cask is filled from the third full one, then there remains in the third only $\frac{1}{4}$ of the wine; but if we wish to fill the third empty cask from the first full one there will not be enough by 50 gallons. What did each one hold? *Ans.* 1st 70, 2d 90, 3d 120 gallons.

101. A capitalist derives from the sum he has at interest a yearly income of $2940. Four fifths of it bear 4 per cent. and $\frac{1}{5}$ five per cent. How much money has he out at interest? *Ans.* $70,000.

102. A servant received from his master $40 wages yearly, and a suit of livery. After he had served five months he asked for his discharge and received for this time the livery and $$6\frac{1}{6}$ in money. How much did the livery cost? *Ans.* $18.

103. A merchant increases his capital yearly by 20 per cent. but takes from it annually $1000 for the support of himself and family. After he had carried on his business in this manner for three years he finds after deducting the usual sum of $1000 that his capital has increased $200 more then $\frac{3}{5}$ of the original sum. What was his original capital? *Ans.* $30,000.

104. On an approaching war, three towns A, B and C are to furnish their complements of 594 men; the division is to be

made in proportion to the population. Now the population of A is to that of B as 3 to 5; whilst the population of B is to that of C as 8 to 7. How many men must each town furnish? *Ans.* A 144, B 240, C 210.

105. A dog pursues a hare. Before the dog started the hare and made 50 leaps, and this is the distance between them at first. The hare takes 6 leaps to the dog's 5; but 9 of the hare's leaps are equal to 7 of the dog's. How many leaps can the hare take before the dog overtakes her? *Ans.* 700.

106. A spendthrift lends his fortune at 4 per cent. interest. After he had let it remain 2 years, he took out the fourth part of the amount and allowed the remainder to stand 7 months. At the expiration of this time he took once more the fourth part of the remainder, and allowed the capital thus diminished to stand 13 months, when he demanded his remaining fortune. In the space of 44 months he had drawn no less than $6291.-96¼ of interest money. What was his fortune at first?

Ans. $50,000.

107. A person has two large pieces of iron, whose weight is required. It is known that $\frac{2}{5}$ of the first piece weighs 96 lbs. less than $\frac{3}{4}$ of the other piece; and that $\frac{5}{8}$ of the second piece weighs exactly as much as $\frac{4}{9}$ of the first. How much did each piece weigh? *Ans.* 720 and 512 lbs.

108. Five brothers in the space of 9 months squandered a capital of $4800, together with the interest of it for the whole time. At the same rate of expenditure two other persons squandered $3320, with interest in 16 months. The rate of interest was in both instances the same. How much did each spend monthly? *Ans.* 110\frac{2}{3}$.

109. Three brothers, A, B and C, held among them $1760 in different amounts, which they agreed to divide equally. The first gave one half of his in equal shares to the others; the second gave one third of the amount he then had to the other two; and the third gave $160 to each of the others. They then found that each held a third of the whole sum. What had each at first? *Ans.* A $533⅓, B $480, C $746⅔.

110. What numbers are those that when added make 25, and when you halve one and double the other, the results are equal? *Ans.* 20 and 5.

111. If 7 gallons of Madeira wine cost as much as 9 gallons of Port, and 9 gallons of Port as much as 12 gallons of Sherry, and the price of three gallons of these, taking one of each kind, is $8.50, what is each worth per gallon?

Ans. Madeira $3.60, Port $2.80, and Sherry $2.10.

112. A merchant bought several yards of silk for $30, out of which he reserved 10 yards, and sold the remainder for $28, gaining one sixth as many cents on a yard as one yard cost him. How many yards did he buy, and at what price?

Ans. 50 yards at 60 cents.

113. A and B jointly have a fortune of $9800. A invests a sixth part of his property in business, and B the fifth part of his, when it appears that each has an equal sum remaining. How much had each at first? *Ans.* A $4800, B $5000.

114. A has 10 acres of pasture, B 8 acres, and C 3 acres, into which they agree to put an equal number of cattle to graze, and C agrees to pay A and B $24. How much should each receive on final adjustment?

Ans. A receives $18, and B $6.

115. A constable is pursuing a thief at a uniform speed, but finds on inquiry that the thief is going a mile and a half per hour faster than he is; he therefore doubles his speed after the first four hours, and takes the thief at the end of $6\frac{1}{3}$ hours from his setting out. The thief had a start of one hour and never varied his speed. How far did they travel, and what was the rate of each?

Ans. Distance $71\frac{1}{2}$ miles, and the constable first traveled $8\frac{1}{4}$ miles per hour; while the thief traveled $9\frac{3}{4}$ miles.

116. Before the Mormons left Nauvoo for California, they boasted that their line of march would be 24 miles long; and that a printing establishment should go with the front to issue the orders of the prophets daily for the whole line. Now suppose they advance 24 miles every 12 hours, how many miles must the herald travel to gain the rear, he traveling two miles to the company's one, and each starting at the same time. And then how much must he increase his speed, so as to regain his place at the head of the column by the hour of encamping? And how many miles must he travel daily?

Ans. He will meet the rear 8 miles from their starting; and must return at the rate of 5 miles per hour and must travel 56 miles per day.

117. A, B, C can build a house in 20 days, B, C, D in 24 days, C, D, A in 30 days and A, B, D in 36 days. Suppose they all commence the job together, but that after 10 days A cut his foot and ceased working; two days afterwards B was compelled to leave, and at the end of the third week, (18th day) C was compelled to leave; in what time would D finish the job? *Ans.* In 284 days.

118. A's house is a certain number of miles from Zanes-

ville, and B's is a certain other number, and just half way from Zanesville to B's house is C's. The square of the distance in miles from Zanesville to A's added to the square of the distance from A's to B's is 90, and the difference between the square of the distance from A's to C's, and the square of the distance from C's to B's is equal to the product of the distance A's is from Zanesville, multiplied by the distance B's is from A's. What are their distances from Zanesville?

Ans. A 3, B 12, C 6.

120. A, B and C start on a journey of 40 miles. A can travel only one mile an hour, B two miles, but C has a horse and buggy, and can travel 8 miles; and as they desire to reach their journeys end in the shortest possible time, C takes up A and carries him so far, that going back and taking up B, they all reach their journey's end together. Required the distance each will travel alone, and the whole time consumed in performing the journey.

Ans. A traveled $5\frac{35}{41}$ miles alone, B $13\frac{27}{41}$, C $20\frac{20}{41}$, and the whole journey occupied $10\frac{5}{41}$ hours.

LECTURE XXII.

ARITHMETICAL PRODIGIES, &c.

Having completed our course of investigation into the philosophy of numbers, we shall devote the present lecture to an investigation of the human mind as adapted to the study of this science. In making this announcement, however, we desire not to excite anticipations that are not to be realized; for we have neither space nor inclination to enter into a discussion of the vexed questions of metaphysicians, as to whether the mind is material or immaterial; and how far the external configuration of the head is indicative of the powers of the mind within. It is sufficient for our purpose to advance the fact, that the powers of the human mind, various in all things, seem in this peculiarly unequal, for while the mass of our race require the aid of long continued education to train the mind to the perception of numerical relations, and some seem incapable of reaching a high degree of proficiency, others possess a power, even in untaught childhood, that cannot be reached by the ablest mathematicians.

Some have supposed that mathematical skill, with which they identify the cases referred to, implies a high development of the reasoning faculties; but though the exercise of the mind in the study of mathematical science, has been always admitted to be an excellent mode of discipline, it by no means follows as a legitimate consequence, that the possession of extraordinary perceptive faculties in regard to the powers and relations of numbers, or of quantities, implies extraordinary reasoning faculties. An astonishing degree of perceptive power, in regard to numbers, has been found indeed to exist in minds but slightly removed from downright idiocy: a fact that would seem unaccountable, if the elementary combination of numbers required the aid of the reasoning powers.

A most striking instance of mental imbecility, combined with a high degree of power in regard to numbers, was brought to light in 1844, in the person of a negro slave, named CAP, the property of Mr. P. M'Lemore, of Madison county, Alabama. We cannot present this case more clearly than by giving the following somewhat extensive extract from a letter, dated October 26th, 1844, written by the Rev. John M. Hanner, and subsequently confirmed by the same gentleman, in a letter written in reply to one from us.

"On the 8th of June, 1844, the Rev. John C. Burruss, Mr. T. Brandon and myself went to see him and were amazed. From himself and Mr. M'Lemore, we learned that he has no idea of a God. When asked, "Who made you?" he answered "Nobody." He has never been but a few times half a mile from the place of his birth. He has not mind enough to do the ordinary work of a slave; eats and sleeps in the same house with the white folks, having his own table and bed. He will not ask for any thing, nor touch food, however hungry, unless it be offered to him. He was never known to commence a conversation with any one, nor continue one, farther than merely answering questions in the fewest words. He speaks very low and tardily. He has never been known to utter a falsehood or to steal, and is but little subject to anger. He will not strike, even a dog, but when vexed by his sister, he will take hold of her arm as if to break it with his hands. He cannot be persuaded to taste intoxicating liquors; and manifests no partiality for females. There is nothing remarkable in the configuration of his head or in his countenance, save that his eye is uncommonly convex, and continually rolling about with a wild and glaring expression. His laugh and movements are perfectly idiotic. He does not know a letter or figure. Withal he is in one respect the most extraordinary human being I ever saw. Almost the only manifestation of mind is in relation to *Numbers*. His power over numbers is at once extraordinary and incredible. Take any two numbers under 100, and he will give their products at once, as readily as a school boy would give the product of 12 times 12. He multiplies thousands, adds, subtracts and divides, with the same certainty, though with greater mental labor. He has, however, no idea of numbers above the period of millions.

With pencil and paper we made the following calculations, and asked him the questions; thus—

How much is 99 times 99? He answered immediately 9,801. How much is 74 times 86½. He answered 6401. How many nines in 2000? He answered, 222 nines and 2 over. How many fifteens in 3355? He answered, 223 fifteens and 10 over. How many twenty-threes in 4000? He answered, 173 twenty-threes, and 21 over. How much is 321 times 789? He answered after a short pause, 253,269. If you take 21 from 85, how many will be left? He answered, 64. How much is 7 times 9, 22 and 14? He answered, 99. How much is 17 times 17 and 16? He said, 305. If you had given one dollar and a half for a chicken and a half, how much would you have to give for two chickens? He said, *Two Dollars*. If a stake three feet long, standing upright, makes a shadow of five feet; how high will a pole be that makes a shadow of thirty feet? At this he put his hand to his chin, drew himself up, and gave a silly laugh. His master said he did not understand such questions as that.

We then asked him, How much is 3333 times 5555? In this instance, as in some of the others, he looked serious, began to twist about in his chair, to pick his clothes and finger nails, to look at his hands, put the points of his thumbs to his teeth, move his lips a little; and then he seemed to think a little, when his countenance gave signs of mental agony, and thus these symptoms continued.

His master told him to walk about and rest himself. He went into the yard, and appeared to be alternately elated with rapture, and depressed with gloom. He would run, jump up, throw his arms into the air above his head; then stand still, and then drag his foot over the weeds, look up and down; in a word he made all sorts of crazy motions. When we rose from the dining table, we found him on the piazza, sitting perfectly composed. He then stated when asked, the amount to be 18,514,815.

We could get no clue to the mental process by which he ascertained such results. When asked how he did it, his unvarying answer was, "I studies it up." But what do you do first and what next? He merely drawled out "I studies it up." He did not count his fingers, nor any thing external, nor did he seem to count at all; and yet he combined thousands and millions, and played with their combinations, just as others would with units. All the instruction he ever received was from his master, who taught him to count 100, and would amuse himself by asking simple questions, such as the twenties, or the fives, in a hundred."

Mr. Hanner saw him a few days afterwards, and found he perfectly recollected the numbers that had been given him on the former occasion; as well as his own answers.

We have since conversed with persons who have seen the above negro, and find Mr. Hanner's account fully confirmed. We might give numerous other answers equally wonderful for one of so little intellect, but we desire not to consume too much space; though we wish to give such description as will enable the mental philosopher to understand the case. In body, Cap weighs nearly 200 pounds, and all agree as to his idiocy. A person who saw him in 1845 says, "Though only 19, he has the appearance of being 30. He does not know a letter or figure, or any other representative of numbers or ideas. He speaks to no one, except when spoken to. His forehead is low, and covered with hair, within an inch and a half above his eyebrows. But the volume from temple to temple is great beyond comparison. I noticed that even numbers were more easily solved by him than odd ones, but could find no clue to his mode of solution. Such is the Alabama Negro, the wonderful being of *one idea!*" Had Mr. Hanner been a phrenologist the shape of the forehead would not have passed unnoticed.

Though other instances of mental imbecility in such calculators, have been found, the above seems to be far the most remarkable. There was a white man, living near Metuchin in New Jersey, some years ago; for whom a guardian to take care of his property was necessary, though himself the wonder of his acquaintance, for his powers of calculation. We have been unable to learn the particulars with sufficient accuracy for publication. Fowler, in his Practical Phrenology, speaks of meeting with a case in 1837, at Fairhaven, Massachusetts, in which the calculating power was combined with a great degree of imbecility; but we have been unable to learn reliable particulars.

The state of Vermont has furnished two of the most remarkable cases on record; Zerah Colburn, and Truman H. Safford. The former died at the age of 35, the latter is now at Cambridge, Massachusetts, receiving the full benefit of a collegiate education. As these cases differ from each other as well as from those we have before alluded to, we shall give a pretty full account of both; and we hope the reader will bear in mind the points of resemblance and difference.

ZERAH COLBURN, was born at Cabot, in the state of Vermont, September 1, 1804; and it is said in his memoirs that of the seven brothers and sisters who formed the family, he was regarded in infancy as the dullest in intellect. The first exposition of the peculiar powers of his mind in combining numbers, was in August, 1810, when he was about a month under six years of age. While his father was at work at a joiner's bench, ZERAH was playing amongst the chips, when his father was surprised to hear him saying to himself, "5 times 7 are 35, 6 times 8 are 48," &c., evidently amusing himself by the process of calculation. He had then been at the district school about six weeks, and his father supposed he might have caught these expressions there; but on examination he found him perfect in the numbers of the common multiplication table, and on proposing other numbers he was found equally accurate. He inquired the product of 97 by 13, when ZERAH promptly answered 1261. News of his wonderful powers soon spread through the neighborhood, and many called upon him to satisfy their reasonable incredulity, who going away more than satisfied, spread the tale of wonder, with additions of what they had themselves seen. The account soon found its way into the public papers, and was spread throughout Europe and America. The boy was taken to the seat of government of his native state, where his powers were more fully tried.

Questions in multiplication of two or three places of figures, were answered with much greater rapidity than they could be solved on paper. Questions involving an application of this rule, as in Reduction, Rule of Three, and Practice, seemed to be perfectly adapted to his mind. The Extraction of the Roots of exact Squares and Cubes was done with very little effort; and what has been considered by the Mathematicians of Europe, an operation for which no rule existed, viz; finding the factors of numbers, was performed by him; and in the course of time, he was able to point out his method of obtaining them. Questions in Addition, Subtraction, and Division, were done with less facility, on account of the more complicated and continued effort of the memory. In regard to the higher branches of Arithmetic, he had no rules peculiar to himself; but if the common process was pointed out as laid down in the books, he could carry on this process very readily in his head.

That such calculations should be made by the power of mind alone, even in a person of mature age, and who had disciplined himself by opportunity and study, would be surprising, because far exceeding the common attainments of mankind;—that they should be made by a child six years old, unable to read, and ignorant of the name or properties of one figure traced on paper, without any previous effort to train him to such a task, will not diminish the surprise.

The project of educating him thoroughly was very early suggested, and many propositions were made to his father who traveled with the boy, but though anxious to effect the same object, he seems to have been of an unhappy, suspicious disposition, always fearful of being defrauded or imposed upon; and hence though he traveled with his son through the United States and Europe, and many efforts were made in both countries to aid him, he succeeded but partially in effecting his object; indeed it would have been infinitely better for the son, had he been alone, for the waywardness of the father kept both poor, and prevented the friends of science from effecting their wishes in the profound education of the youth. Mr. Colburn and son embarked for England, April 3, 1812, and took up their residence in London, where Zerah was visited by thousands, among whom were many of the first men of the kingdom. Some who saw him engaged in calculation, speak of his agitation, comparing it to St. Vitus' dance. The following extract from his Memoir, page 37, may show the kind of exercise to which his mind was subjected:

"Among other questions, the Duke of York asked the number of seconds in the time elapsed since the commencement of the Christian Era, 1813 years, 7 months, 27 days. The answer was correctly given: 57,234,384,000. At a meeting of his friends which was held for the purpose of concerting the best method of promoting the interest of the child by an education suited to his turn of mind, he undertook and succeeded in raising the number 8 to the sixteenth power, and gave the answer correctly in the last result, viz; 281,474,976,710,656. He was then tried as to other numbers, consisting of one figure, all of which he raised as high as the tenth power, with so much facility and despatch that the person appointed to take down the results was obliged to enjoin him not to be too rapid.

With respect to numbers consisting of two figures, he would raise some of them to the sixth, seventh and eighth power, but not always with equal facility: for the larger the products became, the more difficult he found it to proceed. He was asked the square root of 106,929, and before the number could be written down, he immediately answered 327. He was then requested to name the cube root of 268,336,125, and with equal facility and promptness he replied 625.

Various other questions of a similar nature respecting the roots and powers of very high numbers, were proposed by several of the gentlemen present; to all of which satisfactory answers were given. One of the party requested him to name the factors which produced the number 247,483, which he did by mentioning 941 and 263, which indeed are the only two factors that will produce it. Another of them proposed 171,395, and he named the following factors as the only ones, viz:

5×34279, 7×24485, 59×2905, 83×2065, 35×4897, 295×581, 413×415.

He was then asked to give the factors of 36,083, but he immediately replied that it had none; which in fact was the case, as 36,083 is a prime number."

"It had been asserted and maintained by the French mathematicians that 4294967297 ($=2^{32}+1$) was a prime number; but the celebrated Euler detected the error by discovering that it was equal to $641 \times 6,700,417$. The same number was proposed to this child, who found out the factors by the mere operation of his mind. On another occasion, he was requested to give the square of 999,999; he said he could not do this, but he accomplished it by multiplying 37037 by itself, and that product twice by 27. *Ans.* 999,998,000,001. He then said he could multiply that by 49, which he did: *Ans.* 48,999,902,000,049. He again undertook to multiply this number by 49: *Ans.* 2,400,995,198,002,401. And lastly he multiplied this great sum by 25, giving as the final product, 60,024,879,950,060,025. Various efforts were made by the friends of the boy to elicit a disclosure of the methods by which he performed his calculations, but for nearly three years he was unable to satisfy their inquiries. There was, through practice, an increase in his power of computation; when first beginning, he went no farther in multiplying than three places of figures; it afterwards became a common thing with him to multiply four places by four; in some instances five figures by five have been given.

Some persons had very strange ideas as to the manner in which he reckoned; on one occasion, a gentleman came in, and after putting some questions, began to believe that the boy was assisted by some note or hint furnished to him by some one concealed in the room; he doubted so far as actually to request leave to carry him out into the street at a distance from the house, away from his father, to ascertain whether the same readiness of reply would be evinced.

At another time a man came in while the room was full of company, having something wrapped up in a handkerchief under his arm, and taking the father aside, requested leave to propose as his question, "What book had he in his handkerchief?" he manifested considerable dissatisfaction because the question was not allowed."

By this time the child, then between 8 and 9 years old, had at intervals learned to read and write, and he remarks that he was fond of reading as a pastime.

"In the studies to which he subsequently gave his attention, he manifested no uncommon skill or quickness, though his progress was always respectable. The acquirement of language was easy and pleasant; Arithmetic, (in the books,) entertaining; Geometry, plain but dull."—*Mem. p.* 40.

In March 1814, a private instructor was employed to teach the youth Mathematics.

"As might be expected from the nature of his early gift, he ever had a taste for figures. To answer questions by the mere operation of mind, though perfectly easy, was not anything in which he ever took satisfaction; for, unless when questioned, his attention was not engrossed by it at all. The study of Arithmetic was not particularly easy to him, but it afforded a very pleasing employment, and even now, were he in a situation to feel justified in such a course, he should be gratified to spend his time in pursuits of this nature. The faculty which he possessed, as it increased and strengthened by practice, so by giving up exhibition, began speedily to depreciate. This was not as some have supposed, on account of being engaged in study; it is more probable to him that the study of any branch that included the use and practice of figures would have served to keep up the facility and readiness of his mind. The study of Algebra, while he attended to it, was very pleasant but when just entering upon the more abstruse rules of the first part, he was taken away from his books and carried to France."—*Mem. p.* 68.

In 1814, Mr. Colburn and son went to France where the son was fortunate enough to obtain a place in the Lyceum Napoleon, where he spent several months. In 1816 he returned to London, when he obtained a situation at Westminster school where he remained about three years.

In allusion to the study of Geometry, Mr. Colburn remarks:

"Many have inquired if the study of Geometry was easy to him? He never found, that he recollects, any difficulty in understanding the demonstrations laid down by Euclid. Their fitness and adaptation to the various problems or theorems were very evident to his mind, but the study was always dry and devoid of interest. The reason probably was, while studying he did not realize, even in anticipation, the benefits of such a science; had he been engaged in some pursuit that would have required the continual introduction and application of Geometrical principles, the subject would have assumed an interesting appearance, his mind would have been engaged in it, and he would have remembered the principles and arguments laid down."—*Mem. p.* 114.

After leaving Westminster, Mr. Colburn and his son being both without means of support, the stage was suggested and Zerah was placed under the instruction of Charles Kemble; but his dramatic career was brief, and not flattering; and he afterwards spent two or three years in misery through want of means and want of employment. He was then from 15 to 17 years of age and had long ceased to exhibit his peculiar powers for a support. Indeed his extraordinary powers seemed to leave him, as he acquired general education, and ceased to exercise them.

In 1821 he obtained a school and spent several months in that employment, and in aiding Dr. Young, in astronomical calculations. February 14, 1824, his father died, and soon afterwards the son embarked for America. After visiting home he engaged in teaching and afterwards became a Methodist travelling preacher, and continued in that profession until his death; which occurred in 1839. His remains now lie in the town of Norwich, Connecticut, without a stone to mark the spot. *Such is fame.* In heart, he was one of the excellent of the earth.

What might have been effected by the aid of a profound education, cannot be known; but it is fair to presume, from his own account of himself, that though he possessed respectable talents, his mental endowments were not of a superior order; neither did he seem to possess common tact for acquiring a livelihood, and the father appears to have had less than the son; hence they were harassed with want, and instead of helping themselves spent their time in soliciting aid from others. The education of Zerah was respectable, although not what his friends desired. He possessed very considerable mathematical knowledge, and was familiar with the French language, which he learned during his stay in France where he also made some proficiency in the German; and during his attendance at Westminster he must have made very considerable proficiency in the Latin: added to all this was the intercourse he necessarily had with mankind during his travels, which was well adapted to improve his mind, for the individuals with whom he associated were generally of the right kind to induce improvement.

The following are some of the methods of calculation pursued by COLBURN, as explained in his memoirs.

" In extracting the square root, his first object was to ascertain what number squared would give a sum ending with the last two figures of the given square, and then what number squared will come nearest under the first figure in the given square when it consists of five places. If there are six figures in the proposed sum, the nearest square under the two first figures must be sought, which figures combined will give the sum required ;" and the cube root is found by an application of the same principle. The manner in which he proceeded to find the factors of numbers was somewhat similar. He ascertained or rather bore in mind what numbers had certain terminations, and narrowed down his search by the application of established principles bearing upon the case. If for instance the given number was odd, he knew that the factors must be odd, and if it was not divisible by any proposed number, it could not be by any multiple of such number ; thus his process was narrowed down, but still left too wide for the skill of the ordinary mind.

His process of multiplying involved less difficulty, and was something like this : he first divided the factors into their round numbers, thus : Multiply 1675 by 325.

1000	
600	300
70	20
5	5

Then in his mind he multiplied 1000 by 300 and remembered the product - - - - -	300,000
Then 600 by 300, and the product 180,000, added to the other, makes - -	480,000
Then 70 was multiplied by 300, making 21, 000, and being added to 480,000 made -	501,000
To which lastly the product of 5 by 300 being added we have - - - - -	502,500
This disposes of the 300, and we take 20 times 1000=20000, which makes - -	522,500

Then 20 times 600=12000, which added makes - - - - -	534,500
And 20 times 70=1400, which added makes	535,900
Then 20 times 5=100, which added makes	536,000
This is the product by 320, to which we add the products of the several parts by 5, viz: 1000 by 5=5000, making - -	541,000
Then 600 by 5=3000, making - -	544,000
Then 70 by 5=350, making - -	544,350
Then 5 by 5=25, making the whole product	544,375

This process is peculiar in beginning at the highest place instead of the lowest; but it is plain that for mental operation this is far better, as the large numbers are so much more easily remembered from having no low places until almost the last. It is true that the process is prolix, but it is nature's process, and probably was used before our more artificial mode, and it may be profitable to compare it with the common mode, that both may be better understood; for when they are carefully compared, they will be found very much alike. Let us compare the calculations:

```
                   1675
                    325
                   ----
                   8375
                  3350
                 5025
                -------
Product as before 544,375
```

He began by multiplying from left to right, so that his products would stand thus, and a little observation will show that it is in effect the same process we daily use, only that we abridge it by carrying as we proceed from right to left. Compare for instance the products by 5, the last four products in the operation, with the product in a single line, and it will be found substantially the same.

```
   1675
    325
 ------
 300000
 180000
  21000
   1500
  20000
  12000
   1400
    100
   5000
   3000
    350
     25
 ------
544,375
```

Although the foregoing were given by Colburn as his modes of calculation, we are inclined to doubt their accuracy; for some of them seem to pre-suppose a knowledge of figures, which he certainly did not at that time possess. We are rather inclined to the belief that having become, at the time he defined his modes, somewhat acquainted with the use of figures, and being anxious to satisfy the reasonable curiosity of the world to learn his modes of calculation, he deceived himself in using characters while describing a purely mental operation; for as he knew nothing of any representatives of numbers, he must have contemplated numbers themselves.

Perhaps the present will be as favorable a time as any other, to draw a distinction absolutely necessary to be made, between mental calculation by means of figures, and mental calculations without their aid. The latter is what we would understand by the term Mental Arithmetic; but the term is generally applied to all calculations in which neither sensible objects nor figures are used. Pestallozi carried this practice with his pupils to a very great extent; and every child that commences oral exercises before using characters, must study in the same way; but after learning in the usual way, our mental calculations are very similar to our written ones, and this without reference to the question whether we have studied the subject analytically or synthetically. Perhaps it is almost impossible, after becoming familiar with figures as the representatives of numbers, to calculate numbers entirely in the abstract. Our calculations seem naturally to flow into the common form, only carrying on the operation by concentrating the attention and imagining how the quantities would appear if written. Practice and effort will discipline the mind so as to enable it to produce astonishing results; and yet there may be little invention, and nothing whatever peculiar in the mind.

It is said of the celebrated mathematician, EULER, that two of his pupils having differed in the result of a converging series of seventeen terms, at the fiftieth figure of the result, he reviewed their work mentally, and pointed out the proper correction. This was probably in the latter part of his life, as the loss of his sight then compelled him to cultivate mental calculation, and to avail himself of the aid of an amanuensis. He was then able mentally to raise any number less than a hundred to the fifth or sixth power, without difficulty. He had always however cultivated in some degree the habit of mental calculation.

Dr. WALISTON tells us that he himself could in the dark perform multiplication, division, and the extraction of roots to forty decimal places; that he once proposed to himself, while in bed, a number of fifty-three places, and found the square root though extending to twenty-seven figures; and that without writing a single figure, he dictated the result from memory twenty days after.

We sometimes meet with clerks who are able to extend the amounts in bills of items, and to sum up the total, with the apparent rapidity of thought. This may be in part the result of natural quickness, but it is much more dependent on practice and close attention in observing the relations of numbers. For this purpose too it is desirable to be familiar with the products of numbers as far as 20 or 30 at least, instead of 12, the ordinary limit of the multiplication table; and to become familiar with every time and labor saving expedient. The following calculations are said to have been performed by ABRAHAM HAGARMAN, of Brighton, Monroe County, New York, and though they indicate nothing of the peculiar genius of Cap or of Colburn, they show very clearly the power of concentrated attention and long continued practice; for it is said that mathematical studies, and especially the solution of difficult problems, has occupied his chief attention for thirty years, fourteen of which he has been an invalid. The experiment of mental calculation however has been commenced within a few years. We extract the following from his calculations.

1st. 987654×345678=341,410,259,412.

2d. 9753214×2345678=22,877,899,509,092.

3d. 46375619×54625125=2,533,273,984,827,375.

4th. 123456789×123456789=15,241,578,750,190,521.

5th. 9615324516×4256484144=40,927,476,341,768,474,304.

6th. 82527613529 × 49243126216 = 4,063,917,689,313,816,176,264.

7th. 951427523675 × 484324256144 = 460,799,427,678,822,324,209,200.

8th. 831532463519 × 643234375246 = 534,870,264,668,411,251,650,674.

9th. 648728416968 × 421875625125 = 273,682,706,444,726,657,121,000.

The first, second, third and fourth of the above operations he accomplished in from one and a half to two hours. The fifth, sixth, seventh, and eighth, occupied from two to three hours. The ninth he accomplished in less than one hour, owing to the favorable character of the multiplier. This is certainly a great feat to be performed "in the head" alone; and shows very clearly what can be done by persevering effort, with perhaps no peculiarity of mental constitution, except a fondness for such amusements. Close attention is all important, it is the great constituent of inventive powers. Sir Isaac Newton says, "It is that complete retirement of the mind within itself, during which the senses are locked up—that intense meditation on which no extraneous idea can intrude—that firm, straight forward progress of thought, deviating into no irregular sally, which can alone place mathematical objects in a light sufficiently strong to illuminate them fully, and preserve the perceptions of the mind's eye in the same order that it moves along."

This power over the attention may be acquired to a great extent, by any one of sound mind, but with very different degrees of readiness, and probably not always to the same extent by different persons. In some this power seems natural, while with others the acquisition costs great labor. The perceptive faculties are very different in different individuals, and this is true in regard to perceiving the relation of numbers, as well as all other mental perceptions. In some, this faculty seems peculiarly obtuse, and they practice calculations with great difficulty. Some men even of fine minds require great effort in order to learn the simplest rules of arithmetic; and Humboldt speaks of the Chaymas, (a people in the Spanish parts of South America,) that have great difficulty in comprehending any thing that belongs to numerical relations; and that the more intelligent count in Spanish, with an air that denotes a great effort of the mind, so far as 30, or perhaps 50. He mentions, as a peculiarity, that the corners of their eyes are turned up towards their temples.

James Garry, who was remarkable for his powers of calculation, resided some years ago at Harper's Ferry, Va., and from J. A. Fitzsimmons, Esq., who was intimate with him, we have obtained the following account.

Mr. Garry was born in the county of Antrim, in Ireland, but immigrated to this country in early life. He was first employed in New York city, at a large salary; and subsequently by Tiffany, Shaw & Co. of Baltimore. While there it was customary for one of the clerks to call over the items of the largest bills of goods, and as rapidly as the clerk could write them down, Garry would give the extension of each line and the footing of the bill; without requiring the clerk to delay a moment, and with *absolute certainty* of being right. He was subsequently employed as a clerk by Messrs Wager & O'Byrne, of Harper's Ferry, Va., Commission Merchants, where we first heard of him; and where Mr. Fitzsimmons was a fellow clerk with him. In a social point of view, he speaks of him as exceedingly warm hearted, though with a tinge of melancholy, that was probably increased by an unfortunate habit of intemperance. An estimate may be formed of the extent of this singular gift, from the fact that while at Harper's Ferry, his former employers at Baltimore offered him $2000 dollars per annum, if he would return and bind himself to be temperate. But he declined. He was not prepossessing in his manners, and though a tolerable penman, was entirely unacquainted with Grammar, Geography, History &c., his great forte being mental calculation; if that can be called calculation, which seemed to be mere perception. Mr. Fitzsimmons says "His powers of calculation were indeed wonderful, and the gift was natural—not acquired. *He was never known to make a mistake*, except when working with pen or pencil to show work; and then but seldom. He could give the sum total of any sum of figures, momentarily, without his ever having been found in error in any case; but he had very little ability for any science except figures. He was almost as prompt in the higher branches of arithmetic as in the elementary; though complex operations evidently cost him thought.

He could give no account of his modes of operation, but said the answer came instantly, and stood right before his eyes, and he had only to read what he mentally saw. He said it seemed to be there as by magic. In speaking of the extension and summing up of a long bill of items, he remarked to a friend that "The items seemed to pass before him like the ghosts in Macbeth, at the same time adding themselves together as they overtook each other in the journey; thus increasing in bulk until the whole were united, and the sum total was at once before him."

In answer to the question whether there seemed to be any process of reasoning, Mr. Fitzsimmons stated that the result seemed to be matter of instantaneous perception, and that Mr. Garry so described it; but stated that in difficult problems there were some little delay, and indications of mental effort, but Mr. Garry seemed to think, as he expressed it, that the operation was the same, "only the ghosts rose a little slower, and moved more solemnly." When intoxicated, his answers were rather less prompt, but still accurate.

About 1837 or '38, he visited St. Louis where he died, aged about 38 years.

Jedediah Buxton, of England was another instance. He was uneducated, and wrought his solutions by his native ingenuity. The following is given as one of his performances. "On being required to multiply 456 by 378, he gave the pro duct in a very short time; and when requested to work the question audibly, so that his process might be known, he multiplied 456 first by 5, which produced 2280; this he again multiplied by 20, and found the product 45,600, which was the multiplicand multiplied by 100; this product he again multiplied by 3, which produced 136,800, the product of the multiplicand by 300. It remained then to multiply by 78, which he effected by multiplying 2280 (the product of the multiplicand by 5) by 15, as 5 times 15 are 75. This product being 34,200, he added to 136,800, which was the product by 300 and the sum 171,000 was 375 times 456. To complete the operation he multiplied 456 by 3, which produced 1368, and having added this number to 171,000, he found the result to be 172,368."

From this it appears that he was so little acquainted with the common rules as to multiply by 5 and then by 20, to find what the mere addition of two ciphers would have given him. In fact the whole operation seems awkwardly adapted to mental calculation; but with him it was probably nature's method, and it produced the sought for result. We have not the full and satisfactory account of his constitution and habits that would be desirable; nor do we know any thing of his subsequent history. In order fully to appreciate such phenomena it is necessary to know more than merely the results produced by them.

We have seen an account of a clerk in the war office, in France, who in six minutes extracted the square root of 20,511,841; and in a quarter of an hour, without any written memoranda, gave the product of 379,625,348 multiplied by itself. But we know nothing of him beyond this performance, and of course cannot class him with any other.

In 1845 a child named Prolongeau, aged about six years, was announced in the city of Paris, that resolved difficult arithmetical problems, and even elementary operations in algebra; and a committee was appointed by the *Academy of Sciences* to report the facts of the case, with his modes of operation, &c. His countenance is spoken of as expressive; but we have been unable to learn further particulars, or that the committee has reported.

George Bidder, a native of Devonshire in England, born in 1805, afforded another instance of extraordinary calculating powers when a mere child; and a number of gentlemen in Edinburg, undertook the charge of his education; with the design of cultivating his powers to the utmost extent. But though he excelled in Numbers, he proved nothing more than common in Geometry; and by no means realized the hopes of his friends. When only eleven years of age, he would solve difficult algebraical problems in a minute or two; but he failed in Geometry.

We might mention other instances noticed in books, but we have not such particulars as would enable us to give a satisfactory account of them; and after having mentioned two or three minor cases in our own country, we shall close with a somewhat detailed account of TRUMAN HENRY SAFFORD, who differs from all the foregoing, and is perhaps the most remarkable character in this respect, known to be in existence.

An individual, named PETER M. DESHONG, has been traversing the United States for several years past, who possesses an astonishing degree of quickness in performing the elementary operations of Arithmetic, and especially in adding numbers; but his knowledge seems limited to the mere elements of the subject. We saw him several years ago and again very recently and have no hesitation in saying that in adding together long columns of numbers, he very far exceeds in rapidity, any other person that we ever saw attempt the operation. His eye catches the numbers with the rapidity of thought, and he gives the result almost at a glance. In multiplying, he uses but a single line, however large the multiplier may be, and in dividing, he uses a mode very similar to short division, the remainders only being set down. He manifests unwillingness to engage in calculations involving intricacy, and we doubt his ability to reason to any considerable extent on the subject.

His practice is to travel from one important point to another, and exhibit his powers of calculation; at the same time offering for five or ten dollars to teach others to perform with equal rapidity. This he asserts he can do in half an hour; and to aid in the imposition he carries with him charts professing to give his modes of operation. Within a few years he has entirely changed his charts, and they are now well adapted to his purpose. Having carefully examined both his old and his revised charts, we have no hesitation in saying that he who expects to derive any thing valuable from them in regard to adding numbers, or from the instruction of their author, will find himself mistaken. His mode of multiplying is ingenious, and might be profitably employed by many; and to some extent the same is true of division; but his power of rapidly adding and otherwise combining simple numbers, is nature's gift as much as ZERAH COLBURN's was, and cannot be bought for money, nor acquired by any ordinary amount of practice. In addition to the peculiarities of nature, Mr. DESHONG's whole time is devoted to these operations, and he has evidently improved by practice. Like all others whose minds are especially adapted to numbers, his memory on the subject is peculiarly retentive and prompt; and thus he is greatly aided in producing results. His bold and positive assertions are calculated to deceive many; but we understand thoroughly his professed modes of operation, and we have seen no one who has profited by his instruction in the addition of numbers. We say his *professed* mode, for we do not for one moment believe that he adds in the manner indicated by his charts. Addition may be thus performed, but the labor would be greater than in the ordinary way, and could not be performed so rapidly, unless when numbers are set down with special reference to that mode of addition. We have no wish to speak uncourteously of Mr. DESHONG, but feel it to be our duty to warn the unwary, without wishing to prevent any one who desires to seek his instruction from doing so. For his mode of multiplying, see page 280.

We have received a detailed account of the peculiar powers of JOHN WINN, formerly of Clark County, Ohio, and as the case differs from such as we have been considering, we shall give pretty free extracts from the letter before us.

"In person he was large, and in the latter part of his life corpulent. The features of his face were prominent, and indicated decision and determination. Whatever he undertook was pursued with ardor; and this remark applies as well to his religious and political opinions, as to his business transactions. He was decided in his friendships and his antipathies. His early education was limited, but as far as it went, was accurate and thorough. He was a good practical surveyor; his written compositions were free from errors in orthography or syntax, and his hand writing unusually neat, compact and uniform. Papers drawn by him were always executed in a business-like manner.

The most remarkable feature of his mind, however, was his facility in calculation. He was for some years engaged in buying and driving cattle and swine to market; and he prided himself on the rapidity and accuracy with which he could ascertain the numbers contained in droves, especially of swine. An opening would be made in a field containing a large drove, and he would sit on horseback, near the gap, and count as the animals were driven through, expressing himself audibly in something like this manner: "Twenty-five—sixty—eighty—rush them on boys!—hundred and twenty," and so on. Notwithstanding the rapidity with which he counted, he rarely ever made a mistake; and his estimate was considered conclusive.

In adding multiplying and dividing numbers, he possessed uncommon facility. Instead of summing up units, tens, hundreds, &c. separately, as is usual in addition, he would run the whole up together with as little apparent trouble as a common operator would feel in adding up a single column. In multiplying or dividing by numbers of two or three places he took the whole together as we do numbers under 12. He was fluent in conversation, possessed a retentive memory; and with early discipline would have been capable of superior attainments.

We have met with a description of GEORGE BLESINS, son of JOHN BLESINS of Nashville, that would seem to rank him amongst the most remarkable prodigies of the present or the past; but we have been unable by writing, to learn any thing further respecting him.

GEORGE BLESINS is described as being about seven years of age, (in 1847) of common statue, in good health, and very interesting in his appearance and manners. His head is unusually large, his countenance one of those speaking ones that tell the fire within; while his whole demeanor is dignified and commanding. Our informant states that on asking him the product of 25 by 25, he answered instantly 625; and on being asked how he knew, he said "20 by 20 is 400; 5 by 20 is 100, and this doubled is 200; 5 by 5 is 25; and then 400 and 200 and 25 make 625."

"He was then asked how many inches there were around the globe. He replied that there is a certain number of inches in a mile, and this number multiplied by 25,000 will give the circumference in inches. While his thoughts were engaged in the calculation, there was considerable merriment among the company, which did not divert his attention the least. Some person spoke to him, to see what effect it would produce upon him. He replied, "Be patient a moment and then I will answer." Nothing could change the current of his thoughts when once put in motion. He in three minutes gave the exact distance round the globe, in inches; and this entirely by a mental process, *for he knew nothing of figures*."

Other instances of his calculations might be given, but we have not room. His powers of mind seem adapted to reasoning generally, and hence he belongs rather to the Safford than the Colburn school.

We shall close the notices of these cases with an account of TRUMAN H. SAFFORD, whom we shall notice somewhat fully.

TRUMAN HENRY SAFFORD, Jr., is the son of TRUMAN H. SAFFORD, Esq., of Royalton, Vermont, where the son was born on the 6th of January, 1836. His frame is slight and his health has always been delicate, though he is represented as now acquiring greater strength. His hair and eyes are dark, and the latter shine with peculiar brilliancy; while his native modesty and kindness of manner, render him peculiarly interesting. His moral and reasoning faculties are astonishingly developed; but we might well say of his body, as a steamboat captain is reported to have exclaimed of JOHN QUINCY ADAMS, while contemplating that wonderful man as he stood, in venerable age, the centre of an admiring group, "O that we could take the engine out of the old Adams, and put it into a new hull!" But he that formed the brilliant machinery of young SAFFORD's frail bark, can give it strength for his purpose in the hour of need.

At twenty months of age he had learned his letters, and already could be seen the workings of faculties that were soon to astonish every beholder. At three years he was familiar with many things seldom noticed by those of twice his age; and already, though but a prattling child whose tongue had but imperfectly learned the legerdemain (excuse the solecism) necessary to shape the words he used, his mind was breaking its fetters and struggling to understand the objects around him. He was sent to school, but the rules of study and of recitation were irksome to him, and he preferred to be at home where he could revel in study without control. In arithmetic he could not confine himself to the dull routine of the common rules and modes of operation. He saw the whole at a glance, and went through with a hop, skip and a jump, where others spent their days and weeks in slowly feeling their way. Instead of the neatly arranged rows of figures and the long columns that gradually step by step brings the result to the light of common minds, he would throw upon his slate a mass of half expressed numbers, in heterogeneous confusion, while his mind leaped beyond, and the conclusion was reached; but by giant strides that his teacher could not follow; and it was very soon concluded to leave him to his own course. His studies embraced every thing and any thing that came in his way—Geography, Chemistry, Grammar, and whatever afforded food for thought; and all were pursued with success.

The subjoined account of this wonderful youth was written in January, 1846, by the Rev. HENRY W. ADAMS, agent of the American Bible Society, and contains as full an account as may be necessary for our purpose. We may add that since Mr. ADAMS' article was written, ample provision has been made for the boy's education at Cambridge University, by the noble generosity of some public spirited friends of science; and he is now receiving every attention that can guard his health carefully, while he is enjoying the benefit of the greatest facilities that books, apparatus and living instructors can furnish. In order that his parents may watch over him, arrangements have been made to justify Mr. SAFFORD in removing with his family to Cambridge and to support them for five years, during which time HENRY is to remain in that institution free of charge. Every precaution is used to protect his health, and for this purpose strangers are not allowed to visit him unless by express permission; while a board of physicians constantly guard against excess of study; and all tests of his powers, for the gratification of visiters are forbidden. From a friend who has the best opportunity of knowing, we learn that his general health and strength are improving under the judicious course pursued, and that he is rapidly advancing in his studies. He entered the University in September, 1846, and the guardians of his education furnish him with books and instruction, for five years at least; so that whatever may be the result, the friends of science will have the consolation of knowing that every effort has been made to foster the talent that now promises so much for the cause of human knowledge. We ought to remark that before the Cambridge arrangement was made, the youth calculated almanacs for 1846 and 1847, both of which were published, as well as an edition for 1847, adapted to the latitude of Cincinnati, and published in that city. He was but little over nine years of age when the almanac for 1846 was calculated, and only ten when those for 1847 were calculated. This was certainly an effort of childhood that has no parallel.

We will now give Mr. ADAMS' account of his interview with the boy in January, 1846.

"Being a few days in the vicinity of Royalton, Vermont, on business connected with my Bible agency, I was induced, by the reports I had often seen in the public prints, of a remarkable boy of that town, to pay him a visit. The name of this precocious youth is TRUMAN HENRY SAFFORD, Jr. At the age of twenty months he learned his letters. Before three years old, he would reckon time upon a clock almost intuitively. He also learned to enumerate according to the Roman method from Webster's spelling book. He commenced going to school when three years old, but this he did not like. Since then he has been but very little, and now goes none at all. His mode of study was perfectly unique. He did not pursue the common circuitous route to the results of study. Probably no college in the United States could instruct him much, if any. When he first began to go to school, his teachers could not comprehend his ways, nor instruct his infant mind. Every branch of study he could master alone, with rapidity and ease.

He commenced Adams' New Arithmetic on Tuesday morning, and finished it completely on Friday night! And when he finishes a book it is done perfectly. He would not fully set down his sums, but cover his slate with a shower of figures, and at once bring out the answer. The teacher would look on in astonishment, unable to keep up with him, or to comprehend his operations, carried on in his mind with the rapidity of lightning, and then dashed upon the slate, no matter which end first. His thirst for all kinds of knowledge is very great. The whole circle of the sciences is as familiar to him as a household word. His father obtained for him Gregory's Dictionary of the Arts and Sciences, in three large volumes. This work, you know, is a vast encyclopedia of knowledge, treating briefly upon all branches of human knowledge. This was just the work he wanted; for an outline of any thing is enough—he can make the rest. It was this book that first gave him a taste for the higher mathematics. Here he found the definition of a logarithm, and from this alone, went on and made almost an entire table of them before ever seeing one. One day he went to his father and told him he wanted to calculate the eclipses and make an almanac! He said he wanted some books and instruments. His father tried to put him off; but the boy followed him into the fields and whithersoever he went, begging for books and instruments, with a most affecting importunity. Finally, his father promised to accompany him to Dartmouth College, and obtain for him, if possible, what he wanted. At this the boy was quite overjoyed; so much so, that when they hove in sight of the college, he cried out in raptures, "O, there is the college! there are the books! there are the instruments!" But they did not find all they wanted. At Norwich, however, they made up their complement. On coming home, the boy took Gummere's Astronomy, opened it *in the middle*, rolling it to and fro, and dashing through its dry and tedious formulas, *went out at both ends*. By the way, this is his usual mode of study. He does not begin any book at the beginning, but always in the middle, and then goes with a rush both ways. I asked him if, when he opened Gummere's Astronomy *in the middle*, he could comprehend those complicated formulas which depended on previous demonstrations. He replied, he could generally, but sometimes he "looked back a little." On arriving home, he projected several eclipses, and also calculated them through all their tedious operations by figures. This, as all mathematicians know, involves a knowledge of the labyrinths of mathematics, and also of formulas and processes most complicated and difficult. He has recently made an almanac for A. D. 1846. Two editions

—the first of seven thousand copies and the second of seventeen thousand—have already been published and nearly all sold. In the almanac are the calculations of two eclipses of the sun, wrought out wholly by its infant author, besides other valuable tables; especially one showing the amount of duties on wool, under all the tariffs since the formation of the government up to the act of 1842. This table the boy calculated alone. And that he calculated, without aid, the two eclipses of the sun, is attested by the published certificates of judges, lawyers, doctors, and clergymen.

Not satisfied with the old, circuitous process of demonstration, and impatient of delay, young Safford is constantly evolving new rules for abridging his work. He has found a new rule by which to calculate eclipses, hitherto unknown, so far as I know, to any mathematician. He told me it would shorten the work nearly one-third. When finding this rule, for two or three days he seemed to be in a sort of trance. One morning, very early, he came rushing down stairs, not stopping to dress himself, poured on to his slate a stream of figures, and soon cried out in the wildness of his joy, "O! father, I have got it! I have got it! it comes! it comes!" I questioned him respecting this rule. He commenced the explanation. His eyes rolled spasmodically in their sockets, and he explained his work with readiness. To hear him talk so rapidly, and yet so technically exact, and so far above the comprehension of all, save the most profound mathematician, put to flight all my doubts, and filled me with utter astonishment. He said he did not know as his new rule would work in all cases, but as yet it had. He also remarked that the nearer noon the eclipse came on, the easier it was to apply his rule. But young SAFFORD's strength does not lie wholly in the mathematics. He has a sort of mental absorption. His infant mind drinks in knowledge as the sponge does water. Chemistry, botany, philosophy, geography and history, are his sport. It does not make much difference what question you ask him, he answers very readily. I spoke to him of some of the recent discoveries in chemistry. He understood them. I spoke to him of the solidification of carbonic acid gas by Professor JOHNSTON, of the Wesleyan University. He said he understood it. Here his eyes flashed fire, and he began to explain the process.

When only four years old, he would surround himself upon the floor with Morse's, Woodbridge's, Olney's, Smith's, and Malte Brun's Geographies, tracing them through and comparing them, noting all their points of difference. His memory, too, is very strong. He has poured over Gregory's Dictionary of the Arts and Sciences so much, that I seriously doubt whether there can be a question, asked him, drawn from either of those immense volumes, that he will not answer instantly.

I saw the volumes and also noticed he had left his marks on almost every page. I asked to see his mathematical works. He sprung into his study and produced me Greenleaf's Arithmetic, Perkins' Algebra, Playfair's Euclid, Pike's Arithmetic, Davies' Algebra, Hutton's Mathematics, Flint's Surveying, the Cambridge Mathematics, Gummere's Astronomy, and several Nautical Almanacs. I asked him if he had mastered them all. He replied that he had. And an examination of him for the space of three hours convinced me he had; and not only so, but that he had far outstripped them. His knowledge is not intuitive. He is a *pure and profound reasoner.* In this he excels all other geniuses of whom I ever read. He can not only reckon figures in his mind with the rapidity of lightning, but he reasons, compares, reflects, and wades at pleasure through all the most abstruse sciences, and comprehends and reduces to his own clear and brief rules the highest mathematical knowledge. His mind is constantly active. No recreation or amusement can avail for any length of time to divert him from mental effort.

Being accompanied by Rev. C. N. Smith, of Randolph, Vt., who was acquainted with Mr. and Mrs. Safford, I had free access to the boy, and ample opportunity for a long and thorough examination. I went firmly expecting to be able to confound him, as I previously prepared myself with various problems for his solution. I did not suppose it possible for a boy of ten years only to be able to play, as with a top, with all the higher branches of mathematics. But in this I was disappointed. Here follow some of the questions I put to him, and his answers. I said, Can you tell me how many seconds old I was last March, the 12th day, when I was twenty-seven years old? He replied, instantly, "852,055,200." Then said I, The hour and minute hands of a clock are exactly together at 12 o'clock: when are they next together? Said he, as quick as thought, "1h. $5\frac{5}{11}$m." And here I will remark, that I had only to read the sum to him *once.* He did not care to see it, but only to hear it announced *once*, no matter how long. Let this fact be remembered in connection with some of the long and blind sums I shall hereafter name, and see if it does not show his amazing power of perception and comprehension. Also, he would perform the sums mentally, and also on a slate, working by the briefest and strictest rules, and hurrying on to the answers with a rapidity outstripping all capacity to keep up with him. The next sum I gave him was this: A man and his wife usually drank out a cask of beer in 12 days; but when the man was from home, it lasted the woman 30 days: how many days would the man alone be drinking it? He whirled about, rolled up his eyes and replied, "20 days." Then said I, what are the values of x in the equation $a^2+b^2-2bx+x^2=\frac{m^2x^2}{n^2}$? He sprung to his slate, and dashed on a few figures, and replied in about a minute, $x=\frac{n}{n^2-m^2}(bn+\sqrt{a^2m^2+b^2m^2-a^2n^2}.)$ He also gave the negative value of x.

Then said I, What number is that which, being divided by the product of its digits, the quotient is 3; and if 18 be added, the digits will be inverted? He flew out of his chair, whirled round, rolled up his wild, flashing eyes, and said, in about a minute, "24." Then said I, Two persons, A and B, departed from different places at the same time, and traveled towards each other. On meeting, it appeared that A had traveled 18 miles more than B; and that A could have gone B's journey in 15¾ days, but B would have been 28 days in performing A's journey. How far did each travel? He flew round the room, round the chairs, writhing his little body as if in agony and in about a minute sprung up to me and said, "A traveled 72 miles and B 54 miles—did'nt they? Yes." Then said I, What two numbers are those whose sum, multiplied by the greater, is equal to 77; and whose difference, multiplied by the less, is equal to 12? He again shot out of his chair like an arrow, flew about the room, his eyes wildly rolling in their sockets, and in about a minute said, "4 and 7." Well, said I, the sum of two numbers is 8, and the sum of their cubes 153. What are the numbers? Said he instantly, "3 and 5." Now in regard to these sums, they are the hardest in Davies' Algebra. I have had classes of one hundred scholars who have not been able to perform several of them. But young Safford, at one reading, comprehended them at a flash, and returned, almost instantly, correct answers. He also gave me correct Algebraic formulas for doing them. Then I took him into Plane Trigonometry. Said I, In order to find the distance between two trees, A and B, which could not be directly measured, because of a pool which occupied the intermediate space, the distance of a third point, C, from each was measured, viz: C A=588 feet and C B=672 feet, and also the contained angle A C B=55° 40 min.; required the distance

A B? He seized his slate, covered it with a group of figures, performed some of it mentally, and brought out the answer in about two minutes, saying, "592.967 feet." I then gave him this in the mensuration of surfaces: What is the area of a trapezoid whose parallel sides are 750 and 1225, and the altitude 1540? He walked rapidly across the floor, and whirled about to and fro, and replied, "1,520,750." Then, said I, if the diameter of the earth be 7921, what is the circumference? He said, instantly, "24,884.6136." To do this, he multiplied 7921 by 3.1416. This he did mentally quicker than I could write the answer. Then I gave him this: How many acres in a circular piece of ground whose circumference is 31.416 miles? He sprung on to his feet, flew round the room, and in a minute said, "50,265.6." Then, said I, required the number of acres of blue sky in an ellipse whose semi-axes are 35 and 25 miles? He began to walk the floor again, twisting his little body, and whirling his eyes spasmodically, and in about a minute said, "1,759,296 acres." How did you do it? said I. Said he, "Multiply the semi-axes together, and that product by 3.1416, and that product by 640." And did you perform the entire operation in your mind so soon? "Yes, sir." Then I took him into the mensuration of solids. Said I, what is the entire surface of a regular pyramid whose slant height is 17 feet, and the base a pentagon, of which each side is 33.5 feet? In about two minutes, after amplifying round the room, as is his custom, he replied, "3354.5558." How did you do it? said I. He answered, "Multiply 33.5 by 5, and that product by 8.5, and add this product to the product obtained by squaring 33.5, and multiplying the square by the tabular area taken from the table corresponding to a pentagon."

Now let it be remembered that this boy is only *ten years old*—that he did this sum for the *first* time in about *two minutes*, almost wholly in his head—and who can account for it? *

* * * * * * * * * *

I asked him to give me the cube root of 3,723,875. He replied quicker than I could write it, and that mentally, "155, is it not?" "Yes." Then said I, What is the cube root of 5,177,717? Said he, "173." Of 7,880,599? He instantly said, "199."

These roots he gave, calculated wholly in his mind, as quick as you could count one. I then asked his parents if I might give him a hard sum to perform *mentally*. They said they did not wish to tax his mind too much, nor often to its full capacity, but were quite willing to let me try him once. Then said I, Multiply, *in your head*, 365,365,365,365,365,365 by 365,365, 365,365,365,365!! He flew round the room like a top, pulled his pantaloons over the top of his boots, bit his hand, rolled his eyes in their sockets, sometimes smiling and talking, and then seeming to be in agony, until, in not more than one minute, said he, "133,491,850,208,566,925,016,658,299,941,583, 225!" The boy's father, Rev. C. N. Smith, and myself, had each a pencil and slate to take down the answer, and he gave it to us in periods of three figures each, as fast as it was possible for us to write them. And what was still more wonderful, he began to multiply at the left hand, and to bring out the answer from left to right, giving first, "133,491," &c. Here, confounded above measure, I gave up the examination. The boy looked pale and said he was tired. He said it was the largest sum he ever did! In conclusion, I am aware that this narrative is almost incredible. But let it be remembered that I went a skeptic, took a good witness with me, examined the boy carefully, and here pledge my sacred honor that all I have here stated is true. Rev. Mr. Smith, of Randolph, Vermont, is a witness to the correctness of this report. Further, if any are disposed to disbelieve my statement, I beg them to make a tour to Royalton, Vermont, where they will find the boy and have an opportunity to examine him for themselves. I was informed that he had been offered $1000 a year to cast interest for a bank not far from his father's. Mr. Safford has received many urgent proposals to permit his wonderful son to be carried round the world for exhibition, but he will not consent. Gentlemen of wealth have offered pecuniary aid to furnish the boy with books, &c.; especially one of Cincinnati—the patron of the distinguished Powers.

HENRY W. ADAMS.

Concord, N. H. Jan. 1846.

In comparing the preceding cases, we find a great diversity of general intellect, and even diversity in the prominent feature, but in some respects a striking similarity exists. In all cases in which the power of calculation exists in an extraordinary degree, the ability to recollect numbers is found to exist also; and it is believed that it will prove true in every phase of mind. We find boys in every school that are dull in this subject, and others that are bright, and we find in regard to the former, that however the memory may be in other matters, it is difficult to cause them to remember rules involving arbitrary numbers, as .7854, 3.1416, &c.; while with the learner that delights in the subject, these numbers are remembered with ease. Something is no doubt to be placed to the difference in the ability to concentrate the attention, but this is not sufficient to account for all. There seems to be a natural difference, and this keeps pace with the power of calculation, from the dullest to the brightest specimen; and like the power itself it is improvable by exercise.

We have already alluded to the distinction that may be drawn between the cases that seem to possess an unsought and apparently unacquired power, and those that are the result of patient practice. The former shows itself most clearly in uneducated persons, who must of necessity contemplate numbers without the aid of figures, by modes of their own invention; while the latter pursue the modes common with others, and hence in former cases the results seem the more astonishing. But if an individual of ordinary powers were faithfully trained by either mode, it would be found that a degree of proficiency might be acquired, that no one would anticipate: and this would increase with the intensity of attention, which again would be proportioned to the proficiency; for we love best and attend most closely to that in which we can excel.

We have seen detailed accounts of a school kept by J. E. Lovell, Esq., at New Haven, Ct., in which mental operations in arithmetic are made a very prominent subject of study, and the power to which the pupils attain is almost incredible. The multiplying of fifteen or twenty places of figures by as many, is not unusual, the numbers being set down and the operation wrought mentally by cross multiplication. The surprise often expressed on witnessing the performances of arithmetical classes, when that science is made the subject of especial study, would cease on a more intimate acquaintance with the powers of the human mind. Carry the calf daily and you may carry it when it becomes an ox. Mental performances, being more out of the usual routine of what is seen than written ones, excite most surprise; for he that is busied with other cares feels how impossible it is for him to turn away from the world and look in upon his own mind with the intense and unbroken gaze, indispensable to success; neither will they who now excite our surprise be able to do so when the cares and perplexities of life come upon them. Even Colburn, the wonder of the world, was unable to do his accustomed performances after other cares began to crowd upon him. In the case of Hagarman and some others to whom we have alluded, we find persons of maturity; but they were men with whom this was a constantly practised hobby. How far the object to be attained will justify the cultivation of this talent in youth, to the exclusion of others, is not the subject of our discussion. The man who appeared before a king of the olden time to show him with what certainty he could throw peas through the eye of a needle, had acquired astonishing dexterity; and the king duly appreciated his enterprise when he gave him a bag of peas for his pains. The thing can be done but is the acquisition worth the time and labor, and the sacrifice of other things involved?

If in other features the human mind presented no anomalies, her freaks in this particular would be more astonishing; but we find scarcely two minds constituted alike, or in which the several faculties are equally balanced. In our physical, mental and moral developments we find continual diversity; and while some manifest great deficiencies in one point, they exhibit perhaps equally astonishing prominence in others. Our time would fail to point out exemplifications, but any one can supply them in abundance.

In what respect are the gifts of Colburn, Bidder, Garry, &c., different from the endowments of the rest of mankind? Are they something distinct, or only extraordinary developments of what belongs to the human mind generally?

We have heard their peculiarities spoken of as *instinctive;* and have seen the term used in print in reference to them: but this cannot be a correct designation. It is doubtful whether instinct ever improves. We imagine that the first cell of the young bee is as perfect a hexagon as it can construct in old age; and we do not see that the young bird mistakes the materials of which its species usually build their nests, or constructs a different form. In every case of computing power to which we have alluded, and in every case in community the power is improved by cultivation, and lost by disuse. We do not for a moment believe that Colburn would have lost his ability, had he continued to cultivate it; and been freed from his pecuniary and other cares and perplexities.

The term *intuitive*, as applied to the case is less exceptionable; if we understand it to mean "Perceived by the mind immediately without the intervention of a train of reasoning or testimony. Perceived by bare inspection." But it is hard to say how far this is true; for the operations of the mind are often so rapid that the steps elude our observation, and we think we see at once, what indeed costs us a train of reasoning. The celebrated Dugald Stewart believed that all the conclusions of Colburn were reached by processes of reasoning, so rapid as to elude his own grasp, and to make no impression upon his memory; and hence he could give no account of them. After examining Colburn, Stewart seemed to attribute much of his peculiar power to Memory and Concentrated Attention; but these would produce rather small results when brought to bear on a mental blank. Yet there is no doubt but that a high degree of both was necessary to enable him to produce results so astonishing to the world.

It must be that a basis of axiomatic truths exists in all minds; and bare perception, intuition if you please, suffices to establish their character. But then the range and extent of these will depend on the ability of the mind to perceive; and the same mind after being cultivated, possesses greater ability to perceive and compare than when in a state of nature. It might be very difficult to decide where mere perception or intuition ceases, and reasoning commences, for they blend by shades so imperceptible, that there is no clearly defined line. They would seem to vary with different minds, and with the same mind under different circumstances. But whether the cases under consideration owe their peculiarity entirely to a perceptive power, beyond their fellow men, or alone to an ability to reason on the subject, with a celerity and accuracy peculiar to themselves, or to both combined they are equally interesting subjects of philosophical investigation.

In the case of Cap we find the power over numbers existing as almost the only representative of mind; while in Colburn we find it in connection with an ordinary development of the other faculties; and in Safford, we find it combined with an astonishing development of the whole mind. We regard these three as the most remarkable cases on record; while the others to which we have alluded, seem to fill the intermediate spaces and to show a gradual ascent from the lowest to the highest.

In the first we have an idiot, with no other faculty of the mind susceptible of education; and as the cares of the world are not likely ever to intrude, we may expect to see this power continue with him, and no doubt it might be increased under proper cultivation. In the second this ability perished or was choked by the growth of harassing cares and perplexities by which its unfortunate and highly sensitive possessor was weighed down, at an early age. He lost the indispensable power of withdrawing his attention from other things and turning it in upon itself. It would indeed have been strange had it been otherwise. Perhaps too the net work of forms thrown around his mental operations, in breaking him into the ordinary routine of study and school discipline, embarrassed the free operation of those modes peculiar to himself; and of course adapted to his own mind. Safford differs from all the others, possessing the natural power, in being able to perceive and announce truths as promptly as they could; and yet to follow his own mental operation and make it intelligible to others. This would favor the belief that however astonishing the aptitude might be, and however rapid the perceptions, they are still analogous to the every day operations of the mind, and differ only in degree from those of the dullest school boy. The talent was not found to exist in Safford, without previous indications of mind; for he reasoned, as well as perceived, from infancy. And it might be matter of doubt, whether the course he pursues in explaining to others, is always the original process of his own mind. In one respect these individuals seem to have resembled each other, and that was in the effect of their mental operations upon their bodies, producing violent contortions, and seeming to rack their whole systems. A similar expression of the eye is also spoken of and an acuteness in moral perception. How far these things may have been true of the others, we are not advised. In the case of Bidder and of Colburn, the ability seems to have been confined to *Numbers*, for neither one, though tried, excelled in *Geometry;* but Safford seems equally at home in either. From the account given of Winn, we think he might have acquired great proficiency as an engineer. His accuracy in estimating objects within the field of physical vision was not necessarily associated with his power of combining numbers; but taken together, they would have been invaluable to an engineer, or a field officer.

Garry seems to differ in some respects from all the others, but not materially so, and we must make allowances for difference in description, by different individuals. Though he thought he saw through no intervening medium, he admitted that in difficult problems "the ghosts moved more slowly and solemnly." To his mental vision, the sums of large numbers, and their various combinations, were as clearly present, as the sum of 3 and 4 would be to an ordinary mind.

But with all their powers, if it were sought to make a profound mathematician it would probably be better to take a subject of ordinary aptitude, with a sound mind in a sound body, and whose reasoning faculties are susceptible of healthful discipline.

It would be a pleasant task to pursue this subject much farther, and for this the *material* is ample; but if what has been hastily brought together shall lead inquiring minds to investigation, the object hoped for will be attained. Though such cases are rare, they are legitimate and important subjects of study. We have done no more on the present occasion than merely to throw out suggestions, which we hope others will improve. These anomalies are invaluable in the study of *mind;* like some species of mania, they exhibit the mental constitution in weak and strong lights, favorable to contemplation. In the well balanced mind, much of the internal working is concealed; but in the cases alluded to, the features of weakness and strength stand prominently out, and invite scrutiny.

CONCLUDING REMARKS.

HAVING concluded the task assigned ourselves at the outset, we shall devote a very few pages to the encouragement of the reader; and more especially the young reader, for we know well that we are too apt to be easily discouraged in early life. First then, we would say to him, let his object be threefold, The Increase of Mental Power—The Acquisition of Knowledge—and Skill in its use.

We have the testimony of the wisest men that have lived, that toil is the price of knowledge. Sir ISAAC NEWTON says that to patient industry he owed whatsoever of knowledge he had acquired; and the present wonder of our country, ELIHU BURRITT, the "learned blacksmith," who at less than forty years of age, has already learned more or less perfectly, some sixty or seventy languages, and studied various branches of science, says "All that I have accomplished, or expect, or hope to accomplish, has been and will be by that plodding, patient, persevering process of accretion which builds the ant heap,—particle by particle, thought by thought, and fact by fact." The Rev. JOHN TODD, in his *Student's Manual*, a work that every seeker of knowledge should read, very appropriately remarks: "Those islands which so beautifully adorn the Pacific, and which but for sin, would seem so many Edens, were reared up from the bed of the ocean by the little coral insect, which deposites one grain of sand at a time, till the whole of those piles are reared up. Just so with human exertions. The greatest results of the mind are produced by small but continued efforts. I have frequently thought of the motto of one of the most distinguished scholars in this country as peculiarly appropriate. As near as I remember, it is the picture of a mountain, with a man at its base, with his hat and coat lying beside him, and a pick axe in his hand; and as he digs, stroke by stroke, his patient look corresponds with his words,—*Peu et peu*—"little by little."

"The river rolling onward its accumulated waters to the ocean, was in its small beginning but an oozing rill, trickling down some moss-covered rock, and winding like a silver thread between the green banks to which it imparted verdure. The tree that sweeps the air with its hundred branches, and mocks at the howling of the tempest, was in its small beginning trodden under foot and unnoticed; then a small shoot that the leaping hare might have forever crushed. It now towers to the heavens."

He who expects by waiting, to rise by some bold stroke, will probably resemble at last the countryman who loitered on the river bank, hoping that the passing stream would exhaust its waters. But the young man who believes that knowledge is worth possessing, and is willing to apply his energies, has much to encourage him. He may point to some of the brightest ornaments of the nation, and of the world, and tell of the time when they were poor and obscure. ROGER SHERMAN was a shoemaker, and was encumbered with the care of his widowed mother and helpless family; yet he became deeply skilled in mathematics; afterwards read law, was appointed a judge, and rose to eminence as a jurist and politician. It has been remarked of him that he never said a foolish thing in his life. General GREENE, the favorite of WASHINGTON, was a blacksmith; and had only the elements of an English education given him. "But to him, an education so limited, was unsatisfactory. With such funds as he was able to raise, he purchased a small but well selected library, and spent his evenings and all the time he could redeem from his father's business, in regular study." BENJAMIN FRANKLIN, it need scarcely be said, was a practical printer; and emphatically the artificer of his own fortune. RITTENHOUSE, who was pronounced second to no Astronomer living, was a farmer in early life; and it is said that when a boy the smooth rocks in the field, and the fences by the way side, were often covered with his arithmetical calculations. He became eminent as an astronomer, and mathematician. NATHANIEL BOWDITCH, the celebrated navigator and scholar, was poor and enjoyed few opportunities in youth to acquire knowledge; all his science and his fame were the fruit of persevering application. Who was FULTON, whose inventions in the applications of steam power, have added millions to the wealth of our country, and especially of the west? And who was WHITNEY, the inventor of the cotton gin, by which the wealth of the south was doubled?

Let the industrious student read his country's history and he will find that these are but very few of the number that have risen to eminence without the inheritance of fortune's favors. And amongst the living he will find laborers, and mechanics, standing conspicuous in our deliberative assemblies; side by side, with the graduates of colleges and universities. Youth should study too that they may make useful and respectable private citizens; for such as seek to store their minds with useful knowledge, and who train their reasoning powers to think efficiently, will rise, notwithstanding the frowns of fortune, if they are true to themselves. When we look around upon our substantial farmers, master mechanics, and prominent citizens, how large a portion were poor boys! while the worthless and dissolute, are often those who commenced life under favorable auspices.

But there are difficulties in the way! or as the wise man of old has it, "There is a lion in the way—a lion is in the street." Thousands would rejoice to be learned, were it not for the toil. They would gladly enjoy the gratifications that intellectual wealth affords, but they are unwilling to labor for the prize. Think you that the men we have named, rose to distinction without effort. Or rather, did they not climb the ascent step by step? BURRITT, to whom we have already alluded, was not merely a blacksmith by profession, but until very recently a daily laborer for eight hours at the anvil. WILLIAM COBBETT was once a common soldier, and afterwards a member of the British Parliament. He says of himself: "I learned grammar, when I was a private soldier on the pay of sixpence a day. The edge of my berth or my guard bed was my seat to study in; my knapsack was my book case, and a bit of board lying on my lap my writing table. I had no money to purchase candles or oil; in winter time it was rarely that I could get any light but that of the fire, and only my turn even of that. To buy a pen or a sheet of paper, I was compelled to forego some portion of food, though in a state of starvation. I had no moment of time that I could call my own; and I had to read and write amid the talking, laughing, singing, whistling and bawling of at least half a score of the most thoughtless of men; and that too, in the hours of freedom from all control. And I say, if I, under circumstances like these, could encounter and overcome the task, is there, can there be, in the whole world, a youth that can find an excuse for the non-performance?"

Youth are apt to err in attributing too much to genius and

favorable opportunities. There certainly are grades of intellect, and with similar opportunities all would not succeed equally; it is likewise true that wealth may buy advantages: but it often brings its disadvantages. It leads youth to rely too much on their opportunities; and the mind lacks the energy which adverse circumstances generally impart. The young man who relies on his genius and college facilities, will not be apt to distinguish himself by his attainments.

If misfortune overtake you—rally again. When the web of the spider is destroyed by the hand of the intruder, it does not waste its time in idle repinings; but forthwith commences the work anew. Shall man do less? Though the labor of years may have been destroyed, can we better repair the loss than by sitting down patiently to the task of restoration? So the student of mathematics will often find, after spending hours or even days in completing some solution, that his plan is wrong, or that some slight error has been committed, by which all his labor is rendered of no avail. Then it is that, with mind disciplined to the task, he must, if he would succeed, meekly sit down to the work of review. He that would succeed must learn to bear the toil of revision. He must try —and try again.

Tradition relates, that when Bruce, King of Scotland, after a succession of defeats, took refuge in the winter of 1306, in the Isle of Rachrin, on the coast of Ireland, and there lay upon his bed, debating in his mind the question whether to continue further the unequal contest, or to leave his country forever, he saw a spider hanging to its thread, and endeavoring, as is its fashion, to swing from one beam to another, for the purpose of fastening its line. Six times did the insect essay with all its apparent power, to carry its point, and failed each time; but gathering all its strength for the seventh effort, it was successful. Encouraged, the king rose up and returned to his scenes of danger; and as he had never before gained a victory, he never afterwards sustained a defeat. May not the despairing student here find a lesson to cheer him in the hour of despondency!

Let him turn his attention to biography. He will there learn what man has done, and he knows our motto. We would recommend to his attention an excellent book entitled "*The Pursuit of Knowledge under Difficulties.*" It has been republished in this country, and the Rev. FRANCIS WAYLAND has promised a volume of American characters. The reader may learn from these books, the difficulties under which others have labored. He will find poverty, sickness, and physical misfortune opposing in vain. SANDERSON and EULER, and MILTON, were blind; but did they waste their time in idle repining? Who can peruse their memoirs without a thrill of admiration!

It is related of a poor gardener who had learned much by solitary application, that on surprise being expressed that he should understand *Newton's Principia*, he exclaimed "Can I not read? And if I have books what more do I want?" That gardener was Edmund Stone, afterwards celebrated as a mathematician. We find even the deaf and dumb becoming profound scholars; what excuse then can he have, who can hold free communion with his fellows? And there is Laura Bridgman, now in the Massachusetts Asylum for the Blind, who can neither *see*, *hear*, nor *speak*, and her sense of *smell* is very imperfect; and yet she has learned the use of language, so as to convey her ideas and learn the wishes of others. It is all nonsense for a young person to think that because he is not healthy, or has not leisure and books, and learned professors to instruct him, that therefore he cannot learn. Too many books are often a great disadvantage, by distracting the attention and preventing close application. A small, *well selected*, library is better for the student than a great variety of books. The text books should be studied very closely, very intently, and perseveringly; it is not sufficient that they be read once, and then thrown aside.

Coleridge says, "That readers may be diversified into four classes:—The first may be compared to an hour glass, their reading being as the sand; it runs in, and runs out, and leaves not a vestige behind. A second class resembles a sponge, imbibing every thing and returning it in the same state, only a little dirtier. The third class is like a jelly bag, which allows all that is pure to pass away, and retains only the refuse and the dregs. The fourth class may be compared to the slave in the diamond mines of Golconda, who casting aside all that are worthless, preserves only the pure gems." We should always endeavor to be of the fourth class. To search for the diamonds of thought, and by patient investigation to make them our own. We should *think*.

But how am I to think? Be assured that to make thinking effectual it must be diligent, undivided, concentrated. One great object of Education, I might say *the* great object, is to acquire this mastery over the mind; the amount of knowledge stored up is a matter of secondary importance. Gathering up the ideas of others may make a person *learned;* but discipline of the mind is *education.* He whose mind is disciplined can at any time add to his stock; he can learn facts and reason upon them; but he who has not learned to think properly, is in danger of exhausting his stock. The mind of one is a living spring, of the other a mere cistern that may be filled and emptied. The mechanic acquires manual dexterity in using tools, by long practice and patient training;—the mind of the thinking man, is the power by which he operates, and the dexterity with which he can use it, depends upon the power he has acquired to concentrate its efforts and direct its energies. The habit of trifling, instead of studying, is fatal to improvement. The student should never forget that while sands make the mountains, moments make the year; and he should bear in mind the school boy's motto, "Play when your work is done."

The American boy has much to encourage him; for the wide field of preferment is open before him, and earlier or later, merit will be rewarded. Let him remember that though but a boy now, a few years will place him in a different relation to those around him; and these years may be idled or improved. If idled, the ignorant boy will be an ignorant man; but if improved, the ignorant boy, may become the intelligent, useful, eminent man. Consider the truth contained in our first motto, "WHAT MAN HAS DONE, MAN MAY DO," and then fearlessly and with full and persevering purpose of heart, practise the resolve "I WILL TRY."

But how shall I try? To what point shall my efforts be directed? These are reasonable questions; for every person, and especially every young person, should ask himself, "What is my aim—my enterprize—my object?" If we commence life with nothing particular in view, we shall generally end it with no acquisition. BURNS, the poet, attributed most of his misfortunes through life to the want of an *aim*. Hence his efforts were ill directed; the beneficial results produced were not such as he anticipated, and with all his genius he fell an early victim to excesses. Adopt, therefore, some plan, fix on some object which is worth attaining; and make it your guiding star. What this object should be, your taste and circumstances may decide; but whatever it be, let your battle cry be "*Onward.*" Let it not be said of you as was said with caustic severity of one in the British Parliament, who taunted a member with having been a cobbler. "Had you, sir, been a cobbler, you would be a cobbler still." Improve by books—improve by thinking—improve by conversation. It is less the facts and ideas which we acquire from others in conversation that are valuable, than the exercise given to one's own mind. The collision of mind with mind, makes youth prompt and self possessed; discriminating and judicious. It is for this reason that teachers of youth should be men of sound sense. Employ a fool for what you please, except to teach children.

The memory, as well as the reasoning powers, must be

cultivated, or all attempts to accumulate knowledge will be but the task of Sisyphus. The leaking vessel is never filled—the spendthrift never becomes wealthy—neither can he who constantly forgets what he has just acquired, ever become learned. It has been sarcastically remarked that "many complain of their memories, though but few of their reasoning faculties." It should be considered that without memory the reasoning powers would be without efficiency; being without material to operate upon. The memory is far too important a faculty to be lightly esteemed or left uncultivated.

"Knowledge is profitable unto all things, therefore get knowledge." If you are a farmer or mechanic, study your profession, and if possible, study the arts and sciences connected with your occupation. You will be more useful, more respected, more successful and more happy. I speak of positive feeling, not the negative happiness of stupor. No youth should be content without the every day branches of education. He should *Read* and *Write* well—practice and care will soon enable him to do both. He should study *Geography*, it is a pleasing study; and no man can even read a newspaper, or converse, to advantage, unless he knows the situation of places and countries. He should study *English Grammar*, for without it he cannot hope to succeed as a speaker or writer; or to acquit himself well in conversation. Are you afraid of the study? How did COBBETT learn it? When COBBETT stood in the house of Parliament, think you he regretted his early efforts? Study *History*, especially the civil and political history of your own country and its institutions; no man can be properly qualified to exercise the privileges of a citizen in our republic, who is ignorant of his rights and obligations. Study yourself, study your fellow *Man*.—"The proper study of mankind is *man*." There is so much sameness in our race, that a critical investigation of a few specimens, will give you a very good idea of the whole. *Geology* and *Botany* are pleasant and useful studies. The former will teach you the nature of the materials which compose the globe we inhabit; the latter, the wonders and beauties of the vegetable world. These furnish a fruitful source of enjoyment in every ramble through town or country: but especially as we view the rich and varied scenery of nature. Study *Astronomy*, that noble and sublime science, which tells of the heavenly bodies and their laws. If you fear the enterprise, turn to the history of FERGUSON, RITTENHOUSE, BOWDITCH, HERSCHEL and others, and see what poor and obscure boys have done. No study can be more gratifying, or better adapted to improve the mind. The general laws and principles involve no intricate calculations; they are easily understood and cannot be forgotten. Study the nature of material bodies, through *Chemistry;* study their laws of being and action through *Natural Philosophy.* Here is a rich and inexhaustible treat; and you will be treading in the footsteps of thousands whose difficulties were great as yours.

I might run through the whole circle of literature and science, but I have perhaps already presented an appalling array. Be not however dismayed—take one subject at a time; difficulties, like hills, look most formidable at a distance; and think not mental application a fruitless task; it brings its reward. At each step the mind becomes better disciplined,—the workman better skilled in the use of his tools. It was said by STEPHEN GIRARD that the acquisition of the first thousand dollars of his immense estate, cost him more trouble than all the rest; and the analogy holds true in the acquisition of knowledge.

We have known teachers who had pursued the occupation for life, and could teach to a certain rule in the Arithmetic, or a certain branch of mathematics, and no farther; like the cobbler who mended shoes all his life, without ever attempting to make one. This any sane man should be ashamed of; for with common sense and reasonable industry, he may fit himself to teach any thing that is demanded. Perhaps there are few situations better adapted for study than that of a teacher. He receives and imparts, but the process of imparting does not exhaust; it only deepens the impression and makes the view clear. Many have studied at night, what they must teach by day; and one at least studied Algebra and Astronomy, under just such circumstances. The advanced rules of Arithmetic are but an application of the four elementary rules; and, in point of difficulty, the step thence to Algebra is not greater than from one arithmetical rule to another. With a familiar treatise on this subject, any expert arithmetician may master the study in a few months; and no one who is a good arithmetician should fail to study Algebra also. Surveying is a simple study, with which any arithmetician may become familiar in a few weeks; and the same remark is true of several other mathematical branches. The great difficulty seems to be in persuading ourselves that we are competent to the task. We survey the mountain, but forgetting the motto, we throw down the pickaxe; and while the diligent workman is picking his way through, we are doing nothing. We are waiting for the stream to exhaust its waters, but it flows on, and must forever flow, for the springs which supply its streams are inexhaustible.

There are few young persons who cannot buy a school treatise on Geography or Philosophy, on Geology or Botany, and what more does any one want? If he were surrounded by the most learned, they could not learn for him: every one must learn for himself. Perhaps there is selfishness in enjoying the thought that riches cannot buy wisdom: that there is no royal road to learning. There is a sort of real democracy in the matter; every one must help himself, or be content to plod through the world in ignorance. He must pick his own way through the mountain—must be his own ferryman across the waters.

"Education," says PHILIPS, "is a companion which no misfortune can depress, no clime destroy, no despotism enslave. At home a friend, abroad an introduction, in solitude a solace, in society an ornament, it chastens vice, it guides virtue, and gives efficiency and ornament to genius."

What is Education? Is it knowledge? Is it the acquisition of knowledge? Not exactly either. If it were possible to discover some mode by which a knowledge of Reading, Writing, Arithmetic, Grammar, &c., could be imparted with the instantaneousness of the electric spark, would that alone constitute education? I think not. Education consists, in training and duly exercising the faculties of the mind; and fitting them for successful application: and whether this is done in the course of acquiring one branch of knowledge or another, is comparatively unimportant. Civilized men educate youth through the means of literature and science, while the savage educates his child by exercising him in the toils of the chase, or of war; and in acquiring the skill which will provide for his wants and protect him in danger. It is this practical education that makes men strong in adversity, and able to rise amidst difficulties appalling to the slothful. It is this that often gives the child of poverty advantages over him whose mind rusts amidst the appliances of wealth. It is probable that many who have been distinguished for their self taught acquirements, and have fought their way against difficulties innumerable, would have lived unknown had they been born to wealth.

But we are not to understand that knowledge is valuable only so far as it enables its possessor to acquire eminence, or accumulate wealth. It is valuable in this way as a means to aid in acquiring something beyond; but it is valuable also for its own sake. It is a perpetual source of gratification to its possessor. Who would part with his intelligence for dollars and cents, and consent to become a stupid dolt? A man may be poor, he may suffer from absolute want, but he would almost as soon think of selling his hope of a better world, as of selling the characteristic that distinguishes him from the inferior animals. Fortune may frown upon him, friends may desert him, he may be despoiled of his goods, but his education cannot be taken from him; it will abide with him and enable him to rise again. I am aware that there are visionaries and bookworms who seem unfitted for the world, but this is from improper education, or from constitutional peculiarity; not excess of education. Correct education never unfits a man for discharging the duties of life. It gives him confidence in his own abilities to do what is within his power, but does not puff him up with self conceit, or lead him to suppose that he has reached the bottom of the ocean, because he has exhausted his own sounding line.

We repeat that an American youth has much to encourage him. Not only are the offices of his country open to all, but there are few family clans to sustain the worthless, or crush the deserving. No proud aristocrat can boast his exclusive privileges; and oppressive taxes, to support pensioned favorites, do not break down the energies of the people. The genius of our government invites to action; and an ample field is open in the various occupations of life. Whether he seeks the secluded paths of science,—engages in the quiet duties of domestic life,—or joins in the strife and turmoil of public service, he has the assurance, that his rights and privileges are equal to the best. Let him then survey his position, let him consult his means and his taste; and then deliberately resolve on what he will effect. Having so resolved let him not look back; and though adversity may cross his path; though storms of opposition may beset him, and difficulties rise up in the way, let his course still be "*Onward.*"

www.ingramcontent.com/pod-product-compliance
Lightning Source LLC
LaVergne TN
LVHW020116110826
845151LV00001B/172